System Lifecycle Management

Martin Eigner

System Lifecycle Management

Engineering Digitalization (Engineering 4.0)

Springer Vieweg

Martin Eigner
EIGNER Engineering
Consultant Baden-Baden, Germany

ISBN 978-3-658-33873-2 ISBN 978-3-658-33874-9 (eBook)
https://doi.org/10.1007/978-3-658-33874-9

Springer Vieweg

Responsible Editor: Eric Blaschke

This Springer Vieweg imprint is published by the registered company Springer Fachmedien Wiesbaden GmbH part of Springer Nature.
The registered company address is: Abraham-Lincoln-Str. 46, 65189 Wiesbaden, Germany

Acknowledgements

I would like to dedicate this book to my daughter, Sarah, and my granddaughters, Alva and Emelie. After her experience as a student in my company and the recognitions she gained from this of the volatility, hectic nature, and complexity of the PDM/PLM business, Sarah decided to study medicine. On the day she was awarded place study position, she came to my office and said to me "Daddy, I'm going to do your stupid job after all," then studied business and ended up in the PLM field. So, it seems there is something about this sector after all ☺.

A further matter is to thank the countless people who have helped me along my long, professional journey and who, through positive and critical discussions, have contributed to us being able to live and implement our vision. This primarily concerns my colleagues at BOSCH, EIGNER+PARTNER / EIGNER Inc., at the institute for Virtual Product Engineering (VPE) at the University of Kaiserslautern and the employees at Aras, Minerva, ProSTEP, SIEMENS, Unity, and XPLM. However, in these thank yous, I would also like to gratefully include the many colleagues in the industry, in consulting, at the SW providers, and the university research and teaching. You are all part of the mosaic which makes up this book. However, I would like to particularly thank the following people by name, as they contributed significantly to the content of the book:

At VPE incl. former and external Ph.D. candidates:
Christo Apostolov, Oliver Bleisinger, Thomas Dickopf, Alexander Detzner, Thomas Eickhoff, Kalle Faißt, Torsten Gilz, Marcellus Menges, Fabrice Mogo Nem, Christian Muggeo, Philipp Pfenning, Thomas Psota, Patrick Schäfer, Philipp Heiner Schmidt, Sebastian Sindermann, Mona Tafvizi, Rajeeth Tharma, Radoslav Zafirov, and Mathias Zagel

At Aras:
Paweł Chądzyński, Craig Currie, David Ewing, Mike Gavlak, Tim Keer, Rolf Laudenbach, Marc Lind, Rob McAveney, Matteo Nicholich, Malcolm Panthaki, Peter Schroer, Ayla Singhal, John Sperling, Lopa Subramanian, and Patrick Willemsen

At Minerva:
Leon Lauritsen, Christoph Golinski, Anthony Ponceot, and Leigh Young

At ProSTEP:
Martin Strietzel und Steven Vettermann

At SIEMENS:
Urban August and Matthias Schmich

At Unity:
Ulli Deppe, Daniel Baldus, Magnus Meier, and Philipp Wibbing

At XPLM:
Robert Huxel, Raoul Markus, Rolf Pfenning, and Bernd Zimmermann

Thanks also to the two PLM bloggers Oleg Shilovitsky[1] and Jos Voskuil[2] who have accompanied me over the last few years and who have enriched this book with their diverse contributions and a very special thank you to Mona Tafvici and my two Michaels (Michael Muschiol and Michael Pfenning), for proofreading the book and for many stimulating and motivating discussions, as well as Juliver Napitupulu, who took care of laying out the images. Last but not least, a special thanks to Linguistic Services Inc. in Boston for the translation ofthe book.

Martin Eigner

[1]http://beyondplm.com

[2]https://virtualdutchman.com

Contents

Foreword

1

Dear reader,

When I published my first PLM book – the terms product data management (PDM) or product lifecycle management (PLM) did not exist and we used the term Engineering Data Base (EDB) – together with my colleagues at EIGNER+PARTNER in 1991, we already had a vision of complete integration into the operational process. However, only the "downstream" integration of the development process to production planning and production were considered [1]. The second PLM book that I wrote together with my colleague from Dresden, Professor Dr.-Ing. Ralf Stelzer, already considered the early or "upstream" phase in its 2nd edition in 2009 and assumes the integration of requirements and function structures. The PLM system was introduced as a central engineering backbone, which practically represented the integrating backbone of all engineering processes from requirements management to the start of production (SOP) [2]. However, the industrial practice looked very different. An estimated 70–80% of my customer base actually used PLM more as a PDM systems. The focus of management was CAD data from the area of mechanical construction, partially purely document-centric, but partially also already 3D design with derivation of E-BOMs,[1] which were then transferred to the MRP[2] system. Limited approval and change management were also already used. Those responsible for the PLM complained of the lack of interest from management. While the MRP system was generally the responsibility of CIO's[3] or CFO's,[4] the PLM system bobbed around on department levels way under the radar of the C level. Standard ROI[5]

[1]BOM Bill of Material, E-BOM Engineering Bill of Material.

[2]MRP Material Resource Planning.

[3]CIO Chief Information Officer.

[4]CFO Chief Finance Officer.

[5]ROI Return on Investment.

M. Eigner, *System Lifecycle Management*, https://doi.org/10.1007/978-3-658-33874-9_1

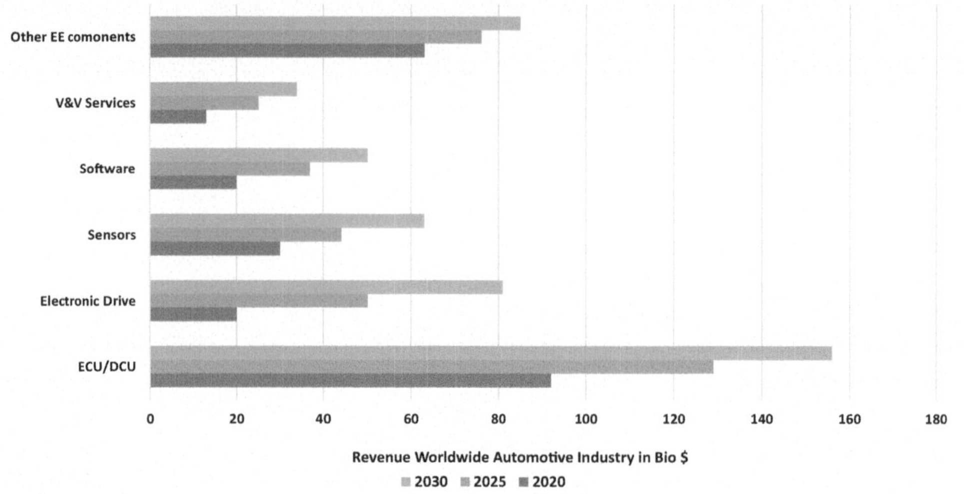

Fig. 1.1 Increasing revenue of worldwide automotive industry in electronic and software. *Source* Statista/McKinsey[7]

considerations did not bring a positive result with the argument that the development/construction only caused about 10% of the overall costs of the company. Meanwhile, the high importance of product development on the subsequent (→ planned) costs determined for the product lifecycle was completely ignored. It is assumed that the early concept and development phases define approx. 70–80% of the subsequent lifecycle costs. The fact that most market-leading PLM system providers ignored the trend toward mechatronics – most PLM providers at the time emerged from business with M-CAD[6] systems – also had a negative effect and they did not dedicate enough attention to electronics and also the software in particular. Monolithic PLM systems with the focus on mechanic design and a subsequently developed WEB client were no longer state of the art. However, the internet was gaining pace. Terms which were already presented at the end of the 90s, such as internet of things (IOT), were extended by internet of services (IOS), Industry 4.0, and the Industrial Internet dominated the literature and global research projects. In 2013, Ulrich Sendler, Professor Manfred Broy and I used the term *System Lifecycle Management (SysLM)* [3] for the first time in the book "Industry 4.0." This was intended to express that more and more highly complex, interdisciplinary cybertronic products and production systems were to have their data and processes managed over the entire product lifecycle using SysLM. The extreme challenge in system architectures to think and act in holistic and digitalized business processes in order to develop and produce interdisciplinary and innovative products is the objective of modern design methodologies. Figure 1.1 shows the need based on current and projected revenue

[6]M-CAD CAD applications from the field of mechanical construction.

[7]Handelsblatt February 19th, 2021.

in the automotive industry, which show the extreme increase in software and electronics. Supporting and implementing this philosophy on a digital platform as an engineering backbone is the task of a SysLM system.

As the term Industry 4.0 was heavily engaged with automation projects in production, the postulation was already made in the series "acatech discussion" in 2012 that an IOT-based production and a new disruptive business model developed from the ideas of the IOS would require smart engineering [4]. The term Engineering 4.0 was later developed from this [5, 6].

Engineering 4.0 was the overarching term for all methods, processes, and software tools which contributed to the digitalization of the engineering. In this phase, in which the BMBF[8] also recognized that Engineering 4.0 was to be promoted as much as Industry 4.0, I had a fantastic opportunity together with my colleagues from the VPE institute at the university of Kaiserslautern to lead and indeed initiate two industrial research projects on the topics of digitalization of the engineering process in automotive construction (MecPro^2)[9] [7] and innovative service-oriented business models in the agricultural industry on the basis of a digital twin (InnoServPro)[10] [8]. The knowledge gained from these and further industrial projects on the topics of the digitalization of engineering, MBSE,[11] service-oriented business models on the basis of digital twins and the role of system lifecycle management in this environment are included in this book. However, I will not only present gray theory on this topic, but also I will also present the underlying functions in the scope of vertical and horizontal integration of the engineering process using the example of a market-leading, modern SysLM system (Sects. 4.2.1 and 4.2.2).

If I may be permitted one personal note. This book was written during the Corona pandemic, which provided the opportunity to finally have enough time to write a book. This crisis has dramatically changed the general opinion about the two current buzzwords: globalization and digitalization. Globalization in economic crises has almost become a taboo and will certainly bring about a drastic change in the supply chains. In contrast, digitalization is experiencing an ever more positive estimation in popular scientific and economic articles. The Corona pandemic mercilessly reveals the current status of a state and has become a stress test for national and international state communities. Within the shortest time, all digital attainments and deficits were clear. This was clear in the fields of work, education, new media, public administration, and infrastructure. However, Corona also showed that many areas not only needed to be more efficient but also more resilient. The benefits of digitalization were also gratefully accepted in the

---

[8]BMBF Federal Ministry of Research and Education.

[9]MecPro^2 Model-based development process of cybertronic products and production systems.

[10]InnoServPro Innovative service products for individualized, availability-oriented business models for capital goods.

[11]MBSE Model-Based Systems Engineering.

private sector. However, we will still experience an effect similar to the 2008/2009 financial crisis. Due to the drastic crash in the world economy, in the next year or two, we will experience, on the one hand, a delay in many digitalization investments, and on the other, also a concentration on economically sensible and technically essential digitalization strategies.

I hope you gain lots of knowledge from reading this and possibly even a few déjà-vu experiences.

Yours,
Martin Eigner

References

1. Eigner, M., Hiller, C., Schindewolf, S., Schmich, M.: Engineering Database – Strategische Komponente in CIM-Konzepten. Carl Hanser, Vienna (1991)
2. Eigner, M., Stelzer, R.: Product Lifecycle Management – Ein Leitfaden für Product Development und Life Cycle Management, 2nd ed. Springer Verlag, Berlin (2009)
3. Sendler, U.: (Publisher): Industrie 4.0 – Beherrschung der industriellen Komplexität mit SysLM, p. 1–20. Springer, Berlin (2013)
4. Anderl, R., Eigner, M., Sendler, U., Stark, R.: (Publisher): Smart Engineering – Interdisziplinäre Produktentstehung, acatech DISKUSSION, 1st edn. Springer Vieweg, Berlin, Heidelberg (2012)
5. Abramovici, M., Ottheim, H.: Engineering im Umfeld von Industrie 4.0 Einschätzungen und Handlungsbedarf (acatech STUDIE). Herbert Utz, Munich (2016)
6. Künzel, M., Schulz, J., Gabriel, P.: Engineering 4.0 – Grundzüge eines Zukunftsmodells, Begleitforschung AUTONOMIK für Industrie 4.0. iit-Institut für Innovation und Technik in der VDI / VDE Innovation + Technik GmbH (2016)
7. Eigner, M., Koch, W., Muggeo, C.: Modellbasierter Entwicklungsprozess cybertronischer Systeme. Springer, Berlin (2017)
8. Aurich, J.C., Koch, W., Kölsch, P., Herder, C.: (Publisher): Entwicklung datenbasierter Produkt-Service Systeme – Ein Ansatz zur Realisierung verfügbarkeitsorientierter Geschäftsmodelle. Springer, Berlin (2019)

2 Forty Years of Product Data Management from PDM via PLM to SysLM

Abstract

This chapter explains the evolutionary development of PDM via PLM to SysLM. The operating conditions have been permanently changing since 1985. Initially, 2D M-CAD was the focus; however, the complexity of data management increased sharply through the 3D M-CAD systems. The next big evolutionary step was triggered by the rapid increase in mechatronic products and the desire of users and providers to extend the application of PDM beyond the core area of development and construction (→ PLM). IOT, IOS, and the resulting globalized interdisciplinary development of cybertronic and cyberphysical products and systems permanently increased software and electronic parts on the product, increased requirements on companies' internal and external collaboration as well as the interdisciplinary engineering processes and methods to be developed parallel to this (→ MBSE, [MBSE Model-Based Systems Engineering] System Thinking, Digital Thread, and Digital Twin) led in recent years to an extension of the PLM approach to System Lifecycle Management (→ SysLM).

In order to be able to understand the structure, functions and application areas of the current software solutions for PLM and SysLM, it is necessary to know the process of their development. Figure 2.1 provides a brief overview of the evolution of PDM,[1] PLM[2] to SysLM[3] during the period from 1980 to 2020. This context shows that the evolution of

[1]PDM Product Data Management.

[2]PLM Product Lifecycle Management.

[3]SysLM System Lifecycle Management.

M. Eigner, *System Lifecycle Management*, https://doi.org/10.1007/978-3-658-33874-9_2

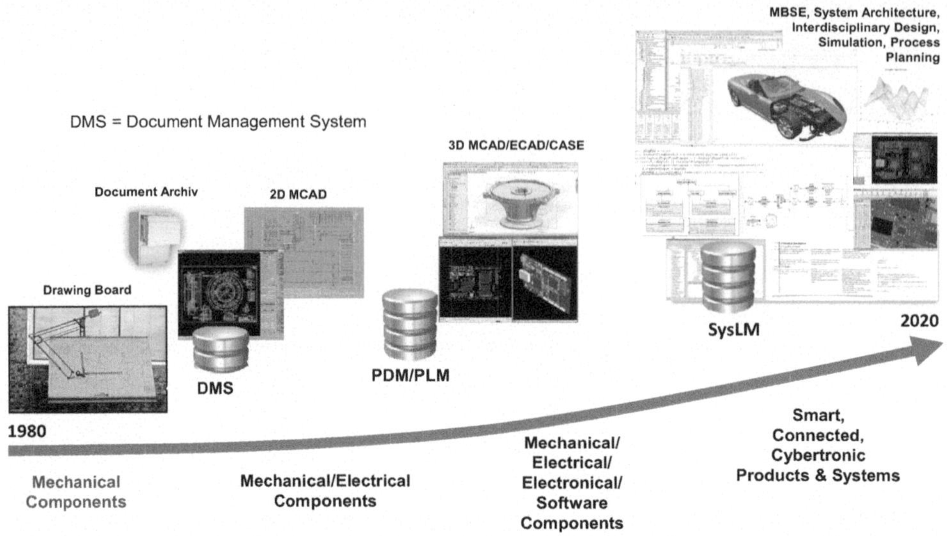

Fig. 2.1 The evolution of PDM via PLM to SysLM

the PLM solutions was formed by the various objectives of CAD,[4] MRP[5] and the independent providers. While CAD and MRP software solutions initially developed independently of one another between 1975 and 1985, in the 1990s, the PLM thought resulted from the connection of both system worlds [15].

Of course, there were no system breakthroughs between the individual phases of PDM, PLM, and SysLM, rather the development was evolutionary and continuous.

2.1 Document Management System (DMS)

At the start of the 80s, there was already increasing development on M-CAD and E-CAD systems, for example, the US company Unigraphics, which later merged into SIEMENS PLM via several stages. Management of product data was developed internally in companies between 1980 and 1985 under various designations, for example, by American Motor Corporation and Rockwell International. The Jeep Grand Cherokee was considered one of the first products that was managed via a product data management system. Parallel to this, the first document management systems (DMS) were also already available on the market. These could manage drawings as well as visualize and plot them on

[4]CAD Computer-Aided Design M-CAD (mechanics) E-CAD (electrics/electronics) CASE (software).

[5]MRP Material Resource Planning.

graphic work stations. For this, traditionally created drawings were transferred to TIFF[6] format through scanning. Drawings which were created using CAD systems could also be saved directly in this format. Back then, this was a big step forward, as at that time, CAD drawings were plotted out, manually signed and then stored in the drawing archive. Manually created drawings were managed via microfilm cards and visualized with reading devices.

The significant application function was the drawing management, sometimes already extended by a simple project management system, in order to simplify access to documents. The advantage of this solution was that, at that time, manually created drawings and 2D CAD documents could be saved and visualized company-wide together in TIFF. The integration functions were very basic. Documents created by IT systems were generally checked in manually. The service functions were already largely complete and contained, for example, applications for visualization, access management, and the "electronic filing cabinet" (→ vault), archiving and back up for safe file handling.

2.2 Product Data Management (PDM)

In 1985, the author and two colleagues founded[7] the first company worldwide for product data management (EIGNER+PARTNER, later EIGNER Inc.). They used the term, Engineering Data Base (EDB) [8]. A year later, the company Sherpa was founded in America. In addition to EDB, the abbreviation EDM, meaning both Engineering Document Management and Engineering Data Management also became widely established. As soon as had the designations were widely accepted, the abbreviation PDM had already become common. The complete functional scope of PDM in comparison to DMS is shown in Fig. 2.2. The area of application was predominately limited to the development and construction of mechanical products.

PDM systems consisted of three basic models:

- The product model, consisting of technical master data, product structures (→ Bill of Material), and documents,
- The process model, which essentially included the Engineering Release and Change Management (ERM/ECM) in development and construction, and
- A limited configuration model which manages temporarily valid products and document configurations resulting from revision via the areas of validity (→ effectivity)

[6]TIFF Tagged Image File Format.

[7]The company founders came from an area of a division of Robert Bosch GmbH, which, in addition to the technical computer center inclusive all engineering applications for the technical organization (Bill of Material, release and change management), was also responsible temporarily for the electronic predevelopment.

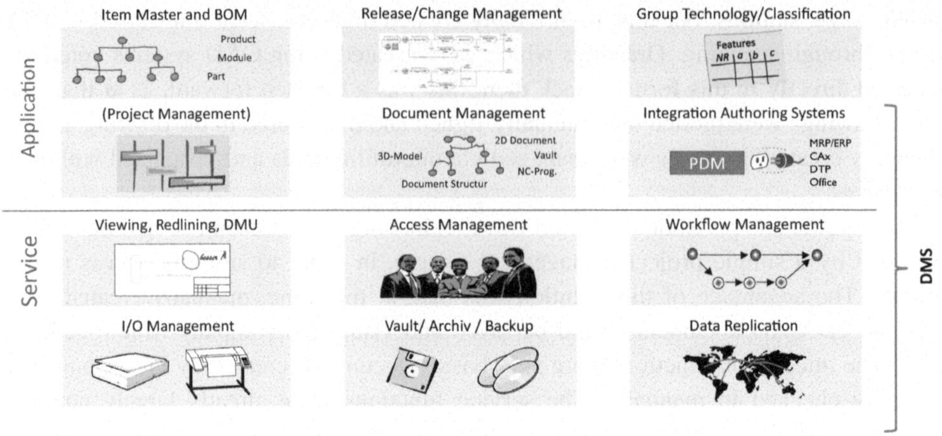

Fig. 2.2 Function scope of a DMS and PDM system (1985–1995)

and is extremely important for reconfiguring the product in the event of damage (→ ISO9001). Generally, only the date (from → to) and revision or version are used to define a configuration from the numerous options. Therefore, company-wide configuration management (CM) was generally excluded.

Several PDM systems were developed in Germany and internationally between 1985 and 1995:

Table 2.1 shows the PDM System developments in various countries. The left column lists the system providers and the right the systems and their updates. (No claim to completeness).

The first four systems came from Germany and still exist today, either directly or via another provider. It was interesting that SAP as a MRP solution provider also entered the PDM market with its own solution, however measured against its market share in the MRP segment, with limited success. The American systems Computervision, Hewlett Packard and MTI, were originally CAD providers. All international systems were later closed down, taken over or will soon be obsolete. iMan is an exception which was later merged into Teamcenter Engineering and now into Teamcenter Unified Architecture (TC-UA). It was a very volatile market that was characterized by lots of take overs. The definition of PDM at the time was relatively simple:

◆ **Product Data Management (PDM):** PDM is the management of the product and process model in development and construction with the objective of creating clear and reproducible product configurations in development and construction.

A typical user interface of a PDM system at that time is shown in Fig. 2.3.

Table 2.1 Overview of the common PDM systems from 1985 to 1995

	Provider	System name
Germany	*CONTACT*	*CIM Database*
	EIGNER + PARTNER	*CADIM/EDB later axalant and e6 (Merge with Agile 2001; taken over by ORACLE 2007)*
	PROCAD	*Pro.File*
	SAP	*SAP PLM*
USA	*Computervision*	*Optegra (taken over by PTC and then closed down)*
	Hewlett Packard	*Workmanager (later CoCreate, then PTC and then closed down)*
	IBM	*Product Manager / (later closed down when IBM temporarily took over the marketing and sales of Dassault´s CAD and PDM/PLM products)*
	MatrixOne	*eMatrix (taken over by Dassault → Enovia)*
	MTI	*Metaphase (later SDRC, then EDS then SIEMENS PLM as Teamcenter Enterprise, 2025 obsolete)*
	SHERPA	*DMS-PIMS (later INSO then PTC and then closed down)*
	Unigraphics	*iMan (later EDS, then UGS, then SIEMENS PLM Teamcenter UA)*
Israel	*SmarTeam Corp*	*SmarTeam (taken over by Dassault, 2025 discontinued)*

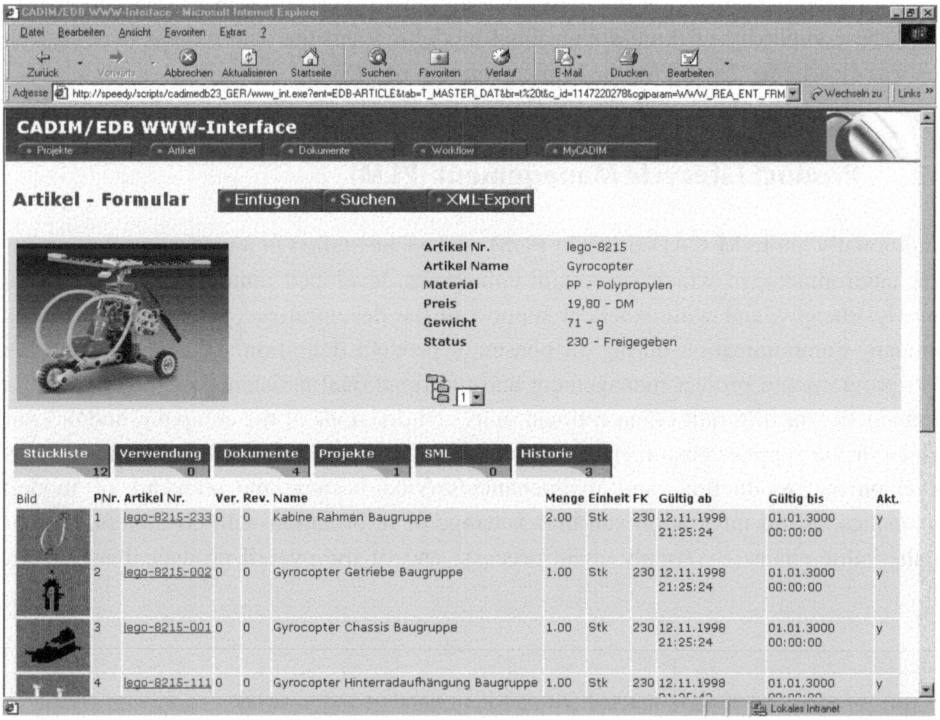

Fig. 2.3 Typical user interface of a PDM system. (axalant HTML Client 1998)

Typical areas of application were in the administration of 2D/3D M-CAD data, partially also E-CAD data already, the derivation of a Bill of Materials (BOM) from the CAD system under the assumption that the CAD structure corresponds to the Engineering BOM (E-BOM), an ERM/ECM limited to the development and construction with restricted possibilities to determine the validity of a component or product as well as the transfer of the master data and BOM to MRP. Some PDM systems had simple project management. The various handling methods for revisioning[8] the part master and BOM's were typical for the interaction between PDM and MRP. The MRP systems were already at least 20 years old and based on the state of the technical organization at the time, so that a 1:1 assignment existed between the part and the document (→ manufacturing drawing) and typically the part and document numbers were identical. Only the document could have a revision or version. The PDM systems had a more open data model, the relationship between parts and document could be n:m, and parts and documents could have their own number circles and both elements could be independently revisioned. The MRP system dominated at this time and so the modern configuration methods of the PDM system could not be utilized in connection with MRP. In the nineties, Product Data Management became an accepted software solution in development and construction for mechanical products and components. The management of the engineering data was necessary with the increasing usage of the three-dimensional CAD models as a fundamental element of product development, at the latest. Manual storage in self-defined directories on the hard drive no longer worked, even in small companies [16]. The complexity of three-dimensional modelling and the data models derived from this was clearly the drivers for the implementation of PDM.

2.3 Product Lifecycle Management (PLM)

Traditionally, more M-CAD-oriented PDM approaches failed in an increasingly competitive environment, in which successful companies developed innovative products more quickly, cheaply, and with excellent support of the decentralized, internal, and external company communication during all phases of product definition. Systems without distinct program and project management and the functional assistance of the cooperation of engineers for different organizational units and divisions of the company, and/or companies in the supplier/customer relationship could not provide this support. Distributed development, production, and maintenance/service became the standard of modern companies, as did integration and the exchange of information with customers. Support of the value chain (→ supply chain process) and of the overall product lifecycle was

[8] A part or document item is marked with an item number and a revision or version. These are international definitions.

demanded by the market. The PDM providers reacted almost overnight at the end of the 90s. A new term, product lifecycle management (PLM), was born. It was interesting that, initially, the providers and systems remained unchanged. When PDM became practically "state of the art," the appetite grew on the customer side to also use data from other areas of the value chain, and on the provider side, the appetite grew for new applications in other disciplines and other departments beyond engineering. Why should not marketing, customers, suppliers, process planning, and assembly already be able to work with the models if they were available via central data management? Initially, this only concerned the geometric data of mechanical construction, which still included a significant part of industrial innovation in the nineties. However, the 3D models were no longer sufficient to depict the behavior of the function of mechatronic products. Thus, PLM stood on the threshold between the isolated, subject-specific IT applications, which were not designed for multi-disciplinary collaboration. The necessity for the integration of mechatronics grew over the years. Thus, the integration of electronics, software, and embedded software was increasingly offered in PLM systems. Of course, PLM itself was not designed for this either and the users in the industry still struggle today with interdisciplinary collaboration. Many customers initially interpreted the renaming more as a marketing move. PDM was at least used extensively with the customers with over 1000 employees and the system providers simply wanted to have a bigger piece of the pie. One of the extremely optimistic definitions of PLM at this time was [11, 12]:

◆ **Product Lifecycle Management (PLM):** PLM is the company-wide information management of product and process-related data. It incorporates planning, management and organization along the product lifecycle and is required for the holistic management of all data, documents, resources, and processes throughout the entire product lifecycle (→ Single Source of Truth SSoT). All people who must solve certain tasks together independently of their location and organizational association work in the system together (→ Engineering Collaboration).

CIMdata [14] also estimated the range of application of PLM with similar optimism in 2005 (Fig. 2.4).

The PLM definitions reflect the visions of the analysts, the university research and the system providers. This optimistic PLM marketing statement in comparison to the actual application areas used in the company is represented in Fig. 2.5.

In contrast to PDM, PLM should provide consistent product and process relevant application functions along the entire product lifecycle and included customers and suppliers via the internet. However, in the beginning, the reality was characterized by more restricted coverage in the direction of process planning and production. The early conceptional and operative phases were omitted completely, the interdisciplinary inclusion of software and electronics, as well as the integration of the simulation both in the early development phase (→ Modelica, Simulink) and in the actual geometric-related

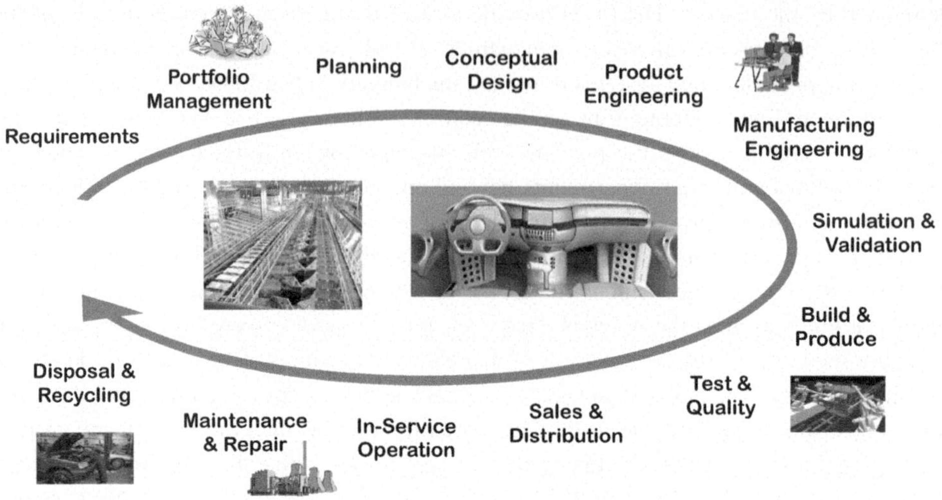

Fig. 2.4 The areas of application of PLM according to CIMdata (2005)

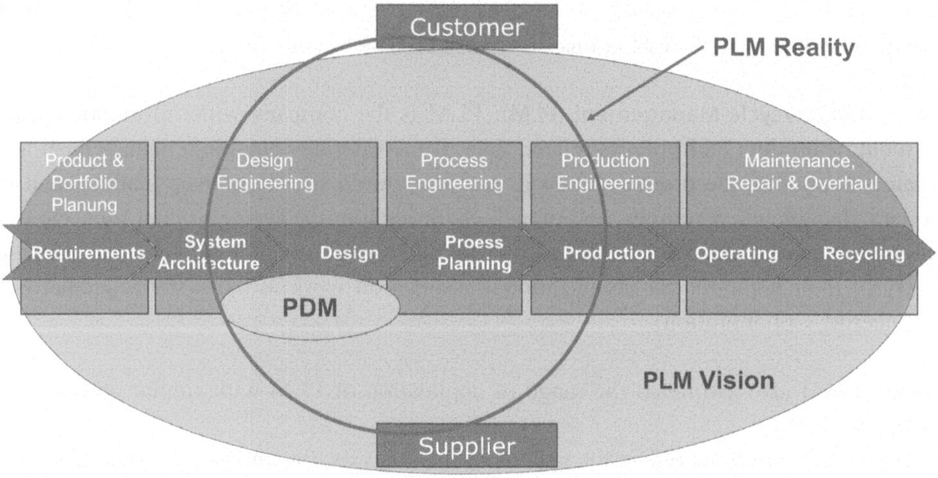

Fig. 2.5 PDM versus PLM—desire and reality

construction phase (FEA,[9] MBS,[10] CFD,[11] …) occurred extremely hesitantly. These applications – if they were used operationally at all – were more fragmented and independent.

[9]FEA Finite Element Analysis.

[10]MBS Multi-Body Systems.

[11]CFD Computational Fluid Dynamic.

Throughout the period until 2008,[12] the PLM systems naturally continued to develop in the direction of their original objectives, so that AMR's[13] definition of PLM from 1999 in the direction of a flexible, distributed application environment, which enabled the provision of the current technical product information along the entire phases of the product lifecycle, was also more and more realistic. A survey of customers by AMR revealed the following ranking of desires for the use of PLM systems [2]:

- Company-wide access to product data (85%),
- Product lifecycle management (45%),
- Configuration management (35%), and
- Financial savings (20%)

From this consideration, AMR derived six competencies for PLM [2]. The first two have since become extremely sophisticated, as they are both functions that form the core of PLM: Product data management (PDM) and engineering collaboration, i.e., the solutions which enable the various internal and external company partners to work together, even in distributed locations. The next two are material sourcing as interface to supply chain support and customer needs management (CNM) [13]. CNM is an aspiring discipline within PLM. It ensures that the customer need and the product requirements are communicated to everyone who is responsible for product development in the course of the lifecycle. CNM should prevent design errors and delays which result from misunderstandings and communication failures, while simultaneously improving productivity, efficiency, and quality. Another important topic inside CNM is the after sales and MRO[14] support. These four key functions are not yet complete. Process planning was offered by some providers over time as further PLM building blocks, started to take over functionality from MRP systems. Thus, PLM solutions departed from the rigid, administrative framework of PDM systems and increasingly offered management support [4]. The latter function is essential to remove PLM from the niche of the technical solution. These extensions—building on the PDM base functionality—are to be taken from Fig. 2.6. The service functions such as vault, archiving, I/O-management, viewing/redlining, workflow, and replications are largely identical with PDM systems.

A further function and service depiction more comparable with Fig. 2.2 is to be taken from Fig. 2.7. The largest common multiplicators of the standard market PLM systems were used as the basis when summing up the functions. The functions in parentheses

[12]2008, the end of the worldwide financial crisis was consciously accepted as a milestone of PDM/PLM/SysLM development.

[13]In 1986, the market research and consultancy company AMR was founded, sold to Gartner in 2009.

[14]MRO Maintenance Repair and Overhaul.

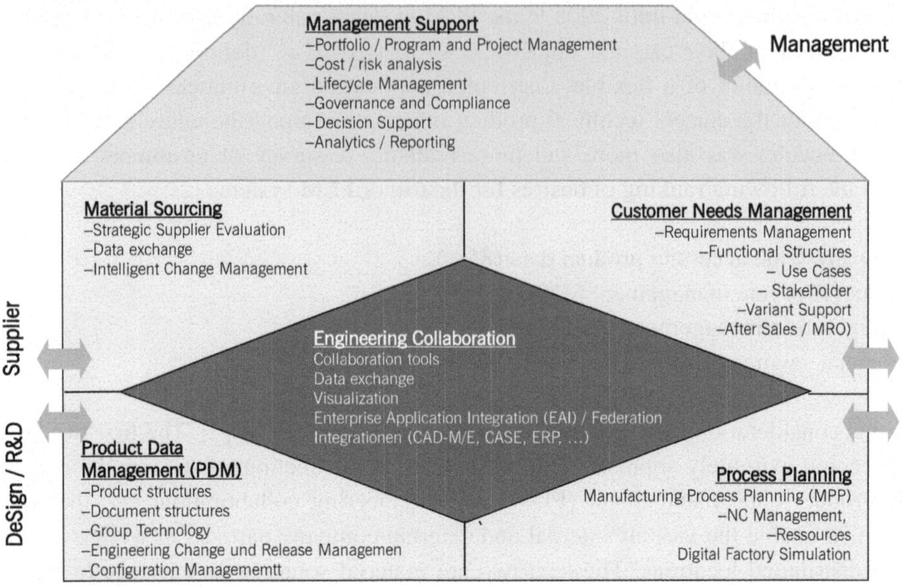

Fig. 2.6 Function groups and functions of PLM (according to AMR 1999 [2])

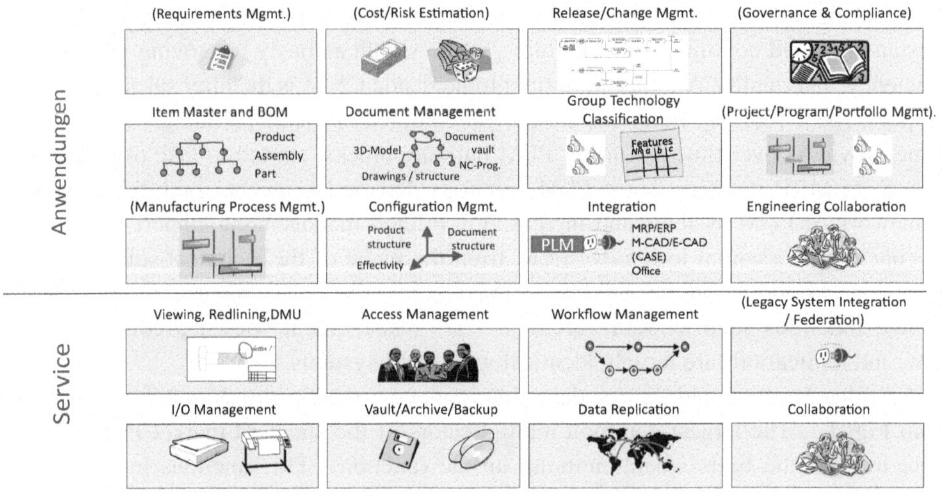

Fig. 2.7 Function scope of PLM systems in comparison to PDM

were somewhat less pronounced at the start of PLM. Federation includes the methodology, processes and IT tools to integrate autonomous applications systems in heterogenous IT application architectures. Methodically it should be based on linked data, technically on REST.[15] But in the 90s, the typical integration technology was based on Application Programming Interface (APIs).

Of course, we have also more or less reduced the initial deficit of PLM solutions over time. We are talking about a PLM maturing process period of over about ten years and we could still determine a difference from the original objective. This difference has four causes [5, 11, 13, 16, 19]:

- The lack of management and user readiness to implement this holistic, overall PLM concept. PLM spans several organizational company units and did not arrive at the C-level (→ CEO, CTO, CIO, CFO) due to the unrecognized strategic relevance. PLM does not have company-wide responsible person, rather it was often domiciled at department levels. Many PLM projects also came to a dead end as the dominance of the MRP fraction was stronger than the technical performance capability of the MRP system for engineering applications. Redundancies and overlapping or incomplete, inconsistent solutions in the approval/review and configuration management were the result.
- The functional incompleteness of some PLM systems in new application areas was evident. There were always providers in functional niches which had offered a better, albeit more difficult to implement, partial solution than the PLM providers could supply. This applied, for example, to applications such as requirements and project management, program and portfolio management, simulation management, process planning, and after sales functionality. Thus, the user had to continuously decide between "Best in Breed" and "Best in Integration."
- Many PLM providers promised the customer an OOTB[16] solution and with this a PLM implementation like a car sale. This statement was generally dubious and awakened completely the wrong expectations in the company. PLM does not concern an OOTB system, and also not an individual application, rather more of a solution strategy, which is partially integrated with several IT systems, whereby the PLM system must generally be heavily adapted (→ configured/customized[17]) as the integrative nucleus.
- As every PLM implementation generally required relatively extensive effort to configure and/or customize the system—as OOTB was not realistic—and also the systems did not further expand this due to out-dated technology (→ adaptation not interactive, no synchronization between PLM data model and user interface). To make matters worse, customizing had to readapt a new version of the PLM basis system for every delivery, which meant approx. 20–30% of the initial customizing.

[15]REST Representational state transfer is a de-facto standard for a software architecture for interactive applications that typically use multiple Web services. Wikipedia (retrieved 1.23.2021).

[16]OOTB Out of the Box.

[17]The term configuration is used in a more interactive adaptation of the system. Customizing is done via programmable interfaces (Application Programming Interface = API).

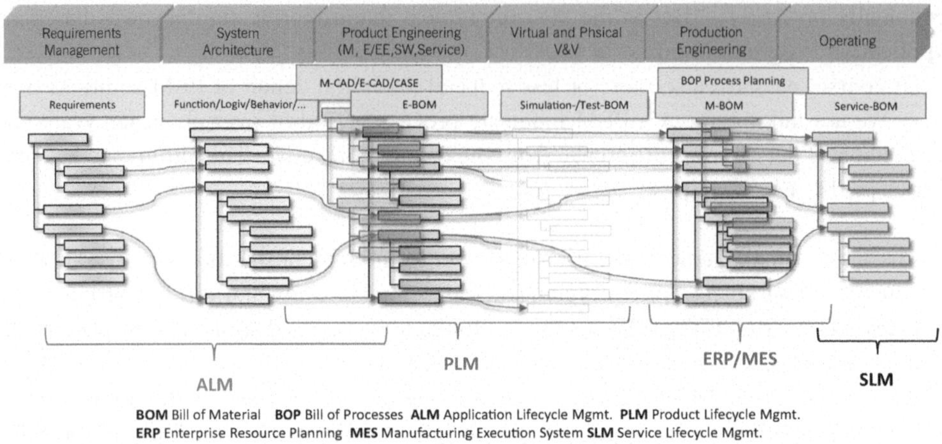

Fig. 2.8 Theoretic PLM data model as a connection of partial models along the product lifecycle and real fragmenting in various partial solutions

Theoretically, all PLM providers agreed what a data model suitable for the requirements mentioned should look like. Every PLM provider had slides such as Fig. 2.8 in their strategic slide set.

Initially, PLM providers reacted hesitantly to integrative objectives and requirements in practical implementation, which led to parallel and fragmented island solutions which balanced out the deficit of common market PLM systems, such as

- Application lifecycle management (ALM) for requirements management, for system architecture development of complex mechatronic systems and for software development (→ IBM Jazz, Integrity, Polarion, codeBeamer, …).
- Interdisciplinary system architecture on the basis of SysML.[18] This interdisciplinary, schematic product description was partly developed as a component of ALM and partly independently developed (→ Cameo, Enterprise Architect, Rhapsody). Open-source solutions were also on the market (→Modelio, Papyrus).
- Simulations PDM[19] (SimPDM) for orchestration and result documentation of simulations (→ Comet, PDTec, Simworx).
- Service lifecycle management (SLM) to support the after sales and MRO[20] processes (→ Impresa, Servigistics, …).

[18]SysML System Modeling Language.

[19]In the European literature, the term team data management (TDM) is used for administration systems close to the application system. In English literature, this is similar to the term PDM today.

[20]MRO Maintenance, Repair, and Overhaul.

Of course, MRP and MES[21] continued to play a significant role as the standard systems for production and logistics and the overlaps with PLM were pre-programmed. So, in reality, the lovely theoretical model from Fig. 2.8 looked more like an overlapping and competitive patchwork quilt. Many of these software solutions and others, for example, on the level of the authoring systems, were later bought by PLM providers and more or less half-heartedly integrated, instead of actually introducing an organic and networked overall data model along the product lifecycle.

Parallel to the continued development of the PLM systems, the provider market consolidated from 2000 to 2008 and the market-relevant international PLM providers to date are:

- ARAS/Innovator (founded in 2001 by former employees of EIGNER Inc. in the USA),
- Dassault /Enovia (acquisition, among other things, of SmarTeam and MatrixOne),
- PTC/Windchill (acquisition, among other things, of Computervision, CoCreate),
- SAP/SAP PLM and
- SIEMENS PLM/Teamcenter Unified Architecture (acquisition, among other things, of SDRC and EDS).

An addition to this, there were country specific providers which largely operated nationally (→ Arena,[22] Contact, ProCAD, Propel, ….)

During the financial crisis of 2007/2008, the PLM market collapsed but then regrew at a rate of 10–15% from 2010 to 2012. However, 2013, the growth rate of PLM investments halved to 4.8%. Was this the result of market saturation, or of new technologies and business models, queried engineering.com in 2014[23]? The analysts saw a combination of all three reasons as being responsible for this development. It was clear that PLM prospered for many years; however, it changed little functionally and when, then only through incompletely integrated acquisitions. Parallel to this, a market saturation resulted, as companies with more than 1000 employees had generally invested in PLM and, due to the economic crisis, had more licenses than users and asked themselves what was the benefit and TCO[24] of a PLM system.

[21]MES Manufacturing Execution System.

[22]December 14, 2020 acquired by PTC.

[23]https://www.engineering.com/PLMERP/ArticleID/8362/PLM-Spending-A-period-of-Digestion-after-two-years-of-explosive-growth.aspx (retrieved 23.10.2020).

[24]TCO Total Cost of Ownership.

2.4 On the Path to System Lifecycle Management

In recent times in particular, customer requirements have changed drastically due to new framework conditions, and the demand for further development becomes stronger. The background to this was the ever stronger influence of IOT/IOS, system thinking, MBSE,[25] engineering collaboration, interdisciplinarity, and artificial intelligence (AI). Moreover, the software technology had permanently developed further. In this sense, national and international ICT[26] research projects were carried out as early as 2004 (e.g., the EU-financed PROMISE project 2004–2008). Here, future PLM solutions were conceived and developed far beyond the standard market PLM systems. Sensor data from the operational phase was integrated in PLM and made it possible to make this information available to various employees across the overall lifecycle, including the operation of an individual product. The term closed-loop lifecycle management (CL2M) in the scope of service-oriented business models was introduced for this closing of the information loop [20]. Further industrial research projects were concerned with model-based system engineering, digital models, digital twin, and digital thread (see Sect. 3.4) as well as overall with the digitalization of the products (see. Sect. 4.1) and the engineering processes (see Sect. 4.2). The term Engineering 4.0 was established as the overall term for all these topics. However, the requirements were not only driven by the academic side, they were taken on in the strategic planning of large companies as early as 2015. Figure 2.9 shows that Engineering 4.0 is relevant for the majority of VW strategic company goals (→ marked yellow).

This was also the period of the BMBF association project mecPro^2 (Fig. 2.10). The aim of the project was to enable companies in automotive construction to efficiently develop cybertronic products and production systems through suitable processes, methods and tools. In mecPro^2, a construction methodology and detailed description schematic (→ Profile) were developed for this (see Sect. 3.6). The interdisciplinary design process recorded in the design methodology describes in detail who uses and generates what information and in which sequence. In contrast, the description systematic defines this information, its connections, and its representation [9].

During the development of the integrated model of mecPro^2, the knowledge grew that more was needed for a model-based development process than a modeling tool on the basis of the language SysML.[27] Rather, all phases of the product lifecycle and specialist disciplines must be consistently transformed in the interdisciplinary, model-based development world. During the project run time, on one hand the acceptance of model-based product and system development has continuously improved, and on the other,

[25]MBSE Model-Based System Engineering.

[26]ICT Information and Communication Technology.

[27]SysML System Modeling Language (Do not confuse it with SysLM (System Lifecycle Management)).

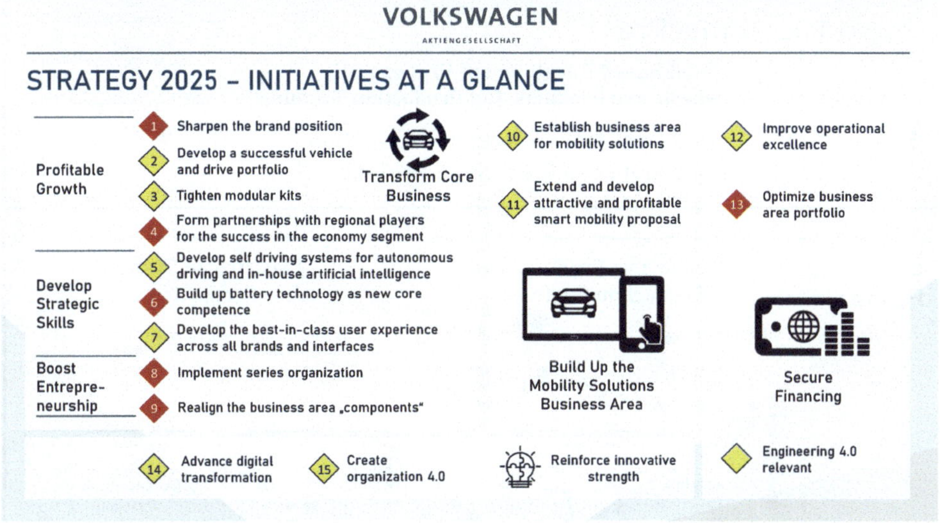

Fig. 2.9 VW Business Strategy for 2025. (VW Homepage/Investor Relationship 2016)

increasing requirements result for the engineering processes based on the increase of the information created and to be managed along the entire product lifecycle, for example, in order to guarantee traceability in the scope of change and configuration management according to EN ISO 10007 (→ traceability) as well as to build up valid quality management along the entire product lifecycle according to EN ISO 9001:2015 or the EFQM[28] model. As strategic component of these changes in the product and process world, *System Lifecycle Management (SysLM)* forms the skeleton for an interdisciplinary, consistently digitalized product lifecycle (PLC), which represents the basis for a digital product, production, service, and process model across distributed locations and along the entire supply chain. Digitalization means a transformation process which rearranges the classic boundaries of a fragmented and competitive IT solution world. An engineering backbone will take over the role of data and process integration between the product lifecycle phases requirements definition, system architecture, product and system development, simulation, production planning, production and service [9]. Current PLM architectures and introduction strategies are already extremely complex and the further increase to be expected from cybertronic products and systems will strengthen this trend. Traceability across the entire product lifecycle, the disciplines, and the supply chain must be guaranteed at all times in order to master and document the engineering process. The term System Lifecycle Management was introduced in the project as an extension of

[28]EFQM European Foundation for Quality Management.

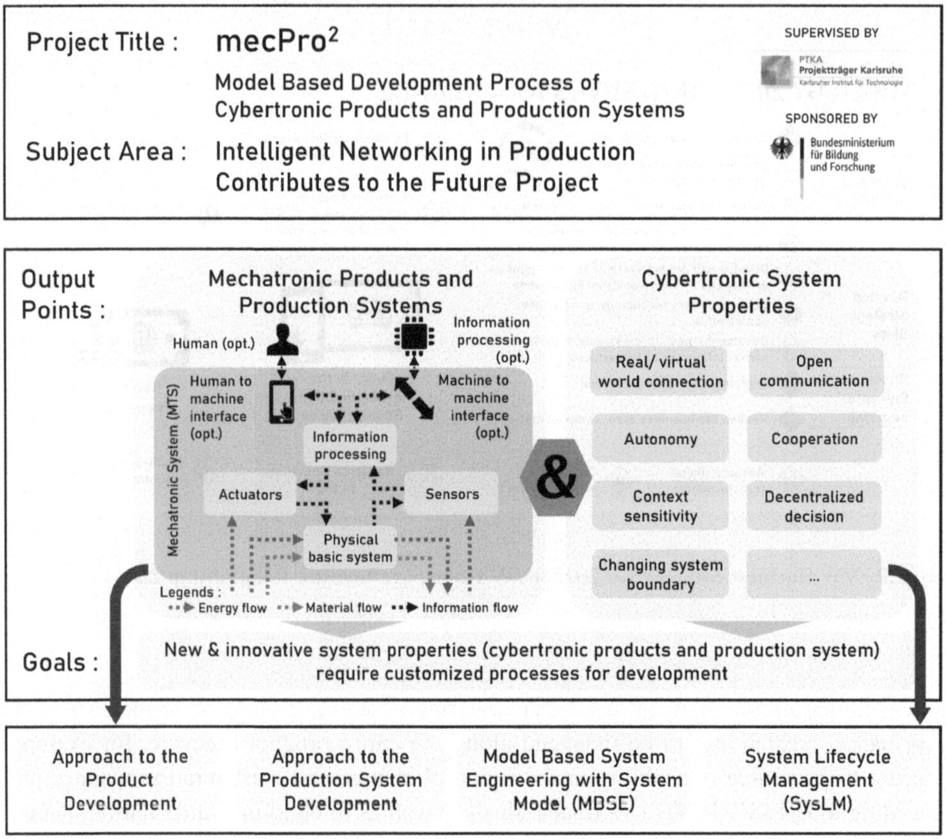

Fig. 2.10 Overview of the industrial research project mecPro² [9]

PLM as the next, consistent evolutionary stage. With this, it should be expressed in the project that SysLM [7, 9, 17]:

- Efficiently supports the development of cybertronic products, production systems and service-oriented business models through interdisciplinary system modeling in the early engineering phase and a high level of support of various discipline-orientated processes in the later phases (see Sect. 4.1),
- Enables the consistent application of the methods and processes of Model-Based Systems Engineering (MBSE) through the creation and further processing of integrating partial digital models connected to one another along the PLC [3],
- Should make vertical integration as redundancy-free as possible through an intelligent function distribution between the authoring system, TDM system and backbone levels (see Sect. 4.2.1). The majority of legacy systems on the authoring system and TDM levels are integrated via modern, WEB-based link technology (→ REST, RDF, linked data, …),

- Integrates (→ horizontal integration) and therefore forms the prerequisite for transparency and agility through the intelligent networking of the engineering process of mechatronic and cybertronic products, production systems and services from the requirements process to system architecture, the detailed development, the simulation and production planning, down to the operation of the respective partial models (see Sect. 4.2.2),
- Guarantees digital traceability through the digital thread, the connection of all information elements throughout the entire PLC (see Sect. 3.4.2),
- Through new, internet-based technologies and architecture (→ WEB services, microservices, linked data, cloud offers), the long valid basis technology of the monolithic program systems can be overcome,
- The draft models are validated and verified through virtual simulation and physical tests in certain PLC phases,
- The gap between engineering and production is reduced through useful movement of MRP functions to PLM (→ M-BOM, MPP[29]) and MES,
- The introduction, administration and synchronization of a Digital Twin lays the foundation for the implementation of service-orientated business models (see Sect. 3.4.1).

The following definitions resulted from the listed requirements and influential factors and from the actual results of the research project [6, 7, 10, 17]:

◆ **System Lifecycle Management (SysLM):** SysLM is interdisciplinary and holistic information and process management which extends product lifecycle management (PLM) with the early phases of system modeling, taking into consideration and integrating all disciplines, including services, in order to support production planning and thereby form bridges to production, as well as extend the operational phase of product operation. The concept is based on the intelligent direct and indirect integration (→ using TDM) of various authoring systems along the entire product lifecycle of a product, production system or the development of a service-orientated business model. As the technical-organizational backbone, System Lifecycle Management solutions are responsible for the horizontal integration of the partial models resulting in the phases of the product lifecycle and for the provision of company-wide engineering processes (→ ECR, ECM, CM). These can be supported by AI. This guarantees clear traceability (→ digital thread, digital twin). In summary, SysLM is a foundation for model-based systems engineering, for the digitalization of the engineering process (→ Engineering 4.0) and for the support of service-orientated business models on the basis of the digital twin.

---

[29]MPP (Manufacturing Process Plan).

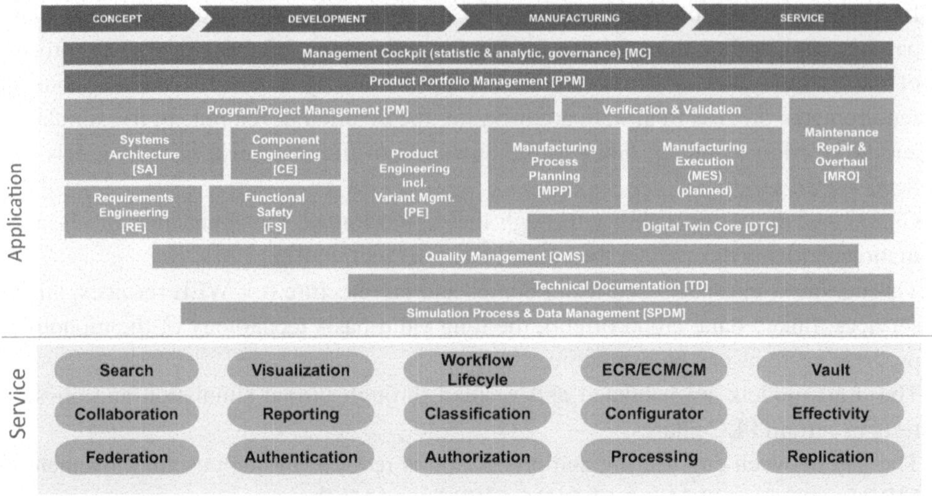

Fig. 2.11 Function overview SysLM. *source* according to Aras

Figure 2.11 shows the typical functions of an SysLM system, sub-divided into application, process support and service. Most of these functions are presented in Sect. 4.2.2.2 in connection with the horizontal integration of the engineering process.

The engineering processes depicted through an SysLM system are based on numerous authoring systems across the various phases of the PLC. These processes must incorporate the various disciplines, as well as the different locations and suppliers and must be connected through a suitable IT architecture in a joint product and process backbone. Five IT levels indicate these concepts, which were determined as an extension of a four-level architecture developed from a VDA[30] work group:

- Authoring systems,
- TDM systems,
- SysLM as an engineering backbone,
- MRP and
- IOT/IOS Enterprise Service Platform.

Details of this multi-level IT architecture are explained in Sect. 3.6, Fig. 3.30. The tasks and function distribution varies between the levels. On the one hand, authoring systems are becoming more intelligent and taking on the functions of the database-oriented management of documents, versioning and cooperation between teams (→ OnShape); on the other hand, the TDM systems would also like to take a bigger slice of the SysLM cake. New business models for producing companies or service providers, for example,

[30]VDA German Automotive Association (Verein der Deutschen Automobilindustrie).

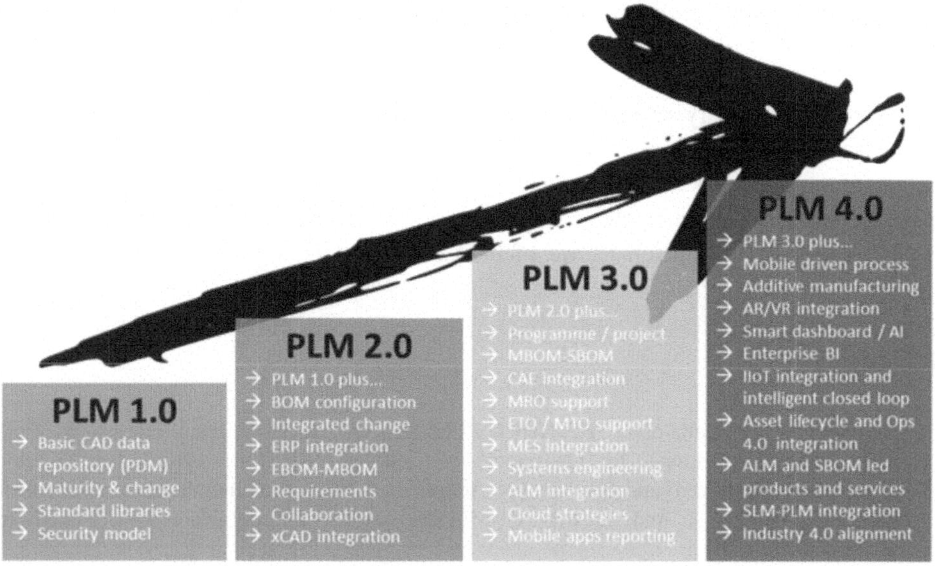

Fig. 2.12 PLM 1.0 to 4.0. *source* Lionel Grealou (https://acqnotes.com/acqnote/careerfields/total-life-cycle-systems-management-tlcsm, (retrieved 11.10.2020)).

predictive maintenance or update over the air, require the embedding of operations in order to permanently report back, evaluate operating data, and visualize to the IOT/IOS Enterprise Service Platform (→ digital and physical twin, big data analytics).

It is interesting that various terms and definitions have developed in parallel. An SOA-based, mechatronic integration platform (mIP) was already presented in 2010 and supports the development of mechatronic systems here[1]. Approaches such as PLM 2.0 (Beyond PLM, Oleg Shilokovsky 2009),[31] "bimodal PLM / IT" (Gartner 2016)[32] "Digital PLM" (Accenture 2016)[33]), Total Lifecycle Systems Management (TLCSM)[34] (2017), Lionel Grealou[35] (PLM 1.0 to 4.0 Fig. 2.12) and integrated lifecycle management (ILM) [18] indicate the necessity for the next evolutionary stage in a similar way to SysML.

[31]http://beyondplm.com/2009/02/26/plm-20-technology-or-facelift/, (retrieved 9.10.2020).

[32]Gartner (Marc Halpern) counts Aras, Propel and OnShape among the bi-modal systems (PDT Europe 2016).

[33]https://www.accenture.com/ro-en/insight-highlights-auto-industrial-using-digital-plm, (retrieved 9.10.2020).

[34]http://acqnotes.com/acqnote/careerfields/total-life-cycle-systems-management-tlcsm, (retrieved 10.10.2020).

[35]http://virtual-digital.com/towards-plm-4-0-hyperconnected-asset-performance-management-Framework, (retrieved 10.10.2020).

As a reader, you're probably now thinking: "Aha, just like PLM, lots of visions again but where are the solutions?" In the year 2020, you are certainly right, however, vision has always been the motivation for engineers to also realize this. We are only at the start of a further evolutionary stage, which will be more disruptive than all the previous ones. Through the strong connection of SysLM with new business models and the operational digitalization strategy, SysLM will not only allow itself to be reduced to the purely technical levels. The participating people, organizations, business, and engineering processes, as well as the value chain, are affected as part of this implementation.

References

1. Abramovici, M., Bellalouna, F.: Neues PLM-Konzept für eine integrierte Entwicklung mechatronischer Systeme. In: Gausemeier, J. (ed.), Entwurf mechatronischer Systeme. Grundlagen, Methoden und Werkzeuge; Adaption, Selbstoptimierung und Verlässlichkeit; Integration Mechanik und Elektronik, Miniaturisierung. Heinz-Nixdorf-Inst, Paderborn. pp. 296–306 (2010)
2. AMR Research Study PLM. Report of AMR Research Inc., Boston (1999)
3. Anderl, R., Eigner, M., Sendler, U., Stark, R.: Smart Engineering—Interdisziplinäre Produktentstehung, acatech DISKUSSION, 1st edn. Springer Vieweg Verlag, Berlin (2012)
4. Bitzer, M., Eigner, M., Langlotz, M., Kasai, C.: Product planning and decision representation and reuse in next generation product lifecycle management solutions. Proceedings PLM08—The 5th International Conference on Product Lifecycle Management, Seoul (2008)
5. Day, M.: What is PLM? CAD Digest. 04.01.2002
6. Eigner, M., Apostolov, H., Dickopf, T., Schäfer, P., Faißt, K.-G.: System Lifecycle Management—am Beispiel einer nachhaltigen Produktentwicklung nach Methoden des Model-Based Systems Engineering. ZWF Zeitschrift für wirtschaftlichen Fabrikbetrieb 109(11), 853–860 (2014)
7. Eigner, M., Faißt, K.-G., Apostolov, H., Schäfer, P.: Kurzer Begriff und Nutzen des System Lifecycle Management—im Kontext von Industrial Internet mit Industrie 4.0 und Internet der Dinge und Dienste. ZWF Zeitschrift für wirtschaftlichen Fabrikbetrieb 110(7–8), 475–478 (2015)
8. Eigner, M., Hiller, C., Schindewolf, S., Schmich, M.: Engineering Database – Strategische Komponente in CIM-Konzepten. Carl Hanser Verlag, Vienna (1991)
9. Eigner, M., Koch, W., Muggeo, C.: Modellbasierter Entwicklungsprozess cybertronischer Systeme—Der PLM-unterstützte Referenzentwicklungsprozess für Produkte und Produktionssysteme. Springer Vieweg, Berlin (2017)
10. Eigner, M., Muggeo, C., Apostolov, H., Schäfer, P.: Kern des System Lifecycle Management—im Kontext von Industrial Internet mit Industrie 4.0 und Internet der Dinge und Dienste. ZWF Zeitschrift für wirtschaftlichen Fabrikbetrieb 111(1–2), 63–68 (2016)
11. Eigner, M., Stelzer, R.: Product lifecycle management—Ein Leitfaden für Product Development und Life Cycle Management, 2nd ed. Springer Vieweg Verlag, Berlin (2009)
12. Grieves, M.: Product Lifecycle Management: Driving the Next Generation of Lean Thinking. McGraw-Hill, New York (2006)
13. Lawrence, S.G.: Additional ABCs About PLM, In automotive design and production, 12.01 (2005)
14. Litjens, G.: What is PLM? PDT Europe 2005, Amsterdam, September 26 (2005)

15. Schuh, G.: Product lifecycle management. In: Gronau, N. (ed.) Enzyklopädie der Wirtschaftsinformatik. Online Ressource, München. http://www.enzyklopaedie-der-wirtschaftsinformatik.de/ (2015). Accessed: 4. Sept. 2020
16. Sendler, U.: PLM und die Zukunft, in PLMportal U. Sendler. https://www.plmportal.org/de/plm-und-die-zukunft.html (2012). Accessed: 4. Oct. 2020
17. Sendler, U.: Ganzheitliche strategie: systems lifecycle management, in PLMportal. https://www.plmportal.org/de/forschung-detail/ganzheitliche-strategie-systems-lifecycle-management-syslm.html (2013). Accessed: 4. Sept. 2020
18. Sherburne, D.G.: Integrated lifecycle management—creating flow for digital transformation. In CIO review PLM special, pp. 25–26, 13.4. https://magazine.cioreview.com/magazines/April2017/PLM/ (2017)
19. Teresko, J.: The PLM Revolution, in Industry Week, 01.02.2004
20. Yoo, M-Y., Grozel, C., Kiritsis, D.: Closed-loop lifecycle management of service and product in the internet of things: semantic framework for knowledge integration, in sensors, 16(7), 7.8. https://www.mdpi.com/1424-8220/16/7/1053/htm (2016). Accessed: 4. Sept. 2016

Engineering 4.0—Foundations of the Digitalization of Engineering

3

When the wind of change blows, some people build walls, others build windmills

German proverb

Abstract

The Internet of things and services (IOT/IOS) as well as Industrial Internet and Industry 4.0 assume networked products, systems, and service in the future. The value proportion of electronics and software will continually increase with these kinds of products and embedded services. When products communicate with one another over the Internet, we refer Cyberphysical Systems or Cybertronic Systems. The development of these new systems will bring several consequences: interdisciplinary, regional, and organizationally distributed and integrated product development, a rethinking of current construction methods, processes, IT solutions, and organizational forms as well as the demand for consistent process chains based on digital models in the requirement definition, system architecture, product development, simulation, product planning, production, and service. Furthermore, planning and design procedures of all disciplines—mechanical, electronic, and software—must be put to the test and their suitability for a new process model for product, system, and service development checked in order to transit them to a common, integrated, and interdisciplinary method, process, and IT solution approach. This approach to the digitalization of product development is called Engineering 4.0. The methodologies of systems engineering (SE), model-based systems engineering (MBSE) and systems thinking form the foundations. Digitalization of products and product development means a transformation process which rearranges the classic limits of a fragmented and competitive IT solution world, a departure from silo thinking to a consistent, integrational solution approach for engineering. A lightweight and federated engineering backbone

M. Eigner, *System Lifecycle Management*, https://doi.org/10.1007/978-3-658-33874-9_3

(→ System Lifecycle Management, SysLM) will take on the role of data and process integration for the entire product lifecycle, including operations. This chapter will present the foundations, framework conditions, and drivers of digitalization and derive an adjusted construction methodology from the results of this for the development of cybertronic products and systems.

3.1 Digitalization—Buzzword or Challenge?

Digitalization, is that just another buzzword? Older readers may even still remember the term CIM[1] from the 80s [93]? "CIM describes the integrated EDP use in all operational areas associated with production. It incorporates the technical information interaction between CAD,[2] CAP,[3] CAM,[4] CAQ,[5] and MRP.[6] Here, the integration of the technical and organizational functions should be achieved for product development, process planning, and production. This is contingent on the common use of all data of an EDP[7] system, also known as a data basis."[8] Comparable to PDM, with CIM an extremely tight interpretation of the integration breadth is also selected concurrently. Some may still remember that the term disappeared as quickly from public discussion as it appeared. This vision hit limits for the most varied reasons, so that many experts also considered it to have failed. Was it not yet time, was management overwhelmed, or were the correct IT tools not yet available? Nonetheless, with a more precise interpretation of the definition, this already concerned the partial digitalization of the process chain process development to production planning and production. After the brief CIM euphoria and the lean wave, with Industry 4.0, Industrial Internet, IOT, and IOS the digitalization and automation of company processes at a new evolutionary stage regain a strong importance. Significantly improved methodological approaches and efficient IT tools 40 years later suddenly push the vision of computer-integrated manufacturing to within reaching distance.[9] However, the dimension of digitalization is dissimilarly larger and more disruptive. It influences every branch, the entire value creation chain from the requirements to various development phases, simulation, production planning and production, operations, and right down

---

[1]CIM Computer integrated manufacturing.

[2]CAD Computer-aided design.

[3]CAP Computer-aided (process) planning.

[4]CAM Computer-aided manufacturing.

[5]CAQ Computer-aided quality management.

[6]MRP Material resource planning.

[7]EDP Electronic data processing.

[8]Definition according to AWF (Ausschuss für wirtschaftliche Fertigung, 1985).

[9]https://www.it-matchmaker.com/news/digitalisierte-auftragsabwicklung-gibt-es-ein-cim-4-0/, Interview Dr. Wiendahl on 08.28.2019 (retrieved on 04.12.2020).

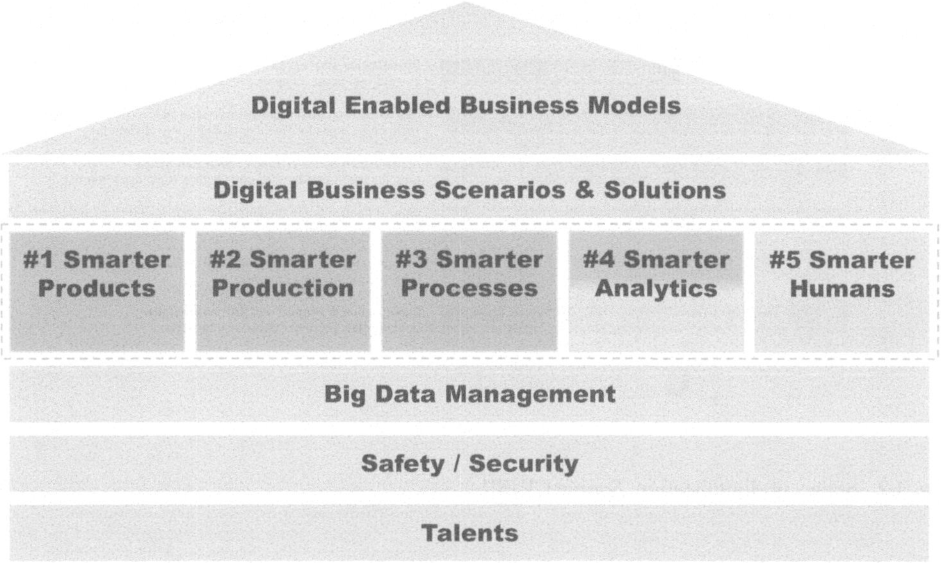

Fig. 3.1 Digital agenda of an automobile supplier. *Source* Schaeffler

to recycling. It relates to innovative products, systems, services, production as well as all business processes. Customers, service providers, and suppliers are to be integrated into these processes. Figure 3.1 shows the uppermost level of a digital agenda of a German company in the automotive supply industry. The yellow marked #1-4 show the strong importance of engineering in the implementation of a digitalization strategy.

Digitalization has become an innovation driver in many branches and market segments. With this technological change, the character of technical, economic, and social processes has changed dramatically [13]. The demand for innovative products and increasingly distributed, interdisciplinary product development, global procurement, and manufacturing put companies under additional pressure and led to more complex product development and the accompanying services. For this reason, current product innovation research concentrates on the design of intelligent cybertronic products and services as well as the digitalization of underlying processes which will change all areas of a company [7, 23, 36, 73].

Many companies and their employees primarily connect digitalization as being "out of their comfort zone," as almost everything centers around innovation, speed, and change [61]. So, it is not surprising that, according to a BITKOM study, 55% of all mid-sized companies in Germany consider digital transformation as a central challenge and opportunity for multiple and consistent optimizations [62]. These are numerous and include increasing turnover or productivity, innovations in value creation, optimized

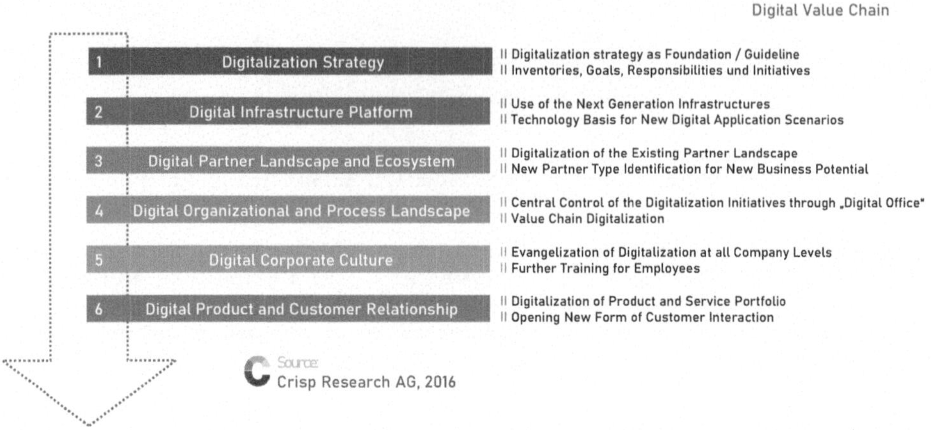

Fig. 3.2 Stages of digitalization. *Source* CRISP

processes, improvement in work systems and skills, and new forms of interaction with customers [97]. Every company must define a new digitalization strategy in advance [28] with the background that digitalization can redesign or replace entire business processes and models (Fig. 3.2) [61].

There are currently supporting methods for the evaluation of the maturity of digitalization measures and their potential for company strategies in development and discussion [80, 104] (Sect. 4.4).

Data as raw materials of a digital world, people as implementers and users of the digital transformation, and the necessary change to the organizational structure due to digitalization are also part of a digital agenda and strategy [83, 84].

Data – The New Resource

In the past, data was mainly generated by internal and external company processes and transactions along the entire value chain: order processing, interactions with suppliers, sales interactions, and customer service visits. Today, intelligent, networked products can generate real-time measurement values which are unprecedented in their diversity and volume. Figure 3.3 shows the unstructured data streams which are provided today in the scope of a service-oriented business model of a product or system for analysis and evaluation.

Data is on the path to becoming the key capital of a company and perhaps the decisive asset [84]. The metaphor "data is the oil of the future" is connected amazingly often with digitalization, particularly in operations and the volume of data occurring from these. Oil not only is the energy source, but also plays a decisive role as a pre-product and is indispensable in many industries. This is precisely the importance that is given to data today: Those who transform intelligent information ideally through smart

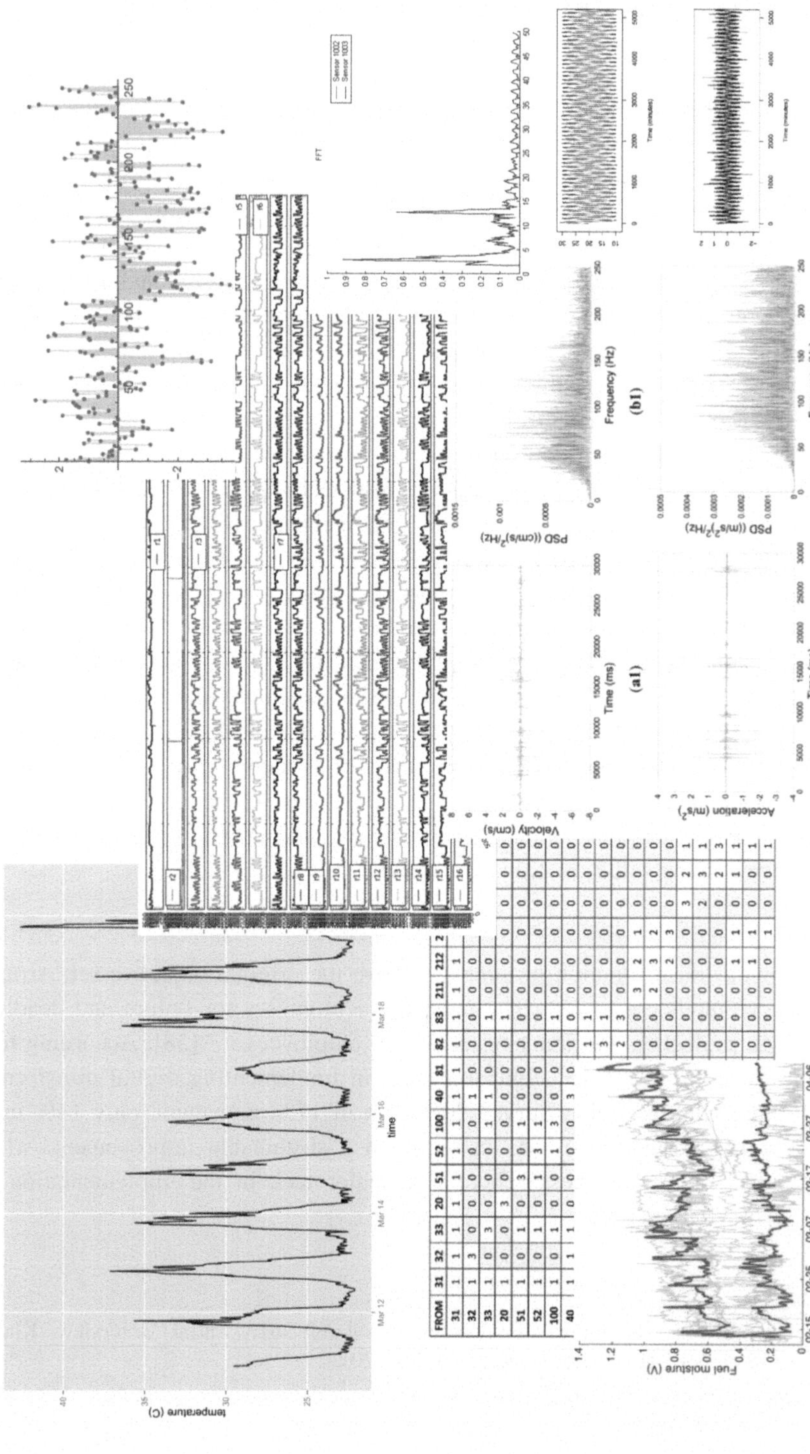

Fig. 3.3 Data streams from products and systems for analysis and evaluation. *Source* Aras

data analysis and evaluation, build new service-orientated business models on this, and thereby upgrade their products through increased value creation will be ahead of the competition.[10] However, only a few companies have been thinking about the economic benefit of data up to now. On the one hand, they can increase their yields through the monetization of their data and, on the other, strengthen their position against competitors in an increasingly competitive market. Data amounts generated from the product and the environment can contribute significantly to value creation. However, only if the company knows how, it can unlock these opportunities. Data collection here is only the first step. The actual value is only created if this information is sensibly analyzed and new recognitions are gained on the basis of which well-founded decisions can be made. This process is called "data mining" or "big data analytics." The latter term, using data models and hypotheses for verification or falsification, has been known in research for many years. With the emergence of IOT and IOS and the possibility that products can report their status from operations, the term data analytics has spread widely for the sensible and selective evaluation of mass data.[11] Big analytics uses techniques to understand this pattern. This challenge lies in that the data from intelligent, networked products is often unstructured. This means there may be many values and formats, for example, GPS coordinates, temperatures, pressure values, vibrations, and many further conceivable sensible measurement values determined and transmitted from sensors from the product or the environment. Standard approaches to data aggregation and analysis, such as table calculations and database tables, are not suitable for the management of many data formats. The new solution is non-relational databases (→ time series databases) and a repository (→ data lake), in which the mass data can be saved in its native format. From here, the data can be investigated with a series of data analysis tools. These tools can be divided into four categories: descriptive, diagnostic, predictive, and normative. Figure 3.4 [84] shows an overview of the value creation from data.

The Human

"Just 5% of digital transformation programs achieve the original objective set. Around 70% only achieve mediocre results. The most frequent causes for failure and deadline delays in change projects is the inner resistance of employees..." [58]. According to a current Forbes headline, "84% of companies fail in implementing digital transformation." And in a McKinsey Global Survey 2018, out of 1733 managers, only 14% indicated that their transformation efforts had created a sustainable improvement, while just 3% mentioned a total success. There is a major lack in the implementation of

[10]https://entwickler.de/leseproben/daten-sind-das-oel-der-zukunft-261377.html, Interview Klaus-Dieter Schulze TDWI, NTT DATA (retrieved on 04.20.2020).

[11]https://www.datenbanken-verstehen.de/business-intelligence/data-analytics-grundlagen/data-analytics/ (retrieved 04.20.2020).

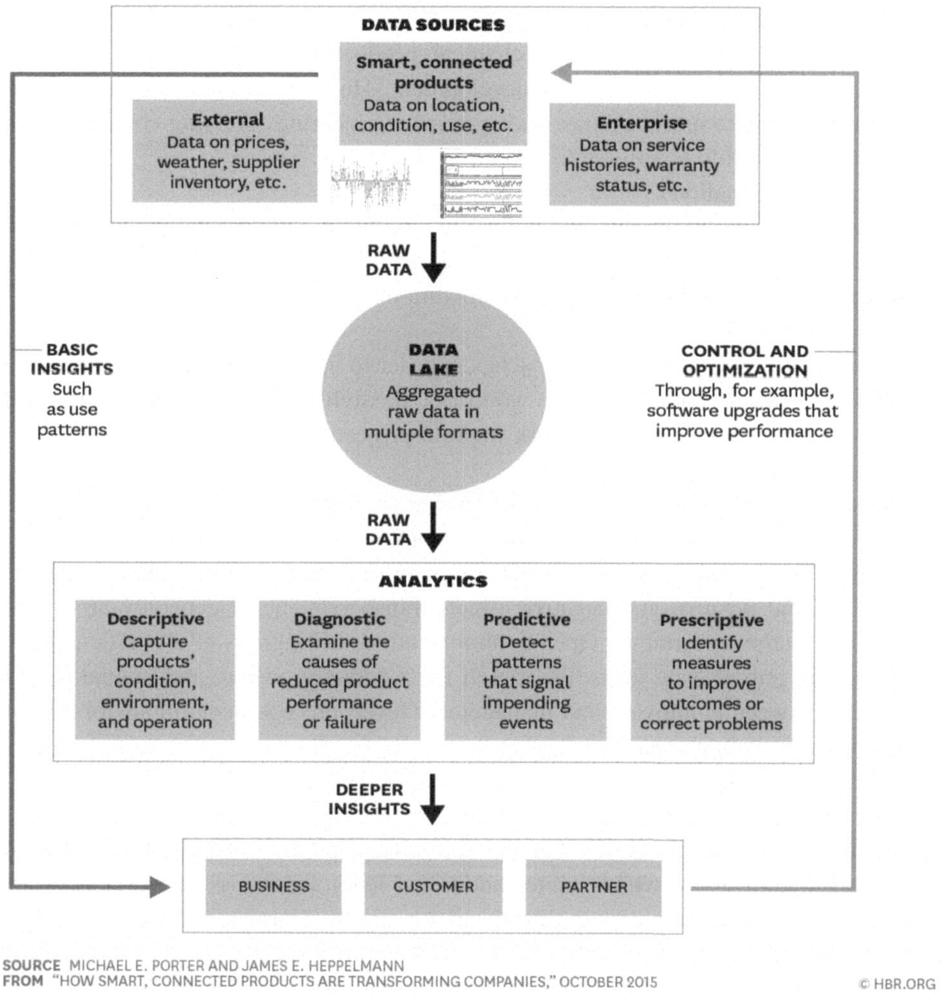

Fig. 3.4 Value creation from data streams [84]

digitalization projects [98]. Only few employees are optimistically embracing these disruptive changes in working conditions, both technically and organizationally. Many behave passively and the rest show inner resistance against any changes. A change is not to be decreed from above. This is why, in the event of a continually introduced disruptive change in a company (→ at departmental, technical, organizational, or social levels), ergonomists demand an accompanying change process by management which ensures that the acceptance and implementation of the project are guaranteed. Options to achieve a positive influence on a digitalization project and thereby "take employees along" are:

- Training and further educational measures
  Particularly in companies with mechanical product backgrounds, new skills are required in electronics and software development. This includes new interdisciplinary skills, for example, mechatronics, and systems engineering. Training courses as system engineers should be offered to interested employees.
- Cultural transformation
  The switch from a traditional, hierarchically structured company to a flat, agile, culture, characterized more by software industry influences, should be seen as an opportunity for a democratization of the work conditions.
- New compensation models
  The companies must find new approaches in order to find, convince, and permanently motivate talent. Advantages such as workplace flexibility, low hierarchies, agile work technology, and leisure time to work on side projects for personal interests are the standard in high-tech companies [84].
- Transparency and communication of the change process
  Digital transformation is only successful if employees' fears and emotions are taken seriously. This is only successful if the company strategy for digitalization makes the objectives and positive and negative effects transparent. Fears generally arise from ignorance and/or uncertainty. Open communication, willingness of all participants to address and rethink new ideas, the search for a common vision, and mutual respect and trust between employees and management levels are good conditions for designing a positive change process.

Managers of such change processes should also be aware of the very controversial sociopolitical discussion being held about the advantages and disadvantages of automation and digitalization, which incites additional fears about loss of workplaces and lower work content. Forerunners in this discussion are the German philosopher Richard David Precht and the Israeli historian Yuval Noah Harari. Precht's thesis in summary is: Industry 4.0 and digital transformation are destroying millions of workplaces in industrialized countries. The Fourth Industrial Revolution affects not only low-wage earners, but also better trained employees such as bankers, insurance employees, and lawyers. New positions are being created, but these are primarily for the few, top-trained IT specialists. "To believe that employment will remain constant or even increase is negligent to insane" [85]. Through the progress in the development of artificial intelligence and digitalization, there is the threat of the creation of a new social "class of useless people," warns Harari [59]. People will lack not only work, but also a sense of purpose. The key theses of Precht and Harari are completely identical in their effects: We are steering toward mass unemployment which could be partially compensated by technical progress, for example, through a basic income, but will take the people's work content and thereby make them "useless."

A commentary from the New York Times on 11.26.2018 is also interesting here [19]: "The redundancies at General Motors are not only incremental, but existential in this sense: It's about accelerating personnel changes which are dictated by the aggressive transition of the company from analog to digital products and from petrol to electricity. Thus, the recent redundancies (and the associated future recruitments) are probably an indication of much more disruptive changes in the automobile sector, that's for sure, but also in companies in the overall economy."

All analysts agree that digitalization and increasing automation – as well as increasing changes to drives (→ electric and hydrogen) – will drastically change the employment market. In 2002, 56 percent of workplaces investigated only required minimal digital skills. Thirty-nine percent of workplaces required medium digital skills and only 5% a high degree of digital skills. A lot has changed. By 2016, the proportion of workplaces which required high digital skills had grown to 23%. The proportion of medium digital skills rose to 48%. And in a major shift, the proportion of workplaces which required low digital skills sank from 56 to 30%[12] [75, 76]. It would be expedient if our education system would foot the bill and evaluate engineering science subjects according to not only scientific depth but also scientific breadth and not run years behind the education requirements of the future. Certainly, engineers with a deep, discipline-orientated training in mechanics, electrics/electronics, and IT will continue to be needed. However, in the future, around 20–30% of engineers should receive a broad, interdisciplinary education and training (→ mechatronics and systems engineering) [98].

Organizational Structure and Culture

Digitalization is a paradigm switch for companies and does not only lead to a rethinking in processes and structures, but also mean a change in organizational development. In a pre-study on change management in a digitalization project, Capgemini defined the attributes of organizational culture for companies which are in this digital transformation process. These attributes can be subdivided into eight dimensions [25]:

- Digital management,
- Digital technologies and processes,
- Autonomous work conditions,
- Agility,
- Entrepreneurship,
- Cooperation,
- Innovation and learning, and
- Customer orientation.

[12] The percentage rates are rounded off and therefore to not equal 100%.

What will the new organizational structure look like? A few important changes will emerge, for example, an ever-deeper cooperation and integration between IT and all organizational units. Furthermore, companies are beginning to form three new types of units [84]:

- Central data and digitalization responsibility (→ CDO chief digitalization officer or chief data officer),
- Development operations groups (or DevOps), and
- Customer success management units.

According to a study by the industry association BITKOM, the CDO in companies is still the exception up to now. According to this, not even every fifth company has a CDO. “A CDO must be able to create bridges and build up internal networks. He must be able to explain, convey and convince, because he may often initially be confronted with resistance to his new ideas” [103]. Fundamentally, there is agreement that DevOps allows for more efficient and smoother cooperation between development, delivery, and operations. John Willis, a veteran in the DevOps movement, describes the framework of the management strategy with five basic principles[13]:

- Culture: Mutual trust, continuous flow of information, and willingness to learn,
- Automation: Automation of certain work processes,
- Lean: Avoiding waste, generating value, transparency, and holistic process optimization,
- Measurement: Standard evaluation criteria (including beyond the application and its components), and
- Sharing: Willingness to share information, learn from one another, and proactively share recognitions.

Customer success is playing an ever-stronger role in the customer–supplier relationship. The principle is simple: The product or service should not only satisfy the customer, but also make them demonstrably more successful. Companies actively support the customer in the optimal use of their offer in order to achieve individually agreed objectives. The customer must permanently recognize the benefit of the product or service.[14] Examples of a customer success manager’s (CSM) tasks are:

[13]Interview Dawn Foster, https://www.linux.com/audience/enterprise/what-devops-john-willis-explains/, 06.28.2016 (retrieved on 04.20.2020).

[14]Galinker, T.: Customer Success—What Does this Term Mean. https://morethandigital.info/customer-success, 01.06.2019 (retrieved on 22.04.2020).

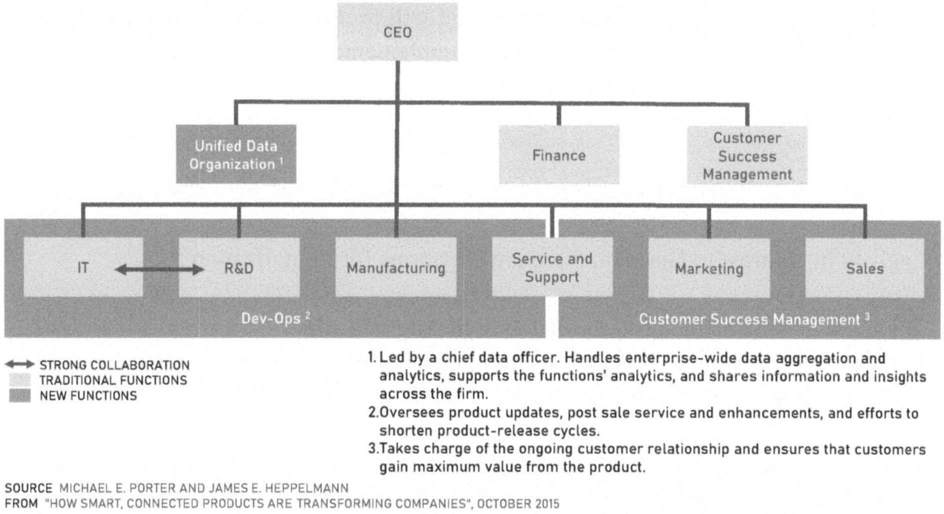

Fig. 3.5 Rearranging the organizational structures in a digital transformation [84]

- Defining individual success criteria together with the customer and determining a time frame,
- Regularly checking success using the agreed criteria,
- Collecting and evaluating feedback and usage data in order to address the problems of each customer,
- Organizing training measures, for example, in the form of webinars, and
- Creating teaching documents.

CSM represents the link between the customer and the company. They are in continuous contact with the customer and ensure that their feedback is also actually implemented[10]. Although it remains unclear which structure will ultimately emerge, practically every traditional function must be rethought and potentially restructured in the face of the dramatic new orientation of tasks and roles (Fig. 3.5).

Just one more note at the end of this chapter: Is digitalization a revolution or an evolution? In recent times, there have been several buzzwords and technical "revolutions"[15]—even IOT was defined as the Fourth Industrial Revolution in Germany. Insofar, the enthusiasm with which the digital change is advertised is nothing new. Together with a promise of progress, which almost always accompanies new technologies, the powerful strength of this discourse is all too familiar. The determination with which digitalization is currently proclaimed as a revolution is certainly exaggerated [18]. Digitalization

[15]The term revolution is especially critical in Germany. Historically, there has never been a single mid-term successful revolution in Germany.

has been increasing for at least 40–50 years and will probably design a more acceptable, honest, and technically successful technical implementation as a multistage evolution and in digital discourse with realistic consideration.

3.2 Digitization, Digitalization, and Digital Transformation

In English, in contrast to German, there are two terms for digitalization[16]:

- Digitization is a change process analog to the digital form.
- Digitalization is the use of digital technologies in order to change and further develop business models and processes.

In the German language, both are subsumed under digitalization. When we refer to digitalization below, we only mean the English term digitalization.

The digital transformation describes a continuous change process which concerns special companies in a technical, organizational, and economic respect. Sudden, upheaval-like changes are called a disruption[17,18]. In a more precise sense, a digital transformation is often described by a change process within a company based on the use of digital technology, processes, and digital business models[19], for example:

- Smart products,
- Digital infrastructures (laptop, smartphone, smartwatch, network),
- Apps,
- Social networks,
- Use of mobile electronic communication technologies,
- Sensors and actuators,
- Big data technologies,
- Cloud computing,
- Artificial intelligence,
- Machine learning,
- Human–computer interaction,

[16]https://www.wortland.com/digitalization-vs-digitization/ 07.23.2019 (retrieved on 04.22.2020).

[17]https://www.gruenderszene.de/lexikon/begriffe/digitale-transformation.

[18]https://www.enzyklopaedie-der-wirtschaftsinformatik.de/lexikon/technologien-methoden/Informatik--Grundlagen/digitalisierung (both retrieved on 04.22.2020).

[19]https://de.wikipedia.org/wiki/Digitale_Transformation (retrieved on 04.14.2020).

- System Lifecycle Management, and
- Intelligent authoring systems with authoring system-close data management (→ team data management, TDM).

Digital transformation in companies takes place on various levels. It equally concerns development[20].

- Digital products,
- Digital processes,
- Digital transformation in customer relationships (→ customer journey),
- Digital transformation in the value chain (→ supply chain), and
- Digital services-oriented business models.

Digital service-oriented business models, such as performance optimization, remote operations, and predicted maintenance, are based on update on the air (OTA) and represent a new form of value creation. They are based on ideas of additionally enriching the product with specific services, in order to achieve customer benefit on the basis of digital technologies for which customers are prepared to pay[21].

The provision of a digital infrastructure counts as one of the foundations of a digital transformation. It plays a decisive role: It integrates the digital technologies in the company. In the hype surrounding the concept, this fundamental "substrate" is often overlooked. While the technological possibilities in the industry are developing rapidly, the quasi-public infrastructure (e.g., fiber glass cables or mobile operated networks of private companies) and internal company IT structures are developing much more slowly. Nonetheless, they are decisive components for the digitalization process [25]. System Lifecycle Management in turn is a significant component of a digital infrastructure. Through its interdisciplinary, holistic, and integrative approach, it allows the administration of mechatronic and cybertronic products, the consistency of the phase-orientated partial product models along the product lifecycle, and also the consistent support of the typical engineering processes, such as the release/change and configuration management. It therefore forms the foundations for model-based systems engineering and for holistic traceability (→ digital thread).

The connection between digitization, digitalization, and digital transformation is shown in Fig. 3.6.

[20]https://www.innolytics.de/was-ist-digitale-transformation/ (retrieved on 04.20.2020).

[21]https://www.enzyklopaedie-der-wirtschaftsinformatik.de/lexikon/technologien-methoden/Informatik--Grundlagen/digitalisierung (retrieved on 04.22.2020).

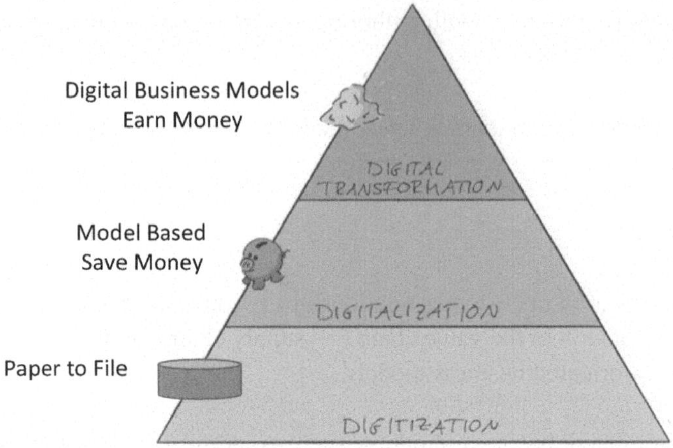

Fig. 3.6 Digitization, digitalization, and digital transformation[22]

3.3 Internet of Things (IOT), Systems of Systems (SOS), Industry 4.0, Industrial Internet, and Internet of Services (IOS)

The Internet of things and services (IOT/IOS) and the national research initiatives and activities of the German BMBF/BMWI[23] (→ Industry 4.0, Platform Industry 4.0[24], Smart Service World [14], Engineering 4.0 [69]) and in the USA (→Industrial Internet[25]) based on these assume intelligent, networked products, systems, and services in the future. The vision of small, embedded computers which humans support imperceptibly can be traced back to Mark Weiser. He first described this in 1991 in his well-known essay "The Computer for the 21st Century." The term *Internet of things* was introduced by Kevin Ashton, the head and founder of the Auto-ID Center. He first used the term in 1999 during a discussion with Procter & Gamble [5]. The term *Cyberphysical System*[26] (CPS) was first mentioned by Helen Gill from National Science

[22]https://medium.com/api-product-management/what-is-digital-transformation-digitalization-and-digitization-c76277ffbdd6 (retrieved on 05.20.2020).

[23]BMBF Ministry for Education and Research, BMWI Ministry for Economy.

[24]https://www.plattform-i40.de/PI40/Navigation/DE/Home/home.html (retrieved on 05.20.2020).

[25]The Industrial Internet Consortium (IIC) was founded on March 27, 2014, by AT&T, Cisco, General Electric, IBM, and Intel now with over 250 international members, https://www.iiconsortium.org/about-us.htm.

[26]The term Cybertronic Systems is used below v43].

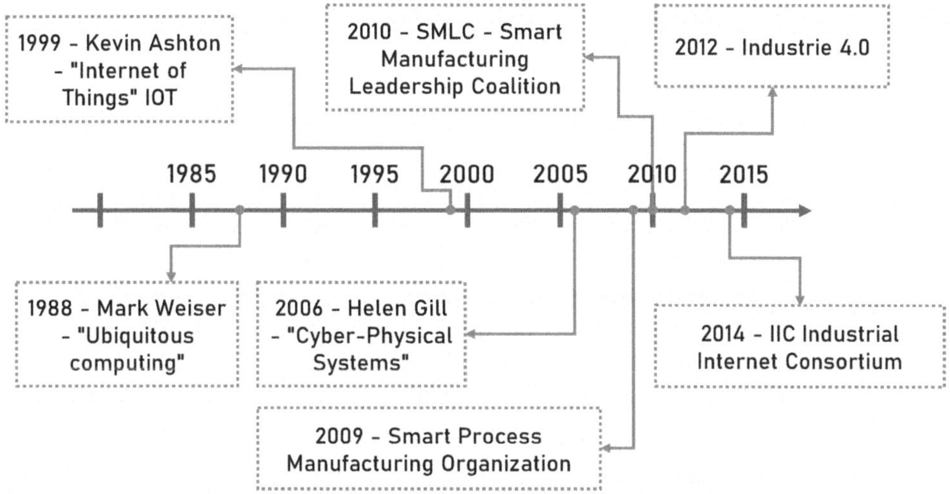

Fig. 3.7 On the path to Industry 4.0 and the Industrial Internet [107]

Foundation in 2006 [54]. This concerns embedded systems, which have a mechanical or electronic basic system and are in a position to communicate with one another via a data infrastructure (e.g., the Internet). The independent, intelligent cooperation of systems which are in a position to make independent decisions is essential. Manfred Broy made this a familiar term in Germany [21]. In 2010, the Smart Manufacturing Leadership Coalition (SMLC) pursued the realization of an intelligent networking of industries with a cloud-based approach. The German *Industrie 4.0* initiative (2012) and *Platform Industry 4.0* (2013)—heavily supported by the Federal Government—came as a reaction to the developments which should lead worldwide to the optimization of the global value chain. In 2014, the *Industrial Internet Consortium (IIC)* was founded in the USA (Fig. 3.7) [107]. Unfortunately, these large initiatives remain more or less national solo efforts.

According to the optimistic prognoses of the implementation from IOT from "Towards Data Science," 32 billion devices should be networked in the year 2020. "BI Intelligence" estimates the number of things networked via the Internet by 24 billion by the same point in time. It is clear that the analysts are really not entirely sure and the assumptions fluctuate heavily.

◆ **Internet of Things:**

The Internet of things (IOT) relates to the vision of connecting every "thing" to the Internet with a standard protocol and naming structure, whereby the reach of the Internet in the physical world will be extended, while its size will be scaled up to thousands of billions of devices[27]. Numerous new application possibilities of data analysis and management result from the increasing intelligence of things.

The research environment Industry 4.0 was originally created over a time horizon of more than ten years and stands for a future project of the German Federal Government which was started in 2011. The so-called Fourth Industrial Revolution is characterized by individualization or hybridization of products and the integration of customer and business partners in business processes[28] [45]. “Plattform Industrie 4.0” was initiated by the Federal Government and founded in 2013 a joint project of the German economic associations BITKOM[29], VDMA[30], and ZVEI[31] for the further development and implementation of high-tech strategies. The focus of the platform work lies in bundling expertise in relation to Industry 4.0 and making these available to German companies, in particular midsized companies[32]. A further implementation strategy was developed in addition to the reference architecture for Industry 4.0. This strategy included a research and innovation roadmap, which determined the most important subject fields for industrial research for the next years from the point of view of the platform. The platform was criticized due to the relatively slow implementation speed compared to the American Pendant IIC[33].

◆ **Industry 4.0:**
“The term Industry 4.0 stands for the Fourth Industrial Revolution, a new stage of organization and management of the entire value chain along the lifecycle of products. This cycle is orientated toward increasingly individualized customer desires and reaches from the idea to the order to the development and manufacturing, delivery of the product to the end customer, right down to recycling, including the associated services.

The basis is the availability of all relevant information in real time through the networking of all instances participating in value creation as well as the ability to derive the optimal value flow from the data at any time. The connection of people, objects, and systems results in dynamic, real-time, optimized, self-organizing, and company-wide value networks which can be optimized according to various criteria, for example, costs, availability, and resource consumption.”

Source: Plattform Industrie 4.0

---

[27]Johann Lukkien (j.j.lukkien@tue.nl).

[28]https://wirtschaftslexikon.gabler.de/definition/industrie-40-54032 (retrieved on 04.20.2020).

[29]BITKOM Bundesverband Informationswirtschaft, Telekommunikation und neue Medien e. v.

[30]VDMA Verband Deutscher Maschinen- und Anlagenbau.

[31]ZVEI Zentralverband Elektrotechnik- und Elektronikindustrie.

[32]https://de.wikipedia.org/wiki/Plattform_Industrie_4.0 (retrieved on 05.11.2020).

[33]https://www.elektroniknet.de/markt-technik/industrie-40-iot/deutschland-hat-die-erste-halbzeit-verloren-116855.html (retrieved on 5.11.2020).

While the initial focus of Industry 4.0 in practice is seen as the optimization of production, one arrives to a somewhat more general approach, by combining the subject fields of the Industrial Internet Consortium with the ideas of Plattform Industrie 4.0 [32]:

- Horizontal digitalization of the entire value chain of a company.
- Vertical digitalization of all internal company processes along the entire lifecycle and all organizational units.
- One focus lies in the digitalization of the engineering process—supported by interactive digital product models along all lifecycle phases (→ Engineering 4.0).
- A further focus lies in the digitalization and further automation of the production and logistics through the integration of innovative technologies in production and logistics with the use of the Internet (→ Manufacturing 4.0).
- Service-orientated business models will have a significant influence on the value creation of a company.
- New social work infrastructures will accompany the process of digitalization.
- Continual development of innovation foundations and cross-sectional and infrastructural technologies.

The networking and integration between areas or departments within a company but also between different companies is a central element of Industry 4.0 and the Industrial Internet. The most complex form of networking is a system of systems (SOS), which has the foundation of a networking of independent Cybertronic Systems (Fig. 3.8) [23, 64]:

- Are target systems whose system elements are in turn systems themselves,
- Solve extensive, interdisciplinary tasks, and
- Consist of many heterogeneous and distributed systems.

These interacting amounts of system elements create a functionality which an individual system could never achieve.

In order to implement all these visions, methods, processes, and IT tools must be further developed for consistent digitalization of the product lifecycle. Building on model-based development, processes for the cross-discipline and integrated development of cybertronic products and systems which contain both products and production and service systems are currently being researched [3, 4, 6, 44, 74, 96]. Software and electronics will further enable numerous new functions in the future and further increase the function complexity of products. On the other hand, development and manufacturing complexity may be partially reduced through the shift of the variance from hardware to software. Service-oriented, scalable cloud platform concepts for predictive quality data analysis can contribute toward creating intelligent, application-specific feedback loops in quality management, such as for product and production data, in order to improve system and process quality for producing companies. A decentralized, self-organizing process is created from central production management [7]. The results are autonomous, smart

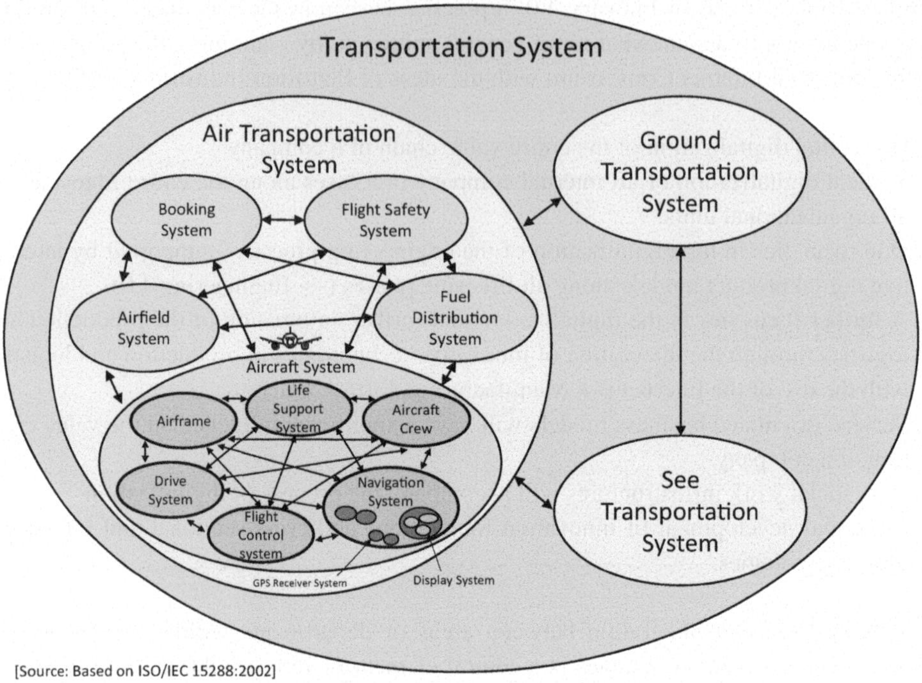

Fig. 3.8 Systems of systems (based on ISO/IEC 15282: 2002)

machines, and communicating products, as well as services built on these. This falls into the area of Internet of services (IOS).

◆ **Internet of Services:**

Internet of services (IOS) is a technology which provides network infrastructures to support a service-orientated business model. The IOS platform offers its users a completely decentralized possibility to provide services and digital goods[34]. This general definition, restricted to the technical solutions, means that services are offered on the basis of communication-capable cybertronic products and can be bundled into completely new, service-orientated business models.

These service-oriented business models cannot yet be implemented full scale today. There is a lack of information and of a transparent evaluation over the entire system during operations with the operator or end customer (e.g., wear of components). These

[34]https://medium.com/@iostoken/the-internet-of-services-intro-to-our-tech-e91abfb13b8c, 2018 (retrieved on 04.24.2020).

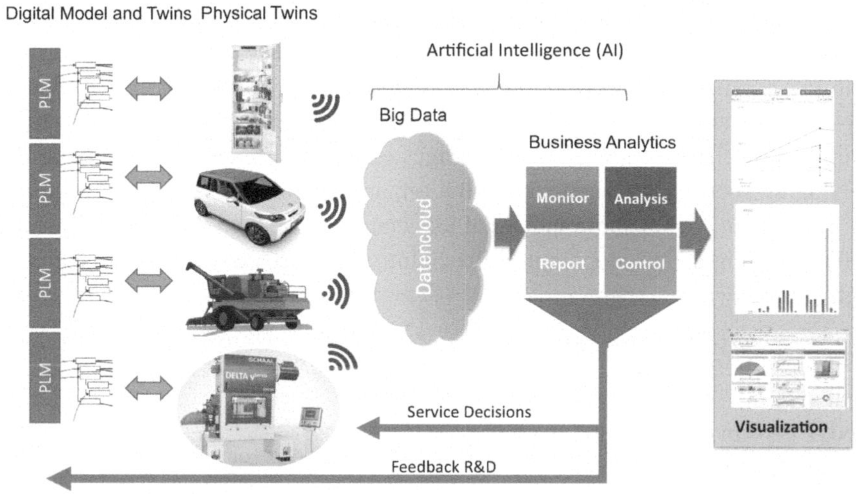

Fig. 3.9 Foundation of service-orientated business models

uncertainties can be eliminated, and the risk of system failure is better estimated with the use and development of closed control loops with intelligent evaluation of field data. Furthermore, it results in new possibilities for the individualization of service products in investment goods. Equally, customer and machine-specific full-service concepts to safeguard the highest availability for improved operational planning, for precise determination of the spare part potential and for the automated trigger of service process, are possible [7, 16]. Figure 3.9 shows a typical example of how both consumer goods and investment goods – based on an active sensor/actuator integration and intelligent data preparation (→ big data, business analytics, artificial intelligence, and visualization)—form the foundation for service-orientated business models [45].

Explicit technologies with sensor and actuator technology as well as embedded intelligence are decisive for the design of the Industrial Internet with the Internet of things and services. Only in this way, is it possible for products to realize the environment and interact with it. Wireless communication, such as broadband mobile connection or radio frequency identification (RFID), is also important. Consequently, the semantic description of services and capabilities which guarantee the interaction of product components and machines in an intelligent way is important. With "smart produce" or "plug and produce," it is possible for machines to automatically recognize their environment and be able to network or interact with other machines. This can enable the exchange of information about orders, capacity, and optimal manufacturing parameters, for example. Building-specific – also cloud-based – backbone concepts, which help on an administrative level to maintain control of a product system along the entire lifecycle, apply as a central element here.

While Germany is on the path to Industry 4.0, Japan is one step further: The government program "Society 5.0" is intended to not only digitalize the economy, but also get started with the population. The Japanese government has in mind nothing less than a new, super-smart society, "Society 5.0." The government program "Society 5.0" formulates solution approaches for the central challenges of Japanese society, in particular the increasingly aging society, environmental pollution, and natural catastrophes. According to a thesis paper from the Japanese industrial association Keidanren, "five walls" which are currently impeding the development of the country from "Society 5.0" must be "broken through" for this[35]:

- Public administration,
- The legal system,
- Knowledge gaps in digitalization matters,
- Lack of specialist workforce, and
- The acceptance of the population.

The common vision of Society 5.0 should attribute a greater significance to the interaction between people and machines and anchor the moral, ethical, and economic aspects of digitalization in society[36]. Infrastructures, platforms, and services form the foundation of Society 5.0. They are based on smart technologies such as artificial intelligence (AI), robotics, the Internet of things (IOT), and block chain, but also on augmented and virtual reality or robotic process automation (RPA). These technologies have now achieved a maturity that enables the largest societal and economic transformation[37].

The many designations in the environment of digitalization—partly with national focusses—are certainly confusing. To end this sub-chapter, Fig. 3.10 represents an attempt at delimitation and containment.

3.4 Digital Twin[38] and Digital Thread

Service-orientated business models are the central factor for company success and pave the way from the Internet of things to the Internet of services. The Digital Model, Digital Thread, and the Digital Twin form the technological basis for these concepts.

[35]http://www.keidanren.or.jp/en/policy/2018/095.html (retrieved on 06.10.2020).

[36]https://e-3.de/society-5-0/ (retrieved on 06.10.2020).

[37]https://www.it-daily.net/it-management/industrie-4-0/21238-die-digitale-zukunft-heisst-society-5-0 (retrieved on 06.10.2020).

[38]The subject digital twin was processed in several industrial research projects together with VPE employees Thomas Eickhoff and Hristo Apostolov as well as the external doctoral candidates at VPE Alexander Detzner, Philipp Heiner Schmidt, and Rajeeth Tharma. The detailed description was published in the journal ZWF [33, 30, 109].

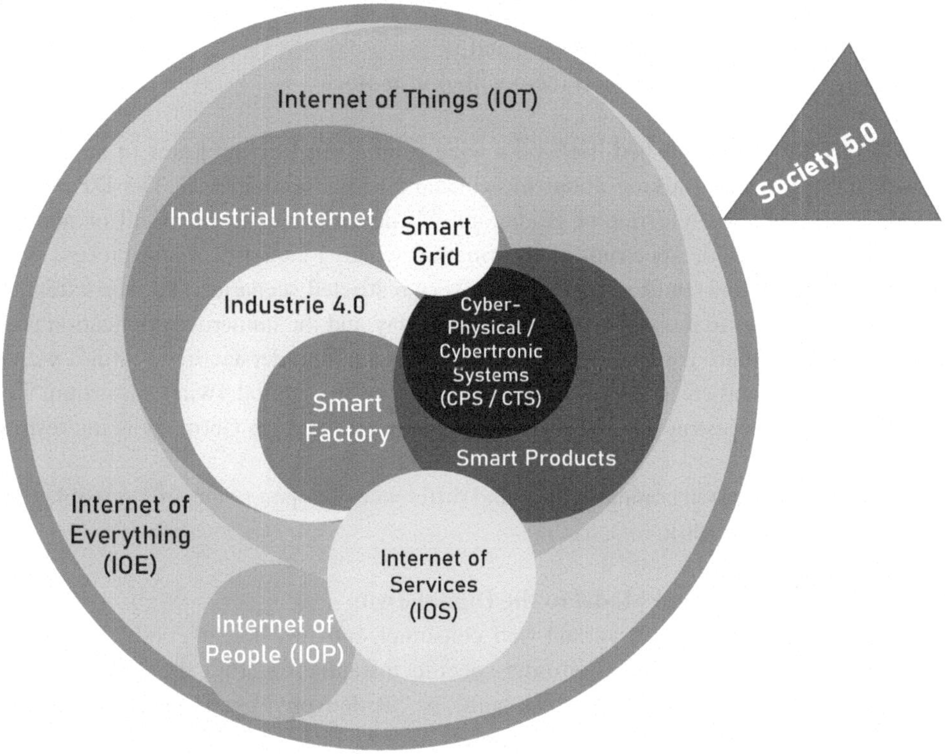

Fig. 3.10 Overview of the terms used in the digitalization environment [79]

3.4.1 Digital Twin

The customer or order-specific instantiation of an individual product is generated from the digital model (see Sect. 3.5.5)—often called the "150% model in Germany[39]"—the basis of the nucleus or core of the *Digital Twin (DT)*. The derivation and management of a DT from the product lifecycle management (PLM)[40] were mentioned for the first time in 2001 by Michael Grieves in a presentation by the University of Michigan for industry [57]. He said that a DT needs three elements:

[39]The term should symbolize that the digital model contains the entirety of all product derivatives and variants, for example, various types of seat for a car (→ normal, comfort, and sport). But the digital model does not only contain the structure data, rather all the information created by authoring systems, for example, CAD models, simulation results, and quality data.

[40]PLM and SysLM systems can already be derived from order-oriented instantiated parts lists today; nonetheless, it should be assumed that the "as-built" parts list comes from the MRP system.

- The physical product on the real market (→ physical twin PT),
- Its digital duplicate in a virtual world, and
- The information which connects them with one another.

The DT has not yet established itself on a wide front. "But I do not know of any industry which isn't at least talking about the concept," explained Grieves. The DT can be introduced in completely different gradations. "It doesn't have to be an all or nothing project. There is a wide spectrum of information which I can collect and process with the twin. Digital twins could also be used in very restricted scenarios and then extended bit by bit" [56]. Due to the still low distribution today and the numerous application possibilities of DTS, there is currently a general fundamental understanding, but not yet an all-encompassing and clear definition. The founding of the Digital Twin Consortium[41] in the USA and the Industrial Digital Twin Association (IDTA)[42] in Germany is interesting in this context.

Below, the recognized commonalities and differences or open points of a possible definition are considered [30]:

Relationship of the Digital Model to the Digital Twin

It is generally known that in variant-rich consumer and investment goods, the digital twin (DT) is based on a customer/order-specific instantiation from the digital model. The core of the DT is based on the "as-built" or "as-delivered" parts list. In the purely theoretical case of a 100% ETO[43] development and production, the digital model would already correspond to the order-specific instance. However, depending on the degree of modularity and reuse of components in the scope of a building block system, reusable digital sub-twins can also be derived on the lower level of the building block, which can be used in several digital twins on the product level, for example, a drive component of a tool machine which is used in several customer-specific derivates. Thus, the more pronounced the building block system of an order manufacturer, the more we can refer to a partially order-neutral digital model here as well. In a good designed, modular, and reusable product construction, a customer order results from new modules in combination with existing product components.

Behavior of a Digital Twin to a Physical Twin

A DT is a digital representation of areal product instance [106]. This could be a tangible or intangible object such as products, systems, services, or processes. The connection of the DT to the PT requires a clear identification. In the simplest case, this occurs via item

[41]https://www.digitaltwinconsortium.org (retrieved on 10.21.2020).

[42]https://www.vdma.org/v2viewer/-/v2article/render/52443301 (retrieved on 10.21.2020).

[43]ETO engineering to order.

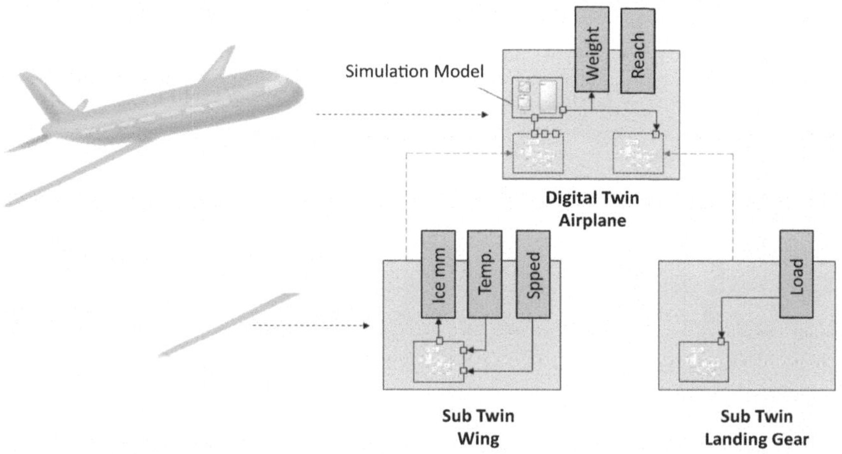

Fig. 3.11 Digital twin with sub-twins using the example of an airplane[45]

number, revision (change index), and potentially the batch from the production process. This scenario is applied if no further after sales business model is put on the DT, for example, with a hand razor. In this case, it is debatable whether we can refer to a DT, as it only consists of the "as-built" status and also does not change over the lifetime of the product. There is no information enrichment and reduction, nor are there any data streams. The relationship of the virtual instance to the physical instance is 1:n. In the case of traceability required by the authorities in the event of damage or in a business model built on the DT, an order number or, ideally for tracing operations, a sequential series number or serial number is used. There must be at least a 1:1 relationship in operations on the top level of the DT and PT. This is not compulsory on the lower modular DT/PT levels. Thus, a certain airplane chassis could be an independent modular DT/PT, but built into many airplanes[44]. These twins should be designated with sub-twins Fig. 3.11.

A DT represents the complete system. Part systems are also represented by hierarchically assigned DTs, the sub-twins. If the sub-twin is a delivery unit, then of course it corresponds to a DT from the perspective of the supplier. DTs can contain simulation models. Sub-twins of composite systems access the simulation models of part

[44] It must be clarified here whether the term digital twin makes any sense for this level of the building block module. However, products result in a mix of top down and bottom up and so the modules of a building block which were developed, simulated, and tested, or are delivered as a whole, are also designated as sub-twins. The prerequisite is that a hierarchical concept is accepted in the DT discussion. The 1:1 assignment between DT and PT then only exists on the top product level.

[45] Dr. Thomas Kuhn, Fraunhofer IESE, Kaiserslautern 2017, https://gi.de/informatiklexikon/digitaler-zwilling/ (retrieved on 05.20.2020).

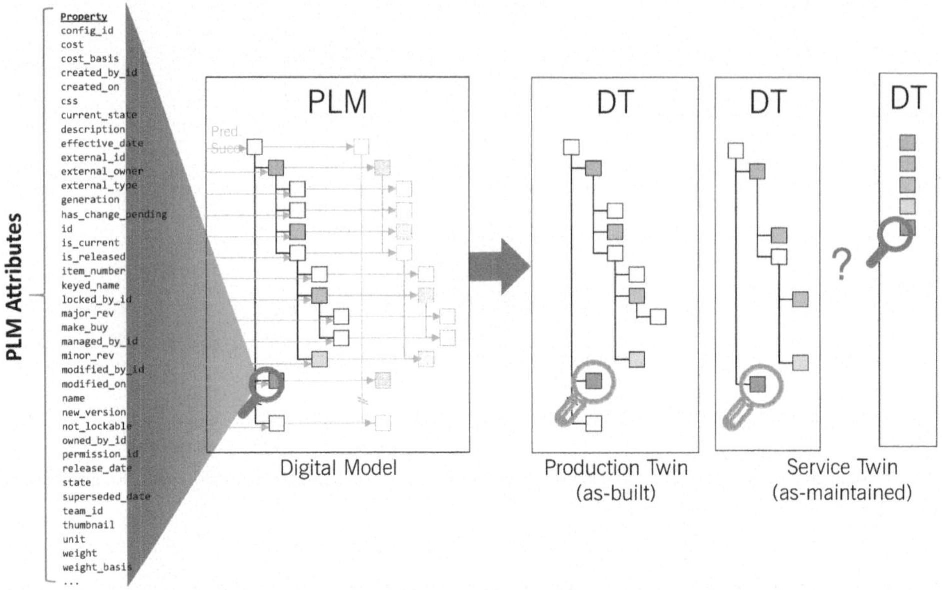

Fig. 3.12 What actually belongs to a Digital Twin, reduction of the Digital Twin[45]

components. Data is provided by means of defined interfaces, and simulated reactions are also returned via defined interfaces[45].

Reduction of the Digital Twin for Operations

A DT contains all information which arises for the description of the development and production until the SOP[46]. In operations, it depends on the business model and the use cases derived as to whether the DT continues to represent a 100% depiction of the "as-built" parts list, or whether it can be drastically reduced and simplified (Fig. 3.12). However, the DT can also be enriched again with additional data from engineering, simulation, production, quality management, and service (see Figs. 3.14 and 3.17). This access to the typical, company-internal IT system occurs through linking.

On the left side, the complexity from the attributes of a PLM system which make the multiple process functions possible (ECM, CM, revisioning, effectivity, etc.) at all is visible and, on the other side, the numerous predecessor/successor relationships, which are not required by an instantiated DT which always represents a temporally valid snapshot, but can always be created again by jumping back into the digital model. This simplification of a reduction of the attributes and the linking of the change history is therefore always possible. A further simplification results from the fact that components which are

[46]SOP Start of Production.

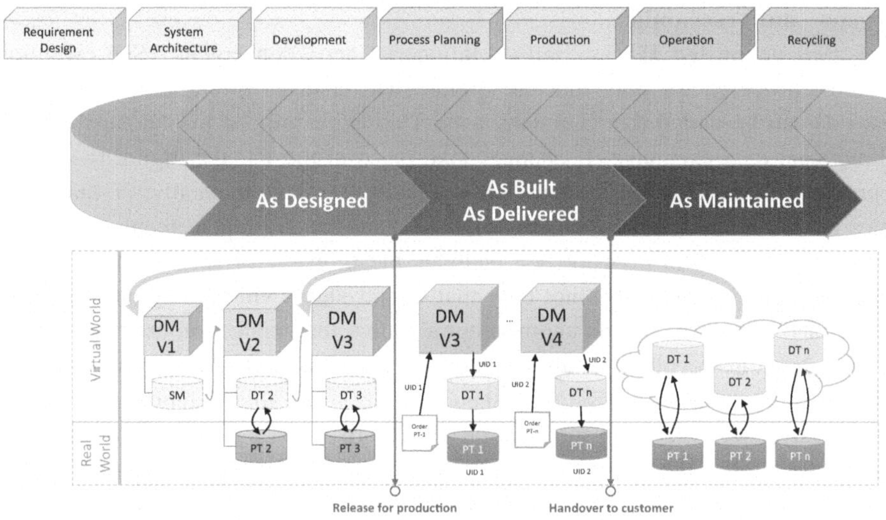

Fig. 3.13 Digital twin along the product lifecycle [33]

not required for the business model are removed from the DT, and even complex parts list structures are potentially to be replaced by flat structures. When questioned, the CIO of Tesla confirmed that the DT only consists of the revised electronic components and the corresponding software modules due to the business model update over the air.

The Digital Twin in the Product Lifecycle

DTs fundamentally have the task of predicting and optimizing the performance features of the physical counterpart through simulation of the behavior. Instantiated models and sub-models are used to simulate, predict, and optimize the product and production system during the entire product lifecycle until delivery, in order to reduce high investments in physical prototypes and resources. Specialists in this area are discussing the more terminological question as to whether a DT can already exist in the early development phase without its physical twin. The majority opinion is that a DT must have a PT as its counterpart. The phases of the digital twin along the product lifecycle are shown in Fig. 3.13. We refer to an instantiated simulation model (SM) during the early development phase without a PT.

For business models based on the DT, communication between the DT and PT is required in operations, among other things to optimize the maintenance and operation of a product. The relationship of DT and PT is also reversed here. Until delivery the DT is dominant. Then, we refer to a bi-directional exchange. The PT feeds both the DT (exchange of components) and the digital model, for example, regarding the likelihood of damages as well as conversely, the specific driver behavior, is assigned to the DT in order to derive consequences for transmission and engine control via algorithms.

Convergence and Traceability

Tao et al. [99] define the DT as "self-evolutionary" and "convergent." Self-evolution describes the property not only that the digital twin changes throughout its life, but also that this can be undertaken autonomously. This is guaranteed by the capability of the bidirectional data exchange. If changes take place in the PT, the digital twin must also reproduce these. This can happen automatically, semiautomatically, or manually. Examples of this are the change of the software status or the replacement of individual components. The term convergence includes the property of self-evolution, but it is further formulated. It describes not only that the physical twin and digital twin must be identical and connected to one another, but also that the connection exists over time between the individual evolution stages. This is necessary in both the physical twin and the DT. Rupp [89] defines traceability as the capability to be able to reproduce connections and dependencies between information which arises during the simulation, analysis, maintenance, and development of a system at any time. "In the context of this consideration, traceability is also mandatory for the DT and is realized in the SysLM system through the so-called digital thread" [33, 109]. This means the DT is always connected with the production-relevant configuration of the digital model.

Even though the Digital Twin can be enriched with data from any source, we propose three major categories of data as integral parts of the Digital Twin: enterprise data, supplemental data, physical twin data, and generated data. Further information is linked to the Digital Twin either from the enterprise IOT platform or sourced from the environment. Figure 3.14 illustrates our definition of the Digital Twin, including the categories of linked data as well as the peripheral systems [33, 109]. For a deeper understanding, the following sections go into the individual components of a Digital Twin, starting with the Digital Twin Core.

Digital Twin Core [109]

The *Digital Twin Core* is derived from the Digital Model. However, closer examination reveals different levels of the Digital Twin Core and thus of the DT. A product instance, for example, is comprised of its components. Consequently, the Digital Twin Core of a product instance can be composed of the Digital Twin Cores of its sub-models. Analogously, these levels of detail for the Digital Twin and its core are described by Lee et al. as they distinguish three levels: component, machine, and fleet levels [110]. The component level contains virtual representations of individual physical parts or components. The machine level contains virtual representations of entire machines or systems. As such, this can also be understood as a system level. Finally, the fleet level consolidates data from several DTs at the machine- or system-level, respectively. In summary, all Digital Twin Cores are derived from their respective Digital Model and serve as an anchor and structuring element for all other information required in a specific use case.

Nevertheless, the relationship between the Digital Model and the Digital Twin Core is not only relevant for instantiation, but it can also be bidirectional. Furthermore, the

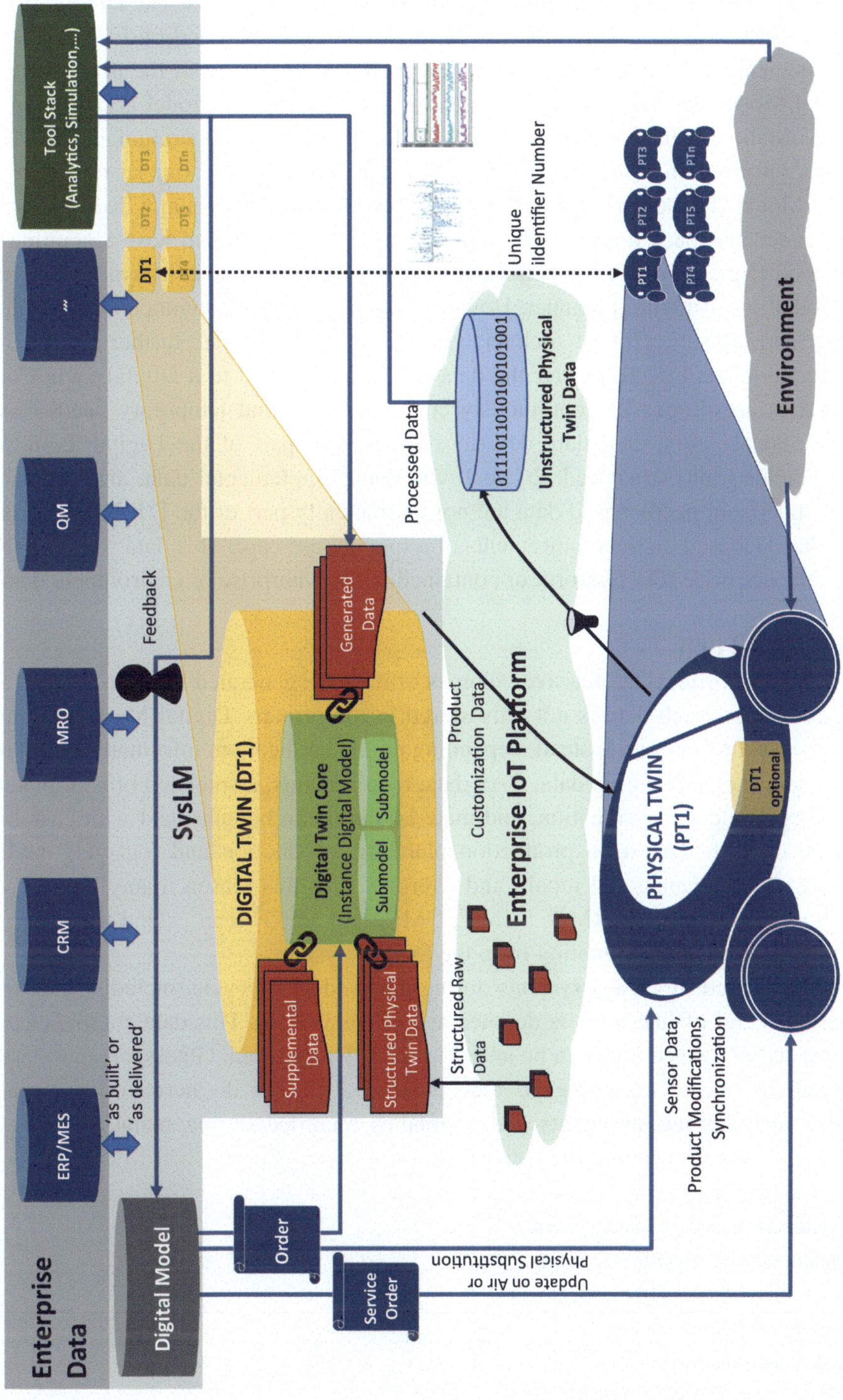

Fig. 3.14 Concept of the digital twin with all relevant data streams [33]

Digital Model can pass on information to existing Digital Twin Cores. The Digital Model is managed in PLM systems, and changes to the Digital Model based on ECM[47] ensure that improvements are implemented in all new products derived from the newly configured Digital Model. However, in general a change in the Digital Model is only applied to existing instances if this change is mandatory for the product's operation and security. Meanwhile, information from one or more DTs can also indirectly impact influence on the Digital Model. The knowledge derived from permanently analyzing a fleet of Digital Twins is termed feedback data, which helps optimize the current and next generation of products. The Digital Twin Core helps interpret the data and makes it available to the Digital Model in a structured manner. However, the information contained in the Digital Twin Core is rarely sufficient for such analyses. As such, if required, further information must be added to the DT. To cluster the data that can be linked to a Digital Twin Core, two categories are proposed: continuously connected data and temporary needs-based data. Continuously connected data describes an integral part of the Digital Twin and can be subdivided into structured physical twin data, supplemental data, and generated data. In comparison, needs-based data are not permanently part of the DT, but rather are linked to the DT based on the requirements of a specific use case. This data will be either stored in the enterprise IOT platform or contained in the enterprise or environment data.

Enterprise Data [109]

During the product lifecycle, massive volumes of data are generated that can be relevant for a Digital Twin. Such data is not only stored in the primary Digital Model administrated by the SysLM system. Instead, depending on the application, information from the SysLM system (i.e., simulation data, CAx data, requirements, as-planned bills of materials, drawings, models, classification, and material data) can be enhanced with customer information or with data from production planning, production and logistics, quality management, maintenance and repair, and overhaul. For this reason, many legacy systems such as MRP/MES[48], CRM[49], SCM[50], QM[51], and MRO[52] can extend the information of the Digital Model originating from the SysLM system.

All data collected by legacy systems during the product lifecycle, including machine parameters and end-of-line tests, is denoted as *enterprise data*. This data comprises part of the extended DT, since they can be added to the DT as needed. Glaessgen and Stargel reveal a similar line of reasoning, as they propose a DT for the aeronautics domain, which also includes measurements and anomalies recorded during manufacturing, in

---

[47]ECM Engineering change management.

[48]MES Manufacturing execution system.

[49]CRM Customer relationship management.

[50]SCM Supply chain management.

[51]QM Quality management.

[52]MRO Maintenance repair and overhaul.

order to predict the remaining operating life more precisely [111]. This idea is confirmed in a recent study by Gartner, where at least 40% of the surveyed companies claimed they had enriched the DT from at least seven sources, primarily consisting of simulation results, MRP, and MRO data. Should traceability of any enterprise data be needed, such as because of a product defect in the PT or a complex engineering change request, then this data can be temporarily linked to the DT. Schuh and Blum went even further and expanded the concept of DTs to the manufacturing domain. To this end, they present an independent DT for manufacturing, which includes the current status and the location of the (semifinished) product on the shop floor [112]. Similarly, for Lee et al, individual machines possess their own DTs, which monitor the parameters and settings of the machine and optimize the manufacturing based on historic data [110]. However, the data contained in a machine's DT can be linked to the DT of a product, which was manufactured using that machine. Consequently, the increasingly connected manufacturing processes enable precisely monitoring products, assemblies, and parts during the manufacturing process. This traceability enables companies to match machine parameters and measurements to their corresponding final product.

Supplemental Data [109]

The majority of data from the PLM system or from enterprise data systems are only added to the Digital Twin on a physical twin data basis. However, some information from these systems must be physically added to the twin's data model as a persistent enrichment of the DT. This data is termed as *supplemental data*. The necessary supplemental data includes information to more precisely specify a Digital Twin and thus its physical counterpart, such as position data of a sensor (e.g., first, second, third, or fourth cylinder). Such data is crucial, as the Digital Twin serves as the anchor and structuring element for other data. In this way, anomalous behavior can be precisely identified via the geometric location (position) of its origin. Furthermore, supplemental data can include additional information about individual parts, including an as-built bill of materials with unique part serial numbers. Based on those serial numbers, access to additional data becomes available. In fact, a supplier might consider their available data regarding a unique part, a DT of their product as well. However, even without such detailed information, the serial numbers still reveal the supplier. For this reason, such information is essential for tracking the root cause of a product defect in case of several suppliers providing identical components.

Physical Twin Data [109]

Almost all definitions of a Digital Twin mention a bidirectional data exchange between the physical product and its digital counterpart. Stark et al. emphasize this aspect by calling the transmitted data *digital shadow*. In their definition, the digital shadow includes operating data, status data, and process data, which is collected with sensors during the usage phase [94, 95]. Dahmen and Rossmann adapt this nomenclature and state that the digital shadow includes all data necessary for a specific purpose [26]. Bracht et al.

also use this term and state that it includes all data sent from the physical product to the DT [20]. Grieves and Vickers mention such data as well, but they do not assign a specific term [56]. However, the term digital shadow is based on further differentiation. Specifically, all information, communicated from the physical twin into the virtual world, is referred as *physical twin data*. Among the physical twin data, there is a differentiation between structured and unstructured data. Structured data includes product modifications during repair and maintenance, as well as preprocessed sensor data. Unstructured data generally requires further analysis in the OEM back end, before being made available to the DT as generated data. As a result, the structured physical twin data can be directly linked to the DT, whereas the unstructured physical twin data is stored in an enterprise IOT platform without immediate linkage to the DT. Within the enterprise IOT platform, the physical twin data is mapped into clusters and can be assigned to a product instance and its corresponding DT, using the product's unique identifier number.

Generated Data [109]

The structured and unstructured physical twin data stored in the enterprise IOT platform is used for data analytics and simulation. On the one hand, information derived from analyzing a fleet of DTs can effect changes to the Digital Model, such as an adjustment of requirements based on actual customer usage profiles. This feedback data helps improve the current and next generation of products. On the other hand, the goal of analytics can be to obtain knowledge about, or for, a specific product instance. We refer to this as *generated data*. A straightforward example of generated data is to extrapolate the remaining life of a product based on individual usage parameters. However, the insights based on data analytics can also be used to manage the settings and behavior of the product, such as to adapt a vehicle's driving parameters to account for attrition of components. This type of generated data can also be called product customization data. In addition to the physical twin data, the enterprise data can be employed for data analytics. Notably, Schuh and Blum employ information from the manufacturing process to serve as the basis for improving manufacturing processes based on data analytics [112]. However, the greatest insights can be achieved by combining physical twin data and enterprise data in order to trace the root cause of a problem throughout the entire product lifecycle. In this context, Dahmen and Rossmann mention a Scenario Library for an Experimental Digital Twin that contains specific testing scenarios [26].

Although physical twin data and enterprise data are essential to the data analytics process, the results can often be improved by incorporating additional environment data. For instance, the vehicle's driving parameters can be adjusted to the current weather. If the weather conditions cannot be determined by a vehicle's own sensors, such environment data can be obtained from an external source, such as dedicated services, based on the vehicle's geolocation. Grieves and Vickers also indicate the importance of external factors for analysis and simulation by introducing the term Digital Twin Environment [56].

In summary, the following definition for the Digital Twin results:

◆ **Digital Twin:**
A Digital Twin is a digital representation of a tangible or intangible product, process, and services from the real world in the digital world. It includes all the necessary information for a specific business model from the digital model, i.e., across all lifecycle phases and disciplines and all relevant IT solutions. The nucleus results, for example, in the prototype construction through the current configuration status in development (→ baseline) and after delivery by the "as-built" or in-plant engineering through the "as-delivered" parts list. The digital twin can be enriched and/or reduced, depending on the requirements of the business model.

3.4.2 Digital Thread

The term *Digital Thread* encapsulates the concept of linking

- All possible Configuration Items[53] (CIs) along the PLC within the generic digital model as well as the instantiated digital twin with one another.
- Analyzing and evaluating the networking of all CIs across all PLC phases, for example, when a part is damaged, which simulations were carried out, which requirements persisted, and how the part is produced. Only in this way can changes be traced forward and backward. In order to ensure this, it is necessary that every type of evolution of the digital model and digital twin is revisioned.
- In order to develop solutions for damage analysis and change and configuration management on this basis.

Such a continuous digital thread allows for the optimization of change management and damage analysis beyond various value processes and the utilization of the most varied forms of digital business models and services offered for products. The digital thread also explains the question "why." If frequent damage is apparent on a component in operation, the cause of the problem must be determined by back-processing of the digital thread. The implementation of the digital thread occurs through SysLM systems and is a significant component of a digitalization of the engineering processes (Fig. 3.15).

The concept of the digital thread was introduced by the airplane industry and their desire to improve the performance of future programs with the use of digital technology.

---

[53]Configuration Items are technical elements along the product lifecycle which are revised according to operational agreement potential in changes and thus are subject to configuration management. The elements actually affected by a change are referred to as an affected or impacted item.

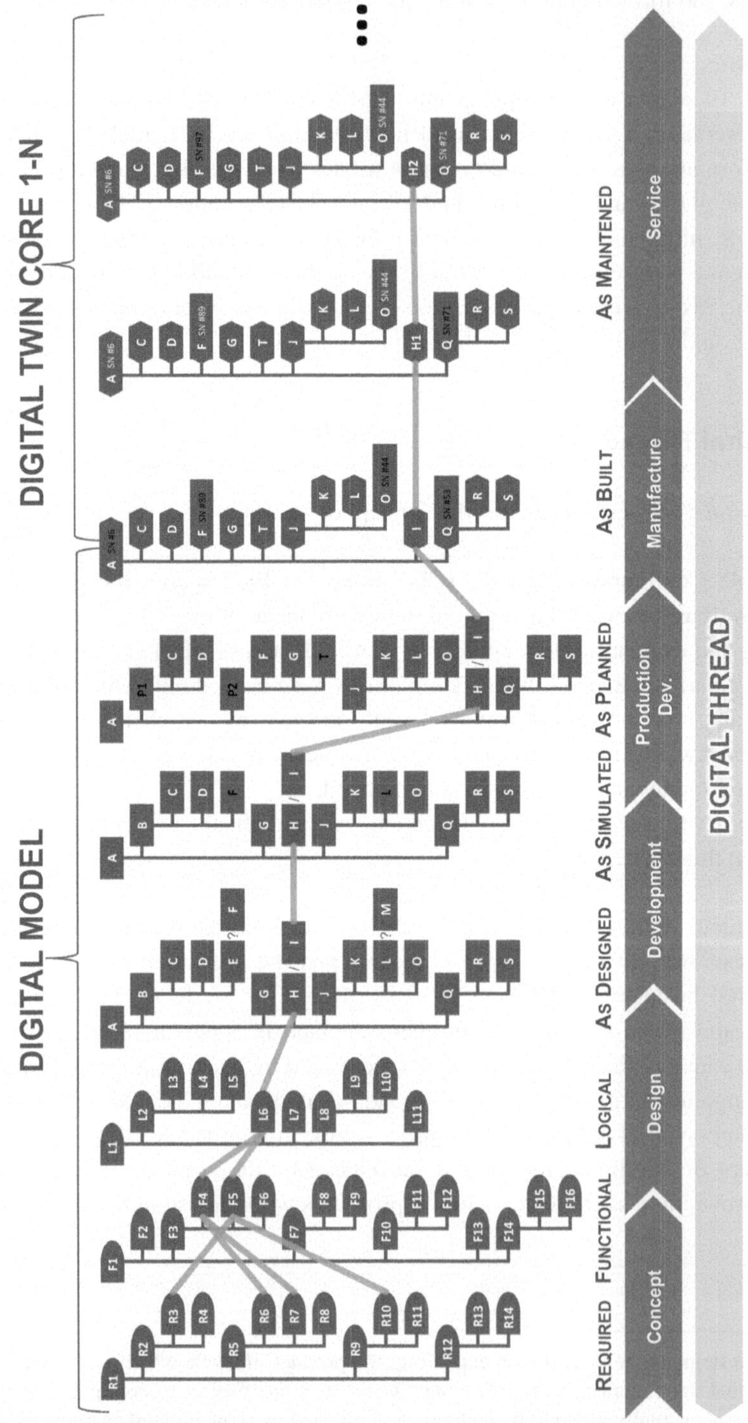

Fig. 3.15 Relationship between digital model, digital twin, and digital thread. *Source* Aras

The objective was to apply acquired recognitions to current and future programs through feedback analysis. However, these concepts and technologies attract interest beyond the aerospace and defense industry and converge with the aims of the digitalization of engineering, in particular in the development of service-orientated business models.[54] Figure 3.15, however, only shows the purely theoretical solution that 100% of the manufacturing of a product occurs in-house. The reality and with it the problem are that tracing a product affects the entire value chain.

◆ **Digital Thread:**
Digital Thread connects the configured items (CIs) of the lifecycle phase-specific partial models of a Digital Model and the Digital Twin, and furthermore also its physical twin, along the entire product lifecycle. In addition to information from SysLM, TDM, and the authoring systems, it can also incorporate further product-relevant information from other IT systems (→ MRP, MES, SCM, QM, and CRM). The objective of the Digital Thread is the continuous traceability of the creation, procurement, and production history of a component. In this way, the digital thread enables

- The company-wide continuity and availability of product and process data,
- The company-wide execution of processes (→ ECM/CM), and
- Complete reconfiguration according to ISO 9001 in the event of damage.

There are different objectives for a Digital Thread, depending on the application:

- Damage event: Traceability of the history resulting in the damaged components and reporting to the responsible organizational unit as feedback,
- Change and configuration management: Analysis of which configured items are directly or indirectly connected with the components to be changed and selection of the actual components affected by the change (→ impacted/affected items).

3.4.3 Implementation of Digital Twin and Digital Thread in Operational Practice

Model-based product development, digital services, and service-orientated business models require continuously knowing the status of a product throughout the entire product lifecycle. In addition to the provision of this data throughout the entire development and production process via SysLM and MRP and in operations via a mixture of SysLM

[54]https://www.ibaset.com/blog/what-is-the-digital-thread/ (retrieved on 05.18.2020).

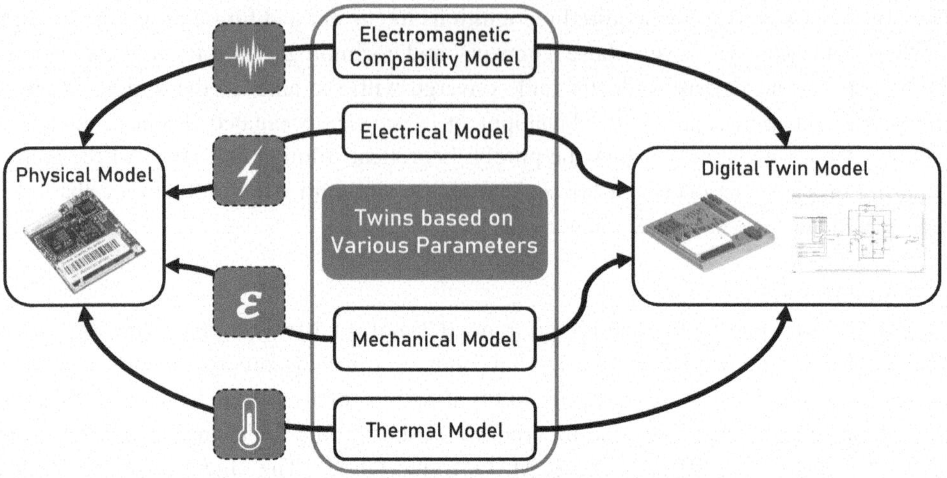

Fig. 3.16 Different views of a DT on a mechatronic product. *Source* Leoni

and Internet-based solutions, an instantiated data management is also necessary. A DT for every individual physical product and a digital thread which connects the DT with all elements of the generic digital model forms the conceptional foundation for this. The analysis and interpretation of the data sent from the PT are enabled by a context formed from the Digital Model, the Digital Twin, and the Digital Thread.

There can be several views of a DT for a product which, respectively, address different applications (Fig. 3.16).

Depending on the application case, information from the System Lifecycle Management (→ simulation data, requirements, parts lists, drawings, models, classification and material data), production planning, production, and logistics (→ MRP/MES), from quality management, service, and many other areas, is linked with the digital twin. According to Gartner, at least 40% of the companies surveyed in regard to the implementation of a digital twin concept indicated that they had enriched it from at least seven sources[55] (Fig. 3.17).

Legal, ethical, and financial questions accompany all the open technical questions, for example, who does the data belongs to, how can personal data protection be guaranteed, and does a maintenance platform operator for the takeover of the delivered DT financially recompense the manufacturer of the original product?

[55]https://www.gartner.com/en/documents/3899678/survey-analysis-digital-twins-are-poised-for-proliferation (retrieved on 06.05.2020).

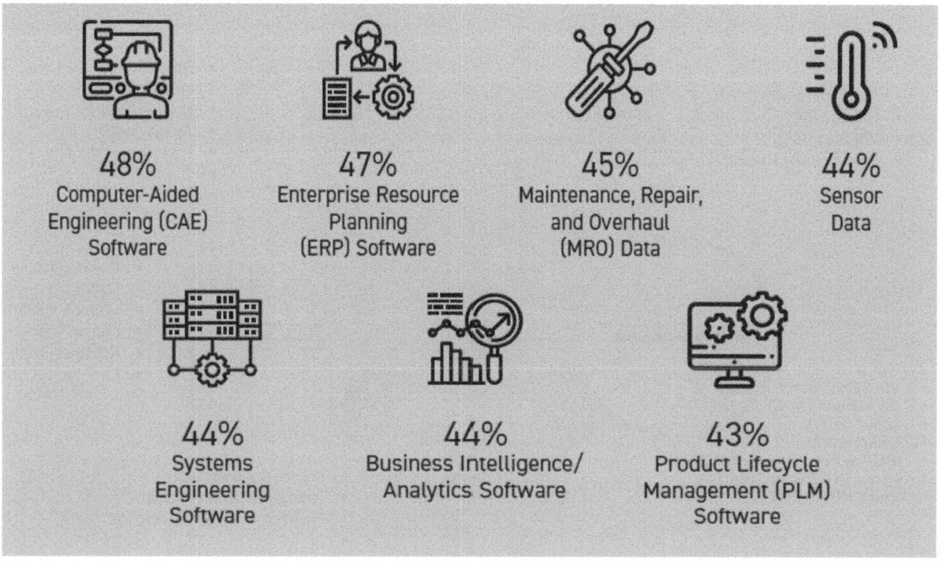

Fig. 3.17 Sources of enrichment of digital twin information. *Source* Gartner[44]

3.5 Model-Based Systems Engineering, System Thinking, and the System Model

3.5.1 Discipline-Oriented Approach in Product Development

Statistics from the last years prove the permanent transformation of the product development process (PDP). The influences result from changed market conditions, new requirements of the product, and the customer perspective. The increase in product complexity results on the one hand from a considerably stronger "multi-market" capable product, derivative, and variant variety and, on the other, from the continuous increase of electronic components and the associated, "embedded software" in the scope of the digitalization of products and services. The value proportion of electronics and software has increased continuously in recent years, and in vehicle construction, for example, it is at approximately 40 percent in value with an increasing tendency. New treatment requirements for methods, processes, and process models for an interdisciplinary PEP are derived from this. This is based on organizationally and technically supporting the engineering product lifecycle, i.e., from the early phase of requirement, to development, simulation, production planning, production, and operations, across all disciplines

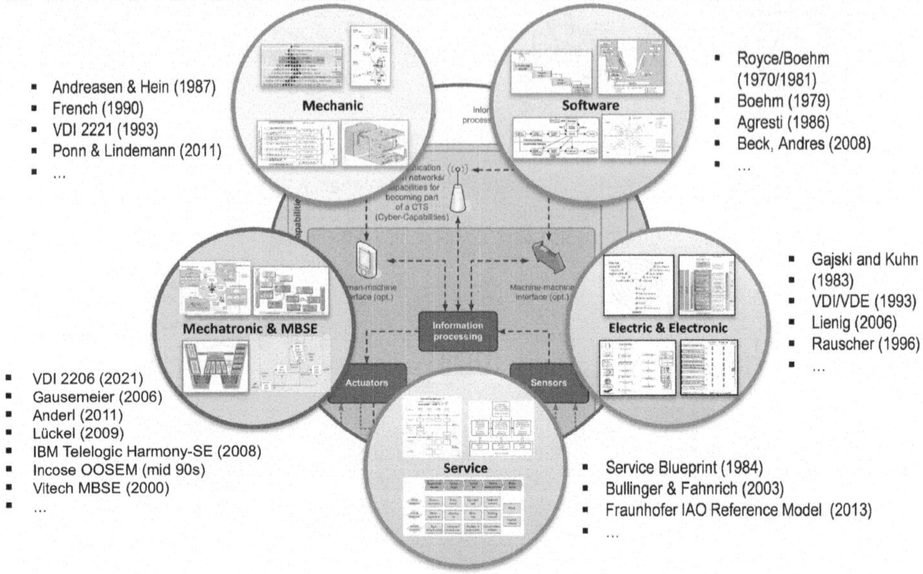

Fig. 3.18 Discipline-oriented education in product development [34]

(→ mechanics, electrics/electronics, software, and services[56]) and across all departments of a company. Today, there is still a lack of established, industrially used methods, processes, and procedure models for the cross-disciplinary development of products and systems. The development methods and processes developed in various disciplines are completely different and led to us educating, thinking, and working in discipline-orientated silos. Figure 3.18 shows an overview of a discipline-orientated education in product development.

Holistic design methodologies of mechanical engineering were suggested in the 60s and 70s, based on a functional approach. Franke, Kesselring, Rodenacker, and in particular Pahl/Beitz were typical representatives [78]. Today, almost most of the established design methods in mechanical engineering are based on the artifact's requirement, function, logic, working principle, and principal solution as well as on four distinct process steps—planning and clarifying the task, conception, execution design, and detailed design. According to Andreasen [1], French [49], Malmquist and Svensson [72], and Ehrlenspiel and Meerkamm [29], the distinct steps of product development are the definition of functions and their realization according to principle solutions. These concepts

[56]In addition to the known disciplines of mechanics, electrics, electronics, and software, a fourth discipline, service, was consciously introduced here. The service proportion of the business model must already be integrated in the early design phase [2].

were also the basis of VDI[57] guideline 2221 [100]. The Munich concretization model from Ponn/Lindemann represented a different approach [82]. Requirements which influence all concretization levels of the solution (function, active, and construction level) play a special role here. In the USA in 1999, Ulrich and Eppinger developed design methodology far beyond the actual construction space. This considered the design and development questions such as the identification of customer requirements, design for manufacturing, prototyping, and industry design. In this way, a series of production development techniques were presented which should bring together the marketing, design, and manufacturing functions of the company [101].

A broader image of design approaches resulted in electrotechnology and electronics (E/E), primarily due to the extremely varied application areas and a rapid technology transformation. A significant classification feature for the respective design model is the degree of technological independence. Typical methods are, for example, the top-down and the bottom-up design [86], as well as the yoyo model preceding this. The method developed by Gajski and Kuhn, also known as the Y-diagram, is one of the best known technologically independent models [52]. It describes the perspectives in hardware design, in particular for the development of integrated circuits in the domains of behavior, structure, and geometry. The domains are represented as the axis of "Y." A further pragmatic approach for the design process of integrated circuits was developed by Lienig [70].

Similarly, to the approach in electronics, the developed methods in the area of software development often show a different pattern – that is, a strong behavior orientation in addition to a functional orientation [2]. Several phase-oriented models were developed under the term software engineering which supported the software design process [81] with the aim of designing software development more productively and efficiently, for example, the waterfall, prototype, or spiral model. In 1979, Boehm introduced a design approach, the concept of which has asserted itself today in the approaches of all engineering domains—the V-model [15]. Due to the extreme effort and difficult reproducibility in the creation and maintenance of complex software, the development of the approaches mentioned takes place in a structured, strictly phase-orientated, and highly formalized process, whereby the development process is subdivided into manageable, temporally, and content limited phases. The Unified Modeling Language (UML) resulted from the development of object-oriented and model-driven concepts in software engineering. Design processes which use the capability of these concepts are the Unified Software Development Process (USDP) [88] and the Rational Unified Process (RUP) [68, 71, 72]. Design approaches which target fast implementation and flexibility during the design process and are not based on specific design procedures are referred to as agile software development approaches [9].

The term mechatronics was first used by Ko Kikuchi [60] in 1969. Initially, the term only related electrotechnical and electronic function extensions of mechanical

[57]VDI Association of the German Engineers (Verein Deutscher Ingenieure).

components and devices. Software only gained its importance in mechatronics much later, and its proportion is continuously growing [2].

3.5.2 Systems Engineering

Mechatronics design today is based on various discipline-oriented methodologies [65], for example, on the mechanical design process of Pahl and Beitz [78] or on variations of the V-model [10, 53, 77]. The established approach in Germany is the design guideline VDI 2206 [102]. Parallel to this, since the 50s, *systems engineering (SE)* was introduced as an interdisciplinary, document-driven approach to the development and implementation of complex, technical systems, particularly in American aerospace and telecommunication companies and in large military projects. This approach was permanently developed from the perspective of the software and electronics industry and, today, offers modeling and simulation support of complex, heavily networked interdisciplinary systems. Systems engineering is based on the assumption that a system is more than the sum of its sub-systems or parts and that, for this reason, it is necessary to consider the overall contexts. According to the requirements of INCOSE[58], systems engineering is a discipline which has the task of creating and implementing an interdisciplinary engineering process which should guarantee that customer and stakeholder requirements are high quality, reliable, and cost-effective, and can be fulfilled across the entire product lifecycle in a prescribed time [64]. In contrast to other approaches of model formation, there is a top-down approach in systems engineering which begins with defining a system limit and then successively continues to refine the system description from rough to fine detail. So-called black boxes are defined for this which represent sub-systems and can only be described in regard to their interfaces with other black boxes. Their content is intellectually omitted in the system consideration in order to keep the model complexity low and thereby be able to concentrate on the most important connections. This system hierarchy thinking is intrinsic to systems engineering and can also be found in model-based systems engineering [79].

◆ **Systems Engineering:**
Systems engineering (SE) is an interdisciplinary approach to enable the development and commissioning of a complex system. The approach aims to define customer needs and the required functionality early in the development process, document the requirements, and then proceed with the system design and the customer's agreement, taking into consideration the problem in its entirety. SE considers both the economic and the technical needs of the customer with the aim of creating a high-quality product and/or system which meets the needs of the user [64].

---

[58] INCOSE International Council of Systems Engineering.

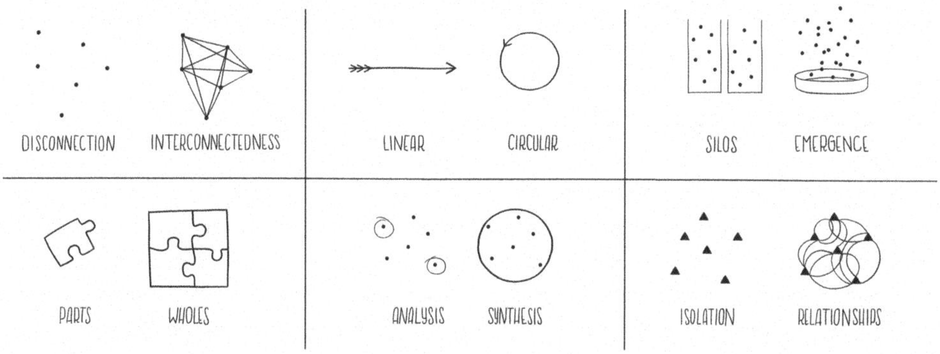

Fig. 3.19 Tools and behavior for a systems thinker[60]

3.5.3 Systems Thinking

Developing cybertronic products and systems means: "VUCA" – volatile, uncertain, complex, ambivalent. That means that we have to make decisions more and more quickly, with less and less reliable prognoses and more and more difficult to predict consequences. We require creative methodologies for this which work as quickly and intuitively as possible. Systems thinking actually already existed in 1956 when Professor Jay Forrester [47] founded the Systems Dynamic Group at the Sloan School of Management of MIT. Systems dynamics is a methodology he developed in the mid-50s, already taking into consideration the arising complexity in aerospace, for the holistic analysis and model simulation of complex and dynamic systems[59]. The term *systems thinking (ST)* further developed from this in connection with systems engineering. The core of this mind-set is: to think in systems but also think laterally beyond the limits of the system in order to analyze the interaction of the system with other systems and the environment. "Systems thinking is what differentiates systems engineering from other types of engineering methods and represents the fundamental capability which is required for system technology" [8]. Systems thinking, according to Senge, 1994 [91], is a discipline to see the whole. Engineering systems thinking is defined as an important, cognitive capability of lateral thinking in regard to engineering core competence, but also with connected fields of competency, such as human resources, cybernetics, economy, and ecology, which enables individual people to successfully execute complex system engineering tasks. As the term systems thinking incorporates such a broad spectrum of application, there is a reduction to the subject area engineering [48]. Systems thinking is inseparably

[59]https://de.wikipedia.org/wiki/System_Dynamics (retrieved on 05.22.2020).

[60]https://medium.com/disruptive-design/tools-for-systems-thinkers-the-6-fundamental-concepts-of-systems-thinking-379cdac3dc6a, Leyla Acaroglu, 2017 (retrieved on 05.22.2020).

connected with the SE philosophy. SE is a process and a method; ST is a mind-set. Figure 3.19 shows a section of tools and behavior for systems thinkers.

◆ **Systems Thinking (ST):**
Systems thinking (ST) is a holistic analysis approach which focuses on the manner of how the components of a system are related to one another and how the systems function over time and in the context of larger systems. The systems thinking approach is in contrast to traditional analysis, which investigates systems by dividing them up into their individual elements. Systems thinking can be used in any research and application area and was used, among other things, in the fields of medicine, ecology, politics, economy, city planning, human resources, and engineering[61].

Moti Frank takes it further with his 16 cognitive capabilities [48]. However, in connection with the target of this book, these capabilities are so significant that they should be listed here:

- Understand the entire system and keep the vision in sight.
- Recognize the connections, relationships, and influences.
- Discover synergies.
- Understand the system from different perspectives.
- Creative and lateral thinking.
- Understand the system without getting bogged down in detail.
- Recognize the effects and changes in the entire systems.
- Quickly recognize new ideas and concepts.
- Recognize analogies and parallelism between the systems.
- Recognize and observe limits.
- Ask good and correct questions.
- Innovators are always curious.
- Observe neighboring areas as well as your own engineering task.
- Set limits regarding teams and tasks.
- Anticipate the future.
- Permanent optimization of function, costs, and time.

Considering many large German projects and vehicle start-ups which are lagging and delayed, or not on budget and function – in particular if there is a high, valuable proportion of software and electronics in the vehicle in addition to the new drive technologies – these capabilities should be the core of every engineer's training.

[61] According to https://searchcio.techtarget.com/definition/systems-thinking (retrieved on 05.22.2020).

3.5.4 Model-Based Engineering and Model-Based Systems Engineering

While the classic methods of systems engineering are document-based, the logical next step on the path to the digitalization of systems engineering is *model-based engineering (MBE)*. This means, engineers are already working with digital models; however, not all the phases or the product lifecycle are covered. The system architecture is still lacking for interdisciplinary work in particular. The scope of the application corresponds to the typical PLM application width [67].

Document-centric work is still heavily characterized by 2D working techniques, from a fragmented, silo-orientated storage of information with the resulting manual or file transfer-based data exchange with the widest variety of formats (→ DWG, STEP, PDF, TIFF, Excel, etc.). *Model-based systems engineering (MBSE)* is essentially the fusion of the thoughts of systems engineering and of model-based engineering. It is a holistic, digital, and model-based procedure model which is now also integrated in an abstract system model for the modeling of the requirements and system architecture phases which are so crucial for interdisciplinary product development (Fig. 3.20) [50].

◆ **Model-Based Systems Engineering:**
Model-based systems engineering (MBSE) is the formal definition of digital partial models which are connected to one another along **all** phases of the product lifecycle and interact in

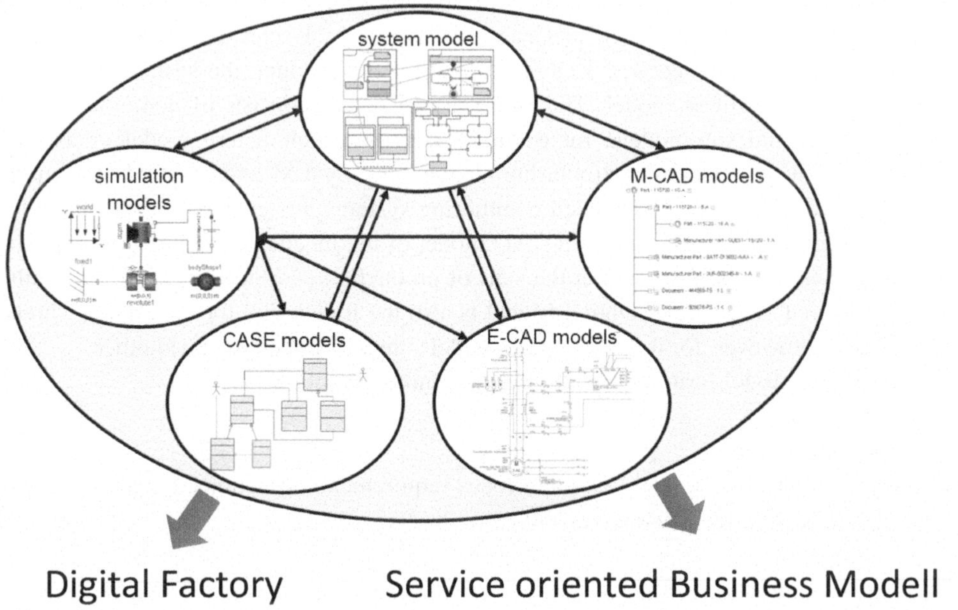

Fig. 3.20 Model-based systems engineering (MBSE) based on Friedenthal [50]

order to support the activities of requirement recording, system architecture, development/construction, verification, and validation through simulation, production planning, and production as well as operations and thereby enable the early optimization of product properties in the sense of a holistic optimization of the entire product lifecycle as well as the drastic reduction of physical prototypes through increasing digitalization of the engineering process. MBSE is the prerequisite for digital traceability based on the Digital Thread.

Often, MBSE is incorrectly understood as a pure concept of the early phase of product development. MBSE, just like SE, has always had the task of covering all phases of the lifecycle.

The presented methods of model-based systems engineering and systems thinking are excellently suited to today's requirements for the interdisciplinary development of cybertronic products and production systems. Nevertheless, MIT has introduced a new term advanced systems engineering (ASE)[62].

3.5.5 The Digital Model

The consistent MBSE approach across the entire product lifecycle is the backbone of the virtual product development and thereby also a significant challenge to the optimization of the PEP for mechatronic and, in particular, cybertronic products and systems. The multidisciplinary approach is based on lifecycle phase-specific, linked, and interactive digital partial models, also partially discipline-orientated, which are integrated along the entire product lifecycle and correspond with the application width of System Lifecycle Management concept introduced in Chapter 2. Figure 3.21 represents this connection. The approach of MBSE creates a *Digital Model* of the product, the system, and/or the service-oriented business model. The entire digital model consists of numerous partial models and partial sub-models, for example, a system architecture model or a CAD model, and always contains a structuring element (→ SysLM and/or TDM) and information derived from the phase-specific authoring systems, for example, a system architecture described in SysML and/or a CAD model of the mechanics and electronics. The core of a Digital Twin is created at the start of production (→ SOP) from the "as-built" or "as-delivered" BOM. The Digital Model is also the foundation for the Digital Thread (Sect. 3.4.2). However, for this, the partial models must be linked to one another.

The Digital Model forming partial models is linked as follows:

- Requirement model.
- System architecture model (→ main focus requirements, functions, behavior, logical structure, stakeholders, use cases, etc.).

[62]http://www.advancedsystemsengineering.org/ (retrieved on 02.01.2021).

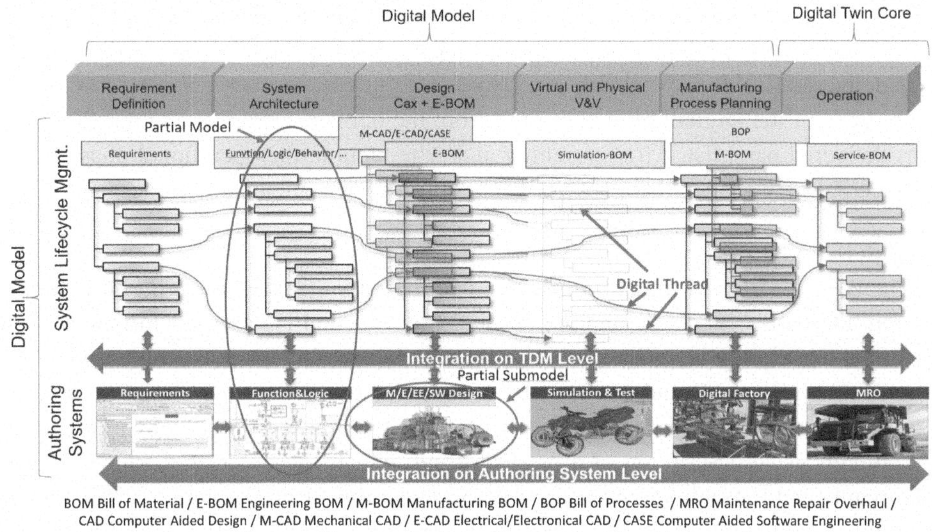

Fig. 3.21 Digital Model along the product lifecycle

- Interdisciplinary development model as a fusion of further discipline-orientated partial models from the authoring systems for mechanical (M-CAD), electrical/electronic (E-CAD), and software development (CASE[63]). The result is the development parts list (E-BOM[64]).
- The simulation model is inserted visually according to the development/construction. In reality, the simulation accompanies the entire product development and planning process. Simplified simulations with Modelica and MATLAB/Simulink can already be carried out during the development of the system architecture. All types of mechanical and electrical/electronic calculations become possible with the concretizing geometry and layout information created by the CA systems used in the development and construction (→ FEA[65], MBS[66], NVH[67], CFD[68], SPICE[69], SystemC, etc.).
- The next phase is production planning in which the manufacturing and assembly appropriate manufacturing BOM (M-BOM) and the BOP[70] derived from this are

[63]CASE Computer-aided software engineering.

[64]BOM Bill of material (E-BOM engineering BOM, M-BOM manufacturing BOM).

[65]FEA Finite element analysis.

[66]MBS Multi-body system.

[67]NVH Noise vibration harshness.

[68]CFD Computational fluid dynamic.

[69]SPICE Simulation Program with Integrated Circuit Emphasis.

[70]BOP Bill of processes.

created. In this phase, we differentiate between complex and temporally extended production still in the as-planned BOM and in the then actually delivered as-built or as-delivered BOM.

- The core of the Digital Twin is created from the delivered product and the as-built or as-delivered parts list. After enriching and/or reduction, the operational phase begins (Sect. 3.4.1), provided there is a corresponding, service-orientated business model.

◆ **Digital Model:**
The Digital Model is the backbone of model-based systems engineering. It is formed from the partial models along the product lifecycle and is based on formal or semi-formal description languages. It has a vertical component (→ authoring system, possibly TDM, SysLM) and a horizontal component (→ linking the partial models and basis for the digital thread and the typical engineering process ECM/CM). It is the comprehensive description of a product group which can be created from derivates, variants, and reusable modules. In order to make this clear, the name "150% model" is also used symbolically. It ends with the order-specific instantiation in the form of the "as-built" or "as-delivered" BOM, which forms the core of the Digital Twin. The Digital Model has a temporal component (when?), which is managed by the configurations resulting over time from the "effectivity" (→ defined by date, revision, serial number, and a view-orientated component (what?). This enables the depiction of the application-related views.

3.5.6 The System Model

The early, system-technical integration of the components of a cybertronic product and system on the abstraction level of a system model is supported by the use of modeling languages for the system architecture, such as SysML[71], for example, by defining the correlations between requirements[72] (R), functions (F), and logical (L) and physical (P) structure. It is often referred to as the so-called RFLP structure. When the systems modeling language (SysML) was published by OMG[73] in 2006, for the first time there was a modeling language which was in a position to support software development as well as model-based systems engineering in the early phase of system modeling. SysML is a UML[74]profile, which means it takes on lots of elements of UML 2.0 on the one hand, but

[71]SysML System modeling language (careful, not to be confused with SysLM).

[72]The requirements are usually defined in a requirement management system and then read into the system architecture via interfaces, for example, ReqIF.

[73]OMG Object management group, int. consortium which develops standards for object-orientated programming.

[74]UML Unified Modeling Language.

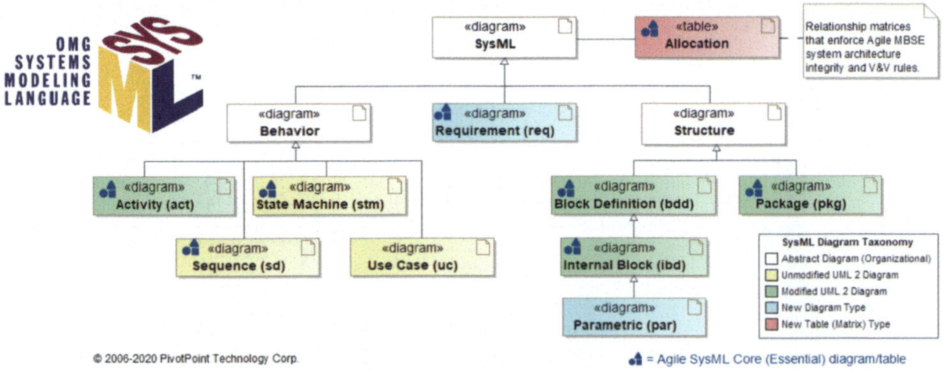

Fig. 3.22 Diagram types in SysML [51]

on the other also extends the UML with concepts which are relevant for the cybertronic product or system design. The possible types of diagram (Fig. 3.22) can be divided into three categories. There are behavior diagrams which describe the temporal processes and system behavior, and structure diagrams, which structure the components of the system and specify flows between various elements. Furthermore, there is the requirement diagram which is used to model requirements.

In addition to modeling in various types of diagrams, relationships with various meanings can also be used for modeling, in order to express that artifacts in different diagrams are related to one another [79].

The concrete implementation of the abstract representation from Fig. 3.20 into a system model described in SysML is made clear in Fig. 3.23.

A subset of requirements (white) and other SysML elements are formulated for an electrically driven golf caddy for easier understanding [55]:

- The drive should have 2 × output of 120 W.
- The maximum speed is 7 km/h.

Both requirements are read in to the system model via an interface and initially connected with the functions (red):

- The caddy must be able to be moved forward.
- The caddy must bear loads.

The hierarchical function structures can optionally form function networks. The functions are then connected with logical elements (blue), the so-called logical breakdown. This is based on:

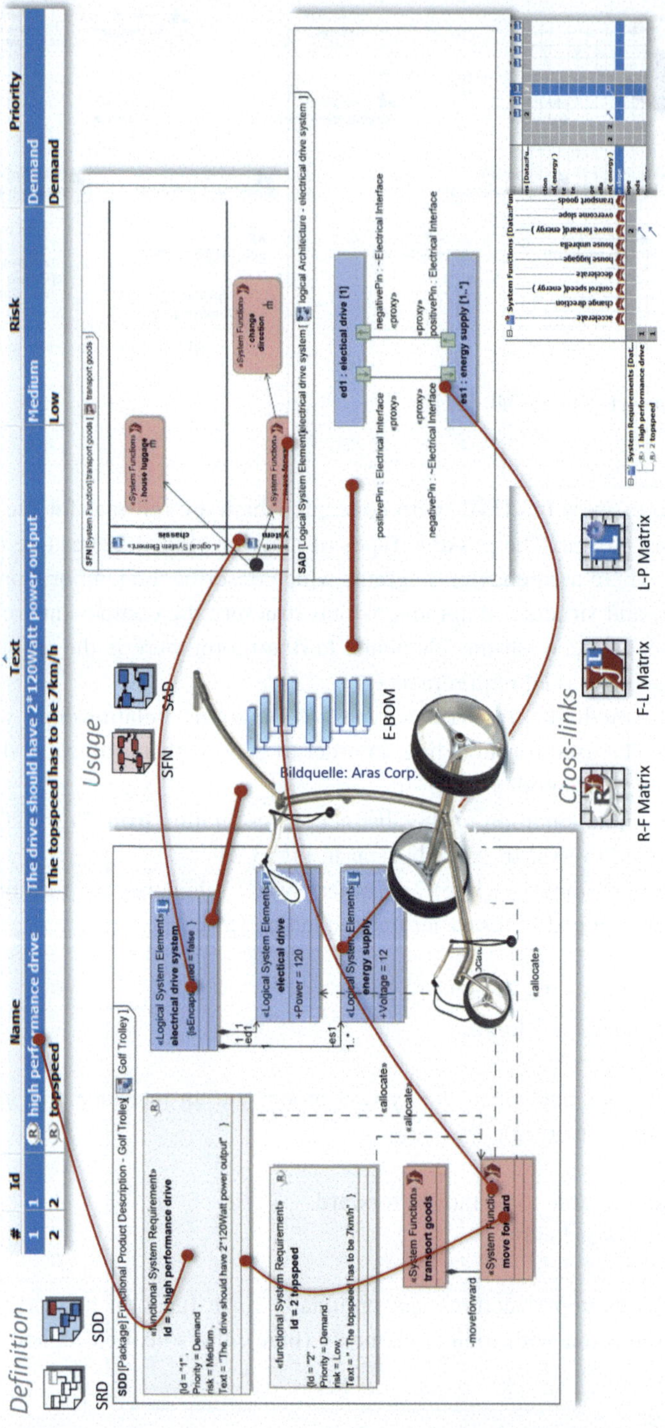

Fig. 3.23 Concrete implementation of a system model with the modeling language SysML [55]

- The electrical drive unit which consists of the following logical components:
  - The electric engine and
  - the battery.

Hierarchically logical elements can also form networks and exchange signals and information via ports. We can see that the connection of various elements of the system model (→ RFL) is documented with the help of matrices.

Figure 3.24 shows a more complex system model from the area of automotive. Further SysML artifacts are visible here, such as use case, stakeholder, test cases, and behavior.

The advantage of the system model is the extensive neutrality in relation to concrete disciplines, at least on the top abstraction level. A function is initially dependent on the implementation of a discipline. Similarly, to Pahl/Beitz [78], a function is only actually implemented with the depiction of the respective working principles. However, before the discipline-specific development is applied, there must be partitioning on various disciplines. The system model, consisting of requirements linked with the system architecture, is the central modeling concept for a digitalized engineering process for interdisciplinary, cybertronic products, and systems as well as services. Figure 3.25 shows the central importance of the system model for the following partially discipline-orientated partial models of the product lifecycle.

The system model forms the foundation for early system verification and validation through the integration of formal simulation models [51, 22, 66]. It is also the foundation for the subsequent discipline-specific design (→ M-CAD, E-CAD, CASE). One of the main advantages of Digital Models is the explicit definition of traceability links between model elements (→ Digital Thread), which make it possible to trace requirements to their respective realization in the modeled and detailed system concept to delivery and in the operation phase and vice versa [11]. Moreover, the use of digital models can contribute toward clearly interpreting system specifications in various disciplines, which leads to a reduction of the realized system complexity and a higher level of implementation security [63].

3.6 A New Interdisciplinary Design Methodology[75]

The term methodology used below—which describes a holistic approach to the development of cybertronic products and systems—is based on the definitions according to Martin [71] and Estefan [46]. Martin defines methodology as a collection of associated

[75]The $^{\mathrm{VPE}}\mathrm{SystemDevelopment}_{\mathrm{Methodology}}$ was developed over several industrial research projects together with the VPE employees Christo Apostolov, Thomas Dickopf [27], Thomas Eickhoff, Karl-Friedrich Faisst, Christian Muggeo, and Michael Pfenning [79]. A detailed description was published in the journal Konstruktion 2018/2019 [34, 35]. The employees mentioned were or are employed at the Institute VPE at the University of Kaiserslautern.

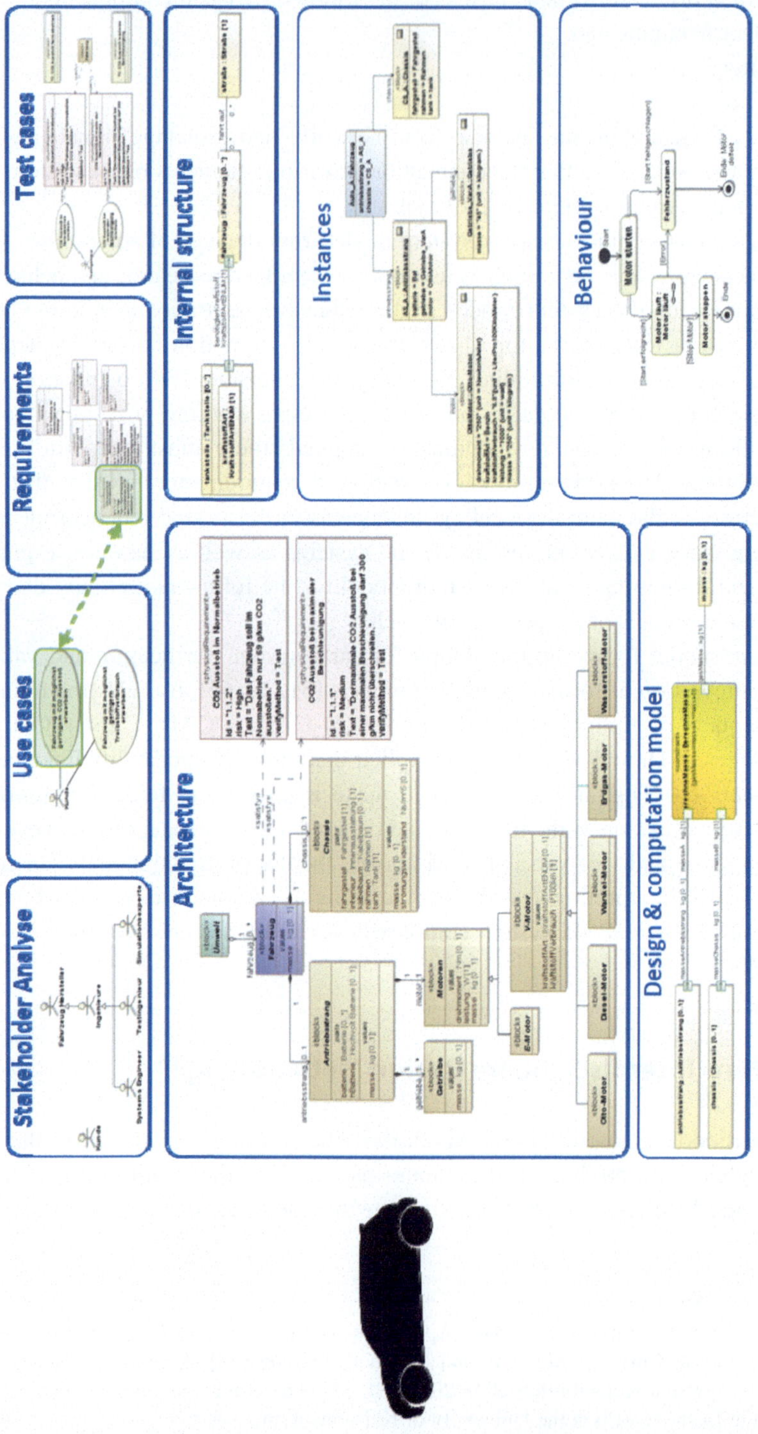

Fig. 3.24 A more complex system model from the automotive industry [17]

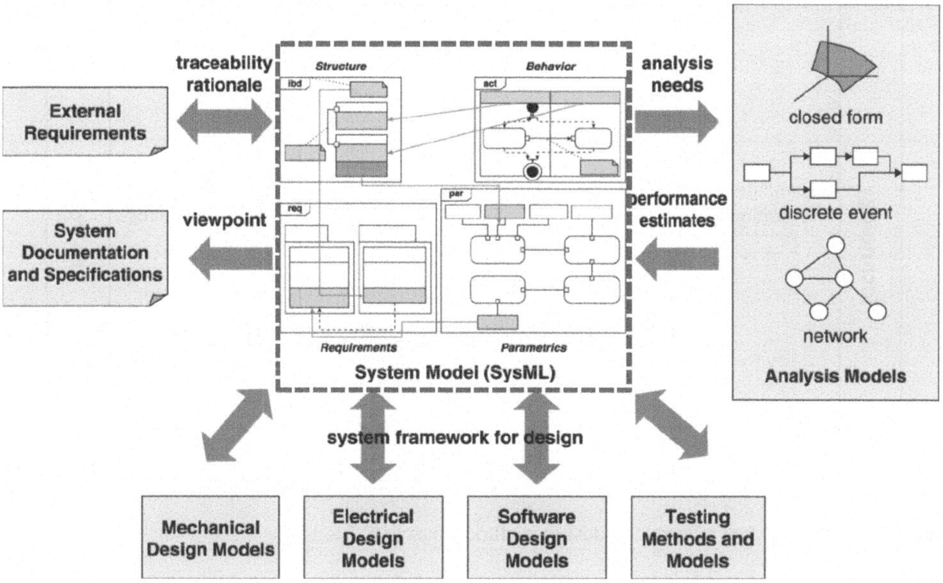

Fig. 3.25 System model as an interdisciplinary meta-model in the concept phase (cf. Fig. 3.20) [64]

processes and tools for the solution of a specific problem. The process here describes a logical sequence of tasks in order to achieve a certain objective. It therefore describes "WHAT" is to be done. The methods, on the other hand, describe the "HOW," i.e., the techniques which should be used in order to execute the tasks of the process. A tool is understood as an instrument which improves the efficiency of a task by being used for the implementation of a certain method. In the context of the book, it is of course an IT solution. Consequently, it improves both the "WHAT" and the "HOW" of the system design. Martin sees the environment as a fourth essential element for a successful system design. It includes all external factors which influence the actions of the developer. Figure 3.26 highlights the connections of the mentioned components of a methodology with additional reference to the topics people and technology.

Estefan's definition of an MBSE methodology is based on the features mentioned by Martin, however adapted to the discipline of the system engineering and the design of a system in a model-based context.

3.6.1 ^{VPE}SystemDevelopment$_{\mathrm{Methodology}}$

The ^{VPE}SystemDevelopment$_{\mathrm{Methodology}}$ describes a holistic macro-methodology for the development of cybertronic products, systems, and production systems in the context of the digitalization of engineering (→ products, processes, services) over the entire PLC.

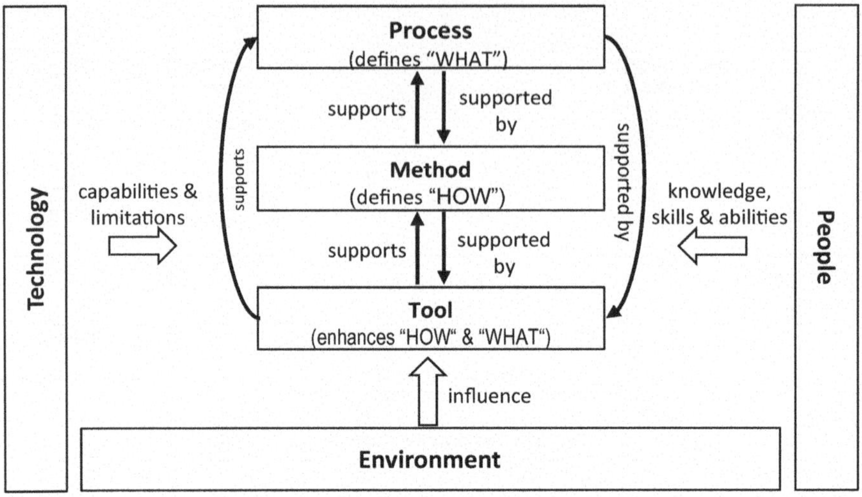

Fig. 3.26 Components of a macro-design methodology [46, 71, 113]

In reference to VDI Guideline 2206, [102] a so-called macro-methodology is understood as a general guideline for determining the procedure to develop cybertronic products and systems. The ^{VPE}SystemDevelopment$_{Methodology}$ here takes on and unifies concepts of the most varied interdisciplinary and discipline-specific development approaches and adapts these to the requirements and framework conditions of the development of cybertronic products and systems. Based on the definitions of a methodology according to Martin and Estefan, the ^{VPE}SystemDevelopment$_{Methodology}$ consists of the following four distinct components (Fig. 3.27):

- The MVPE[76] model developed for the context of the Internet of things (→ process),
- The Kaiserslautern System Concretization Model (KSCM) (→ method),
- A five-level IT architecture concept (→ IT tool), and
- Additionally, the depiction of a macro-methodology on a detailed micro-methodology (Sect. 3.6.2).

The MVPE Process Model

While the MVPE[77] model—with detailed focus on the conceptional design and development—describes the process at the top level of the lifecycle phases of a product or system, the KSCM bundles methods to fulfill the tasks which arise during the phases of

[76]MVPE Multidisciplinary virtual product engineering.

[77]Model-based virtual product development.

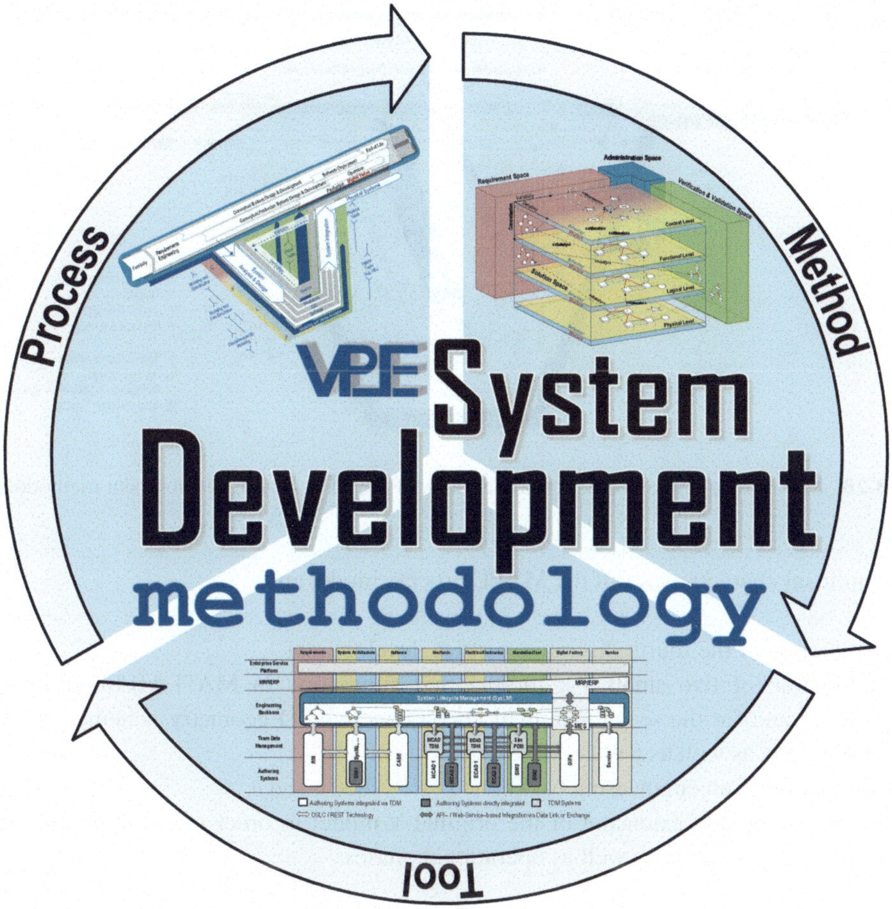

Fig. 3.27 VPE system development methodology

the conceptional design and the discipline-specific development. The five-stage IT architecture concept describes a supplementary approach for tool-based creation and the management of product and system data along the entire lifecycle.

First introduced in 2012, [41] the MVPE model has been adapted and extended many times over the years [36, 37]. However, in regard to the increasing digitalization in the industry and on the basis of recognitions gained from research projects in the context of Industry 4.0, IOT, and Industrial Internet [42, 74, 96], further adaptations and extensions were necessary. These generally concern all phases of the lifecycle, however the early conceptional phases of development in detail (Fig. 3.28).

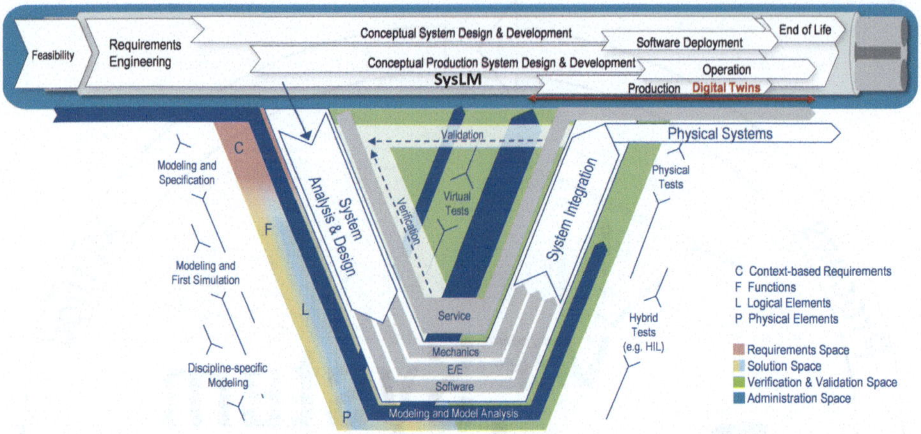

Fig. 3.28 MVPE model as a process building block of the VPE system development methodology

Significant characteristics of the MVPE process model are:

- Introduction of the fourth discipline service development,
- Introduction of two small V's for the first simulation in MATLAB/Simulink and Modelica and for the second simulations on concrete 3D geometry, scheme, and layout drawings as well as SW code,
- Introduction of an engineering backbone (→ SysLM), and
- Left and right side extension of the original V-model in order to cover the portfolio and program planning as well as operational phases.

Roubanov [87], Cadet et al. [24, 113], Sinnwell [92], and Zafirov [108] have shown that the use of MBSE concepts helps not only to overcome the complexity of today's products and systems, but also to describe their production systems. This also enables early data and information exchange between product development and production system development. As a result, the use of MBSE in this context not only leads to an early start of the production system development, but consequently also enables an earlier system production start (SOP). Here, MBSE helps to understand the complex dependencies between various disciplines and departments and to improve the previously mentioned exchange of information and data between these departments.

As previously mentioned, several times, the increasing product complexity of today's systems results, among other things, from the continuous increase of electronic components and the connected embedded software. This also brings many advantages, in particular in consideration of the topics IOT and Industrial Internet. In this way, the states and behavior of today's systems (and their elements) can be monitored and analyzed during their use. This enables immediate or faster reaction to errors, malfunctions, or new requirements which arise by triggering services, providing new software or software-based functions, and adapting or replacing hardware components. In regard to the early phases of the development process the left side of the "V" (Fig. 3.28), the previous RFLP approach was modified to a CFLP approach. This does not mean that requirements have lost their relevance in the development process. Rather, it highlights that the context (C) in which the system finds itself as part of a CTS or CPS (e.g., an automobile in the context of autonomous parking or as part of an autonomous traffic management system) during its operation or already during its production becomes more and more important and requirements are a part of this context. Thus, other components of the context are the people involved (→ stakeholders) and the use case. In other words, the system requirements are considered in relation to the context in which the system acts. The major importance of requirement engineering along the entire system lifecycle shows the high relevance for successful product, system, and service development.

In relation to $^{\mathrm{VPE}}\mathrm{SystemDevelopment}_{\mathrm{Methodology}}$, the MVPE model forms as its central process building blocks the actual basic framework of the methodology, as it dictates what is to be fulfilled at what time and by who. A detailed description of the individual phases and the detailed interdisciplinary process for the development of Cybertronic Systems results in the scope of the research project mecPro² [42].

Kaiserslautern System Concretization Model

The Kaiserslautern System Concretization Model (KSCM), which has developed over several research works, describes a product model for the systematic development of cybertronic products and systems (Fig. 3.29) and supports the MVPE model as a methodological component of the $^{\mathrm{VPE}}\mathrm{SystemDevelopment}_{\mathrm{Methodology}}$ [38, 39, 40, 27, 80]. The axis of concretization here lies parallel to the procedural sequence in the MVPE model.

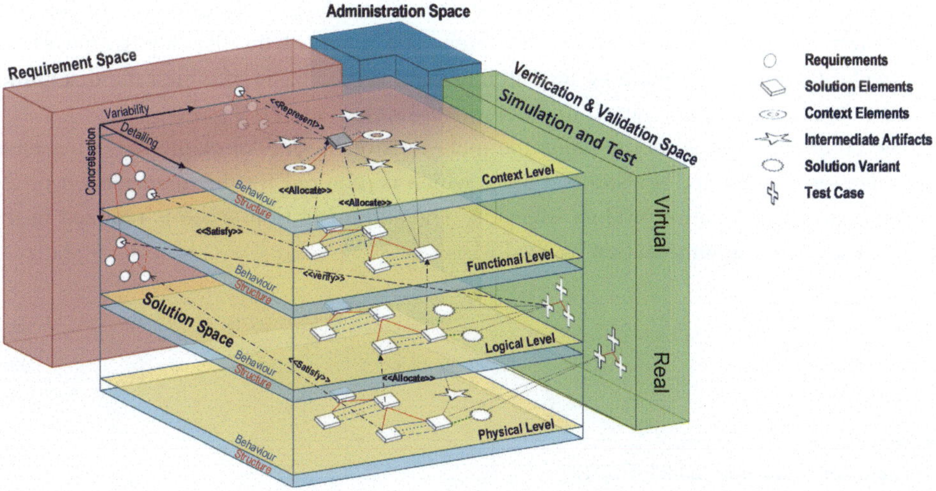

Fig. 3.29 KSCM as a method building block of the VPE system development methodology [27]

The basic structure of the Kaiserslautern System Concretization Model is based on the Munich Product Concretization Model according to Ponn and Lindemann [82] and the SCM[78] according to Pfenning [79] and was supplemented by the distinct concepts of the mecPro^2 model framework. In general, the KSCM can be subdivided into the following four spaces [27]:

- Requirement space,
- Solution space (context, functional, logical, and physical level),
- Verification and validation space, and
- Administration space.

The requirement space contains the natural language customer and system requirements which are translated into higher formality in requirement models. These requirements are related to the elements of various concretization levels of the solution space, the fulfillment which is checked by the simulations and tests of the solutions which takes place in the verification and validation space.

The transfer from the requirement space to the solution space via the context level takes place in the development process. The contexts of the system to be developed—in the sense of system limits, external elements with which the system should integrate in the respective context and through the expected and externally perceptible behavior of the system—are defined on this. Moreover, the context levels can already contain a basis architecture of the system to be developed. The internal functions and the connections between these in the sense of material, energy, and signal flows are identified on the function level. The logical level concerns the logical, concrete architecture of the system through which the function structure should be implemented. The definition of this architecture takes place in a two-stage process in which first, an individual solution variant is identified, analyzed, and evaluated, followed by the analysis and evaluation of the possible alternative logical solution architectures. The result of the development work on the logical level is a complete system architecture, on the basis of which the discipline-specific development can connect to the physical level.

The administration space contains the meta-data of the model elements with a clear identification (→ ID and revision/version), which is required for the management of the elements across the entire product lifecycle. The administration space is realized in the system-technical implementation by SysLM.

[78]System concretization model (SCM).

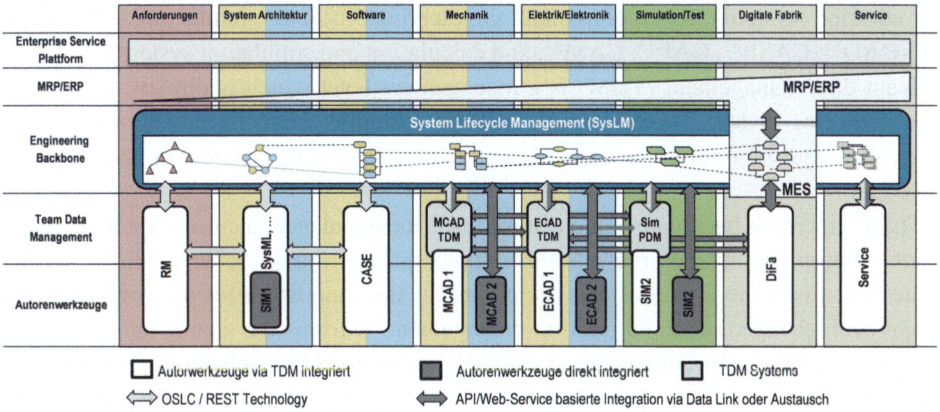

Fig. 3.30 Five-level IT architecture concept as an IT building block of the VPE system development methodology

Five-Level IT Architecture Concept

For an interdisciplinary development of today's systems, the integration and consistency of the system data and information which is created during the entire lifecycle are just as important as the IT tools in which the data is created, processed, or managed. While the MVPE model and the KSCM describe when which data and information are required and how it should be created, there is a lack of a concept to enable the creation, processing, and management of system data and information. As already mentioned, the ^{VPE}SystemDevelopment$_{Methodology}$ describes a macro-approach for the development of complex, interdisciplinary products and systems with the focus on the conceptual system design and development, but taking into consideration all the phases of the system lifecycle. Subsequently, the support of process and methods cannot be managed by a single IT tool; rather, a holistic and cross-lifecycle IT architecture concept is required which integrates all IT solutions and guarantees consistency.

In order to fulfill the requirements of a suitable tool implementation for the entire lifecycle, the ^{VPE}SystemDevelopment$_{Methodology}$ contains a five-stage IT architecture concept [12, 31, 90], which is based on the four-stage model of the VDA[79] [44] and extends it in regard to the development of systems which act in the context of IOT and the Industrial Internet.

As presented in Fig. 3.30, the IT architecture concept is formed from the following five levels:

[79]VDA German Automotive Association.

- Authoring systems for requirement engineering, system architecture, M-CAD[80], E-CAD[81], CASE[82], CAP[83], CAM[84], and calculation and simulation systems.
- Team data management (TDM) is a management level which is directly assigned to an authoring system. The objective of a TDM system is the management of the resulting data and information in a usually native data format determined by the authoring system.
- The engineering backbone is the central level for the management of system data and information. It includes the interdisciplinary product structure with all the associated documents in a neutral data format and is the foundation for the company-wide approval, change, and configuration management. It forms the significant component of every SysLM concept and can optionally be extended by application lifecycle management (ALM) and service lifecycle management (SLM) solutions, if they are not integrated part of the SysLM solution.
- MRP/ERP[85] management levels for the coordination of company-wide finances and resources, production, and logistic-relevant planning processes and production.
- Enterprise service platform and software solutions for recording and analyzing system and service data from PT during the operational phase of a system.

If you compare the two models (→ MVPE and KSCM), it becomes apparent that the KSCM covers the entire MVPE model – despite its focus on the early phases of system development. As indicated in Figs. 3.28, 3.29, and 3.30 by a colored background, the requirement (red) and solution (yellow/blue) spaces cover the entire left "wing" of the "V" along the CFLP approach. As already mentioned, the verification and validation space serve for checking whether the system has been correctly built or whether the correct system has been built. As the check can be applied to part systems and components, the verification and validation space (green) consider both the right wing of the "V" for system integration and the space of the virtual test between the two wings. The administration space is the binding link between data creation and data management. Consequently, the administration space (blue) covers the entire "V," ensures the manageability of the data created in the SysLM backbone and makes them available to the engineering processes.

[80]M-CAD CAD for applications in mechanical construction.

[81]E-CAD CAD for applications in the electrical and electronic construction.

[82]CASE Computer-aided software engineering (software development).

[83]CAP Computer-aided planning (process planning and manufacturing planning).

[84]CAM Computer-aided manufacturing (production).

[85]ERP Enterprise resource planning (MRP + financial/asset/personnel management = ERP).

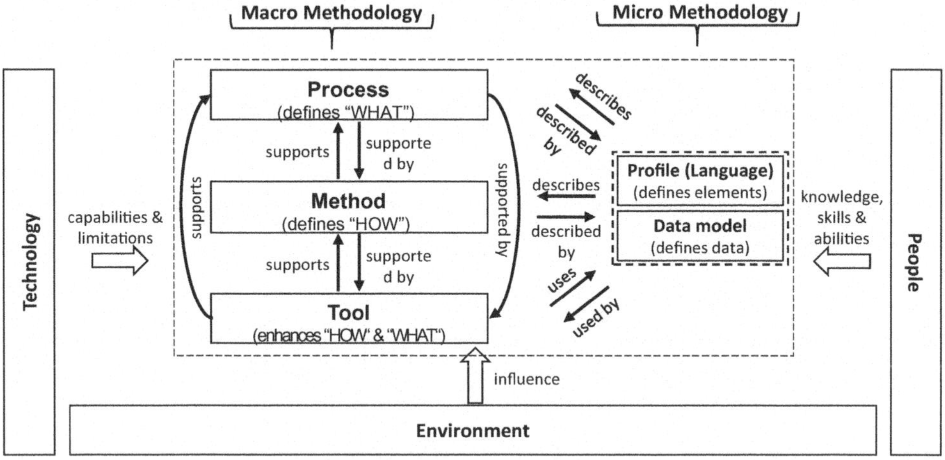

Fig. 3.31 Micro-methodology as extension of the macro-methodology [113]

3.6.2 Detailed Implementation of the System Model at a Micro-methodology Level[86]

The micro-methodology describes detailed features for the conceptional system design of the macro-methodology through a general problem solution cycle specifically for system modeling (Fig. 3.31).

A detailed procedure model on the micro-methodology level exists in all phases of the product lifecycle, for example, requirements, system architecture, M-CAD, E-CAD, CASE, simulation, process planning, and digital factory (see Sect. 4.2.1 Vertical Integration and Digitalization). The phase of system modeling is highlighted here in particular due to the complexity and the novelty of this phase. According to the definition of a micro-cycle by the VDI Guideline 2206 [102], the micro-methodology system engineers should provide support here to work on predictable and consequently plannable sub-tasks. The connection between the macro-level and the micro-level of the construction methodology is to be taken from Fig. 3.32.

The system model was introduced in Sect. 3.5.6 and macro-methodology in Sect. 3.6.1 as the abstract depiction of a cybertronic product or system to be developed. It primarily serves for the conception of the system architecture in to which all relevant requirements (R) are assigned to the corresponding system components (FLP). Here, the

[86]The VPE micro-methodology was realized across several industrial research projects together with the employees Thomas Dickopf, [27] Torsten Gilz [55], and Michael Pfenning [79] at the Institute for Virtual Product Engineering (VPE) at the University of Kaiserslautern. Two solutions resulted from this, the SE-VPE methodology (2014 [55]) and the mecPro^2 methodology which was built on the SE-VPE from the research project of the same name (2017 [27]).

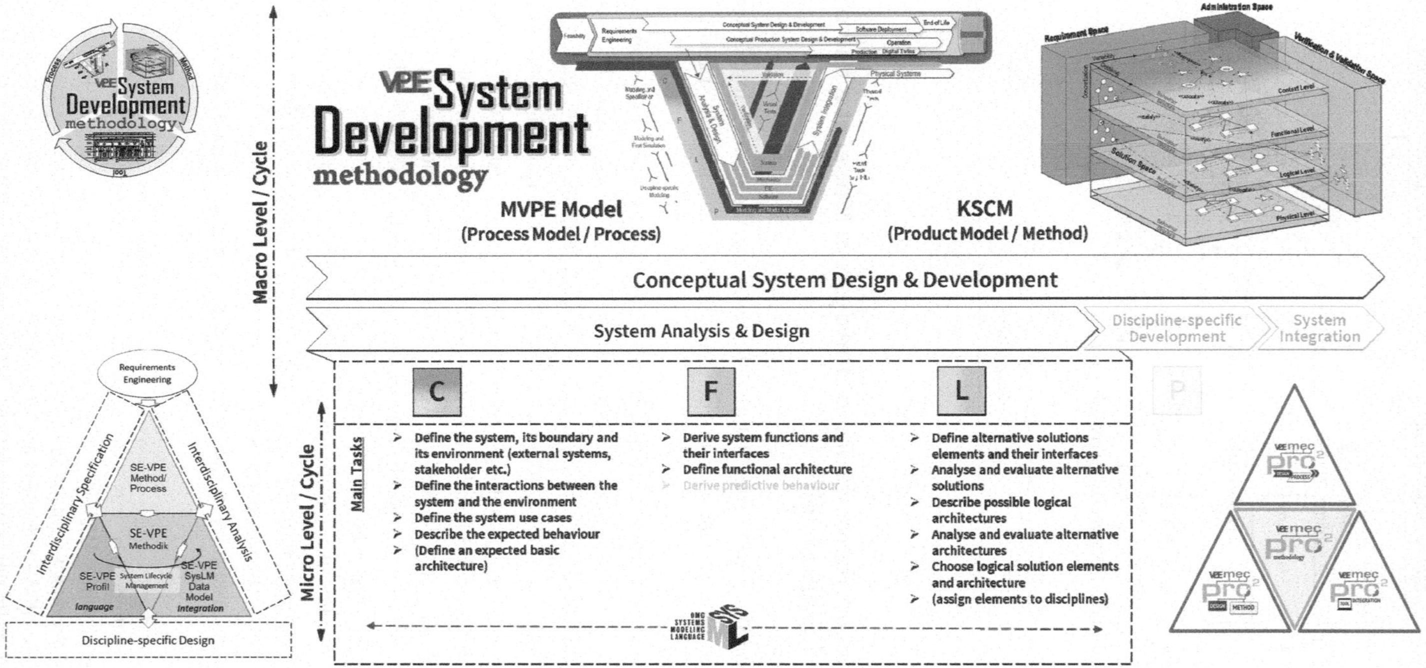

Fig. 3.32 Transformation between macro- and micro-methodology levels (according to [27])

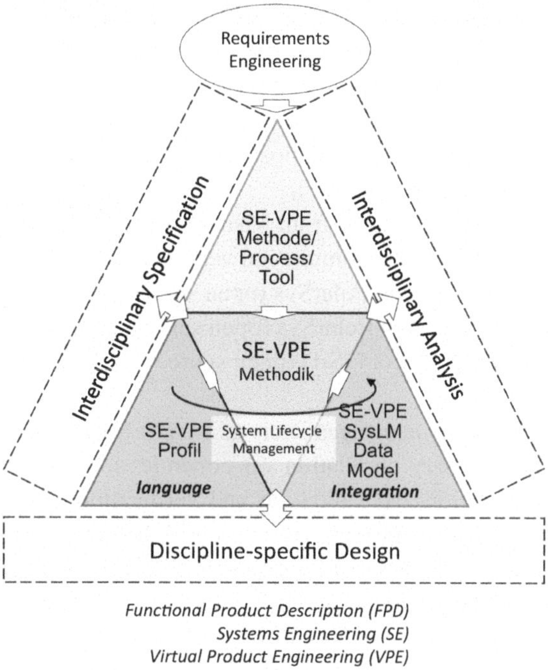

Fig. 3.33 SE-VPE procedure model at the micro-methodology level of system modeling [55]

system components are recorded with their functions, their behavior, physical properties, interfaces, and interactions among one another [42]. According to Gilz [55] and Dickopf [27], for the creation of a system model integrated into SysLM at a micro-methodology level, three building blocks are required in addition to the actual modeling language and the modeling tool (Fig. 3.33):

- A profile which limits the power of the modeling language and also extends it in some points,
- Modeling methodology (methods, process, tool) accordingly to ^{VPE}SystemDevelopment$_{Methodology,}$ and
- A SysLM data model which is in a position to take up and integrate artifacts from the system model in SysLM and apply the typical processes ERM, ECM, and CM to them.

Modeling Language and Tool

System modeling languages such as SysML and in particular the software development UML are used as an entry in the early phases, and these represent a tool for an interdisciplinary description of the system model (Sect. 3.5.6). In order to generate a system model, a wide pallet of commercial and non-commercial modeling tools—mostly based on SysML/UML, partially building on these and partially with their own modeling language—is available in the market:

- Cameo Systems Modeler (CSM) or MagicDraw von NoMagic (now Dassault),
- Enterprise Architect (EA) from Sparx Systems,
- Integrity Modeler from PTC,
- Rational Rhapsody from IBM,
- Modelio from Modeliosoft (only UML in the open-source version, in the commercial version also SysML),
- Visual Paradigm from Visual Paradigm International (only UML in the open-source version in the commercial version also SysML),
- Papyrus from PolarSys (open source),
- Capella from PolarSys (open source, similar to SysML), and
- iQUAVIS from iSiD (open source).

Simulation languages such as MATLAB/Simulink or Modelica can enable early multidisciplinary simulation on concrete development stages, which allow early concept formulation and validation and verification in connection with system description languages. The optimum integration of the system model with the early simulation is still in academic and industrial research [27]. It is also being considered whether the description of the behavior in SysML could be omitted altogether and it could be executed directly on the MATLAB/Simulink or Modelica level. The advantage of this level is a higher acceptance of these systems among users as well as the direct possibility for simulation. The system model described in SysML is more of a descriptive one and not designed for detailed simulation.

Modeling Profile

In order to define a modeling methodology within a profile, we need to think about the system model artifacts to be formed. The following content is relevant in the early phase from the perspective of all involved disciplines (Fig. 3.34):

- Requirements on the system,
- Logical and functional system elements and the networking of these elements (structure),
- Interactions between system elements and their functions (behavior),
- Performance capability and physical features (parameters),
- Adaption, integration, and communication of the system with the environment (context), and
- Analysis/simulation and test cases (validation and verification).

The language SysML can also be extended or limited with the help of the original profile mechanism which originated from the UML. So-called stereotypes can be formed from SysML elements which allow the use of concepts which are not part of the SysML standard in the system model [79]. For example, the SysML state today does not know the concept function, even if functions and function structures represent a completely

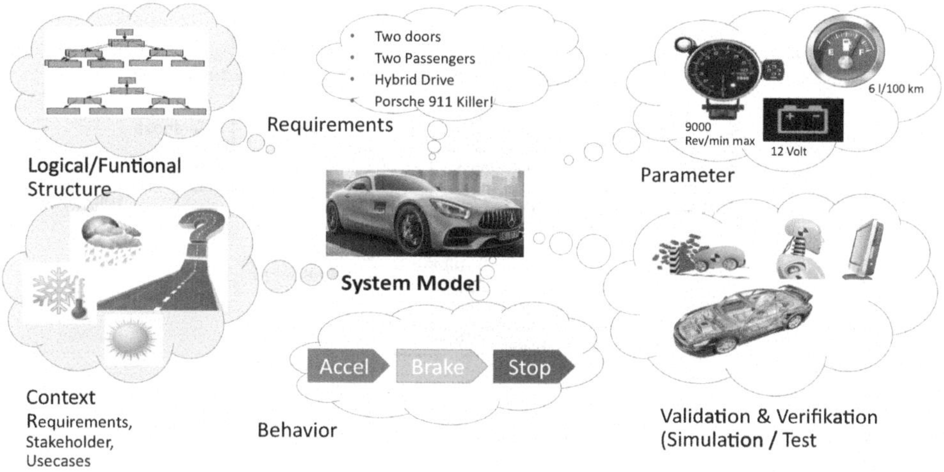

Fig. 3.34 Overview of potential artifacts of the system model

relevant basic concept in mechanics [78]. Based on this, we can also recognize that the language originates from the area of software and electronic development. The function is then anticipated in SysML version 2.0. SysML profiles are mostly an expression of a certain application and a methodology resulting from this. For example, the FAS methodology was developed in a GfSE[87]—working group[88] which introduces the mentioned function term in SysML. Examples for further methodologies are, the SYSMOD[89] [105] suggested by Tim Weilkiens, the OOSEM[90] developed by Sanford Friedenthal [51], the mecPro^2 methodology developed in the scope of mecPro^2 [27], and the SE-VPE methodology introduced by Torsten Gilz [55], which is used in Sect. 4.2.1 in the concrete implementation of vertical integration. An overview and evaluation of significant profiles and methods are contained in [26, 41]. Gilz suggests a restricted SysML profile which has the capability to produce a functional product description (FPD). This includes the

[87]GfSE Society for Systems Engineering.

[88]FAS Functional architecture for system.

[89]SysMOD Systems MODeling.

[90]OOSEM Object-oriented SE method.

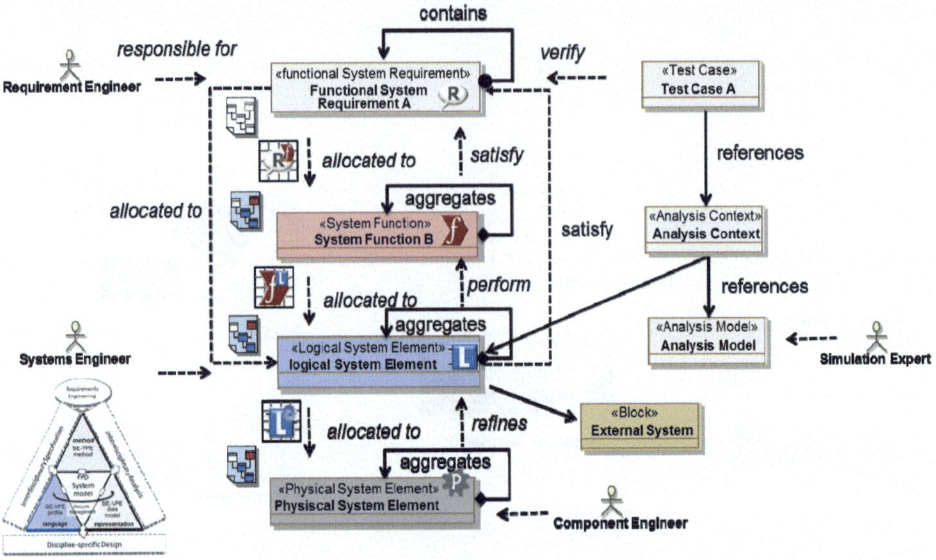

Fig. 3.35 Overview of the elements and relations in the SE-VPE profile [55]

four basic system elements, system requirement (→ functional, non-functional, and imported), system function, logical system element, physical system element, and in addition test case and analysis context (→ connection to V&V[91] and simulation). It is noticeable that system function is also from an activity type—SysML supports the concept of multiple inheritances—thus in this case, it is inherited from block and activity. Thus, system function can be represented hierarchically in a block definition diagram as well as in flows in an activity diagram. Gilz introduces numerous stereotypes, lots of element stereotypes as well as relationship and diagram stereotypes [27, 55, 79]. Figure 3.35 summarizes the elements and relations used.

Modeling Methodology

Gilz introduces five abstraction levels here (Fig. 3.36 compared with MVPE V-model in Fig. 3.28):

- Modeling and specification (R/C[92]),
- Modeling and functional description (F),
- Modeling and logical description (L),
- Simulation on the functional and/or logical level, and
- Discipline-specific model formation on the physical level (P).

[91] V&V Verification and validation.

[92] C Context.

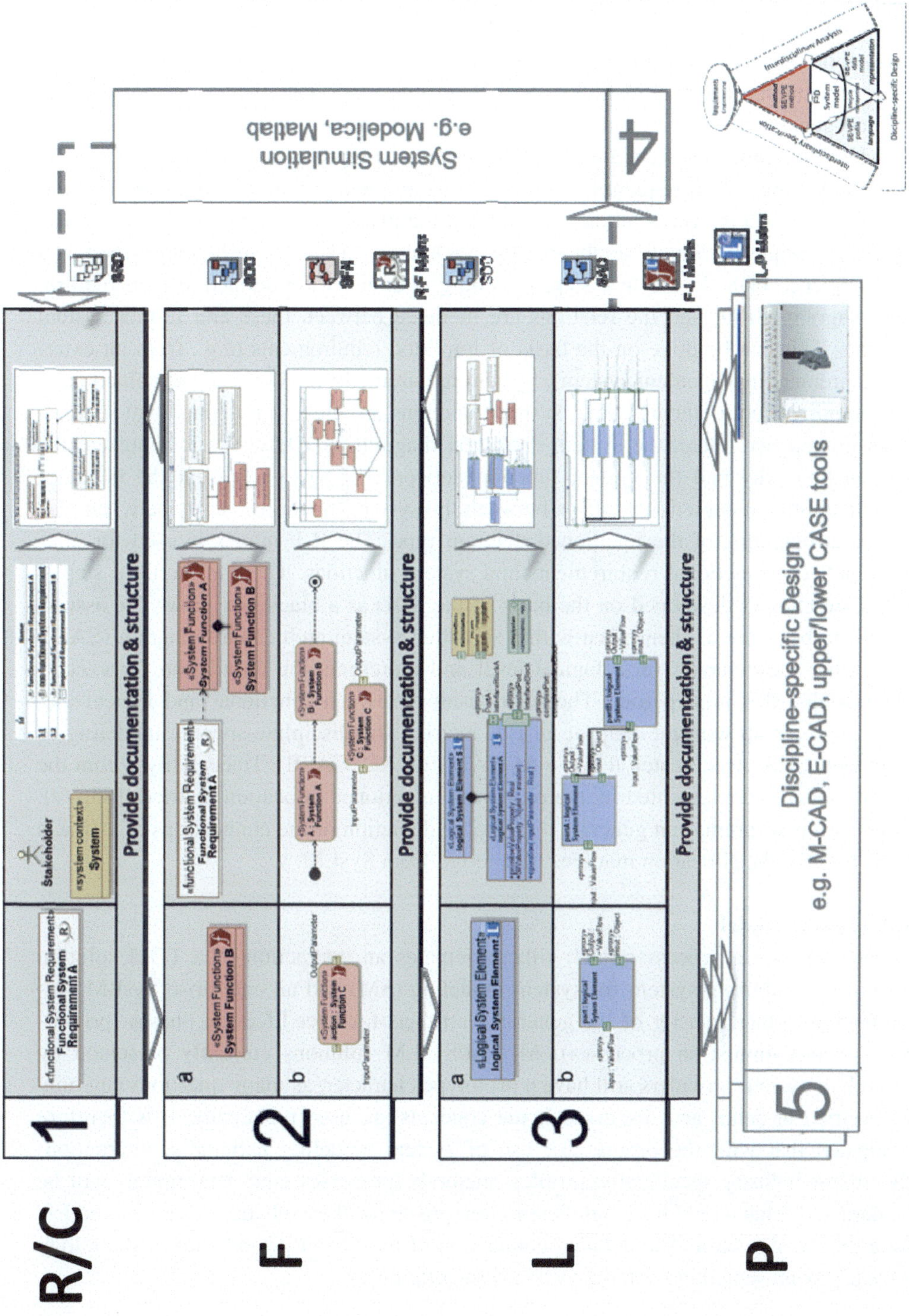

Fig. 3.36 SE-VPE modeling methodology [55]

He concentrates here on the left side of the V-model that means he does introduce test cases, and with these verification and validation methods as a stereotype in the profile, however he only indicates their methodological use. The SE-VPE modeling methodology represents the foundation for the SysML profile used. This determines which artifacts are generated in the modeling process, which relations exist between the artifacts, and which semantic is assigned to these. In addition, it represents a type of process plan for the modeling, i.e., a sequence is suggested for modeling, whereby it should be indicated that this still concerns an incremental and iteratively designed process. The guideline for the sequence is only applicable for orientation and is not to be understood as a strict sequence plan. In the first step, recording is done in the defined system requirement diagram (SRD) and the relations are modeled between these and further system elements. This can be done on the basis of imported requirements (e.g., from an external requirement management system). System requirements are derived and refined from these imported requirements. Next, system functions are created for functional requirements in step two. Here, the system definition diagram (SDD) serves to represent the function hierarchy and the system function network (SFN) to represent the behavior through additional depiction of flows between the system functions. Here, there can also be several diagrams of the mentioned diagram type. The R-F matrix is used for allocation between functional requirements and system functions. Then, in the third step, a logical structure is developed on the basis of the SDD as a black box view. The associated white box view is then given with a so-called system architecture diagram (SAD). It describes the behavior on a logical level and represents the foundation for executable models—this is step four. The relationships between functional and logical elements are built up with the help of an F-L matrix. The discipline-specific modeling of the P-level takes place in step five (→ M-CAD, E-CAD, CASE). Traceability within the system model is implemented by the previously mentioned allocation matrices [55, 79]. Visualization programs can generate the graphic depiction of the connections of artifacts on the basis of the allocation matrices transferred from SysLM.

SysLM Data Model

For support with a process model, Gilz anticipates an interaction of a TDM solution close to the authoring system for system modeling (SM) and an enterprise SysLM system for the further transfer of the generated artifacts for later lifecycle phases (primarily CAx and simulation processes). All SM/SysLM solutions currently presented by research or system providers still have a prototype character, as many questions can only be specified in detail and discussed if the concepts are used practically. It is therefore anticipated that with the progressive use of system modeling nationally, internationally, and/or industry-specifically, profiles, methods /processes, and data models will be standardized. This would be a significant prerequisite for the exchange of system models along the supply chain. The detailed connection of the SE-VPE modeling methodology to SysLM is presented in Sect. 4.2 (Vertical Integration).

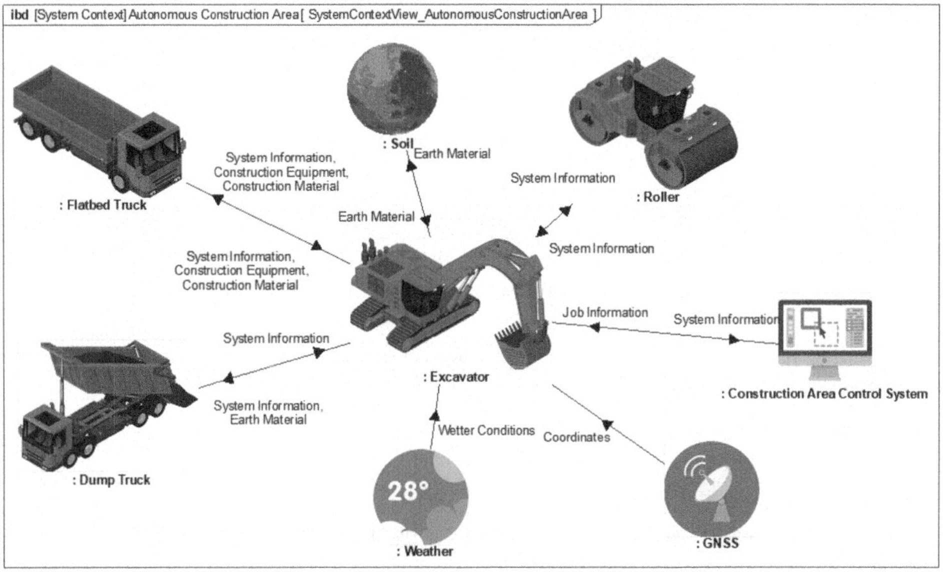

Fig. 3.37 System context of the autonomous construction area (CTS) and its cybertronic element (CTE) [27]

3.6.3 Validation of the ^{VPE}SystemDevelopment$_{Methodology}$

Sect. 3.6 concerns the relatively new approach of system modeling in the scope of an MBSE process model. The ^{VPE}SystemDevelopment$_{Methodology}$ and mecPro^2 methodology were validated on the example of an autonomous construction area [27]. The context of a cybertronic element (CTE) as part of an autonomous building site includes many interactions with the context environment. In addition to the interaction with the GNSS[93] systems, the digger communicates with various construction vehicles (rolling, flat, or tipper trucks) as well as management system for the construction site from which it receives order information and returns status reports. The system must also react to weather conditions, which play a decisive role on construction sites. There is also a passive interaction with the ground surface. Depending on the application scenario, earth is spread or removed and then loaded onto a construction vehicle. Moreover, the digger can help to load or unload heavy building materials and equipment into or out of these vehicles (Fig. 3.37).

The system model was built from numerous diagrams possible in the mecPro^2 profile. This includes:

[93]GNSS Global navigation satellite system.

- Requirements, functions, and logical elements,
- Context (→ use cases and stakeholder),
- System behavior for various states (→ digging, driving, loading, etc.),
- Activities, and
- Parameter model as a foundation of the calculation and management of the CTEs.

The detailed system model is described extensively in [27]. The concept can be subdivided into the five solution building blocks mentioned below, which are connected to the two lifecycle phases "system development" and "system in operation":

- Methodological approach (→ ^{VPE}SystemDevelopment$_{Methodology}$ and mecPro^2 methodology),
- Authoring tool (Cameo System Modeler),
- System Lifecycle Management backbone,
- Real product (physical twin), and
- Digital product as the virtual image of the real product (digital twin).

The first three modules of the demonstrator are assigned to the lifecycle status "system development" and observed the system during its concept development. Through the support of the mecPro^2 methodology along the first three levels of the Kaiserslautern System Concretization Model (Sect. 3.6.1), the authoring tool Cameo was used to create the system model. In relation to the use scenario of a digger which works in the context of an autonomous construction area, the system model includes the behavior of the digger, its system architecture, and one or several validated instances. While the system architecture anticipates the so-called 150% model of the digger with all functions and logical system elements, every instance represents the possible expression of the instantiated system architecture (100% model) and corresponds to the core of a digital twin [27]. Both of the last modules of the demonstrator are assigned to the lifecycle status "system in operation" and represent the "produced system" (→ PT) itself and its digital equivalent (→ DT). The real world represents the digger as a field device which works autonomously in the context of the building area. The 'real world' was built by students at the University of Kaiserslautern in the scope of the research at the VPE institute. In order to collect realistic system data, the digger and its environment were reproduced in the context of the autonomous building area with intelligent Lego Technic models (Fig. 3.38). Through the extension of the original Lego Technic models with sensors and actuators from the LEGO Mindstorms EV3 set and the use of a Raspberry Pi[94,95] as a control device, the construction vehicles are in a position to communicate with one another and send data [27].

[94]The interface between Raspberry Pi and Lego Mindstorms components is realized with BrickPi.

[95]Dexter Industries: BrickPi. https://www.dexterindustries.com/brickpi/ (retrieved on 06.03.2020).

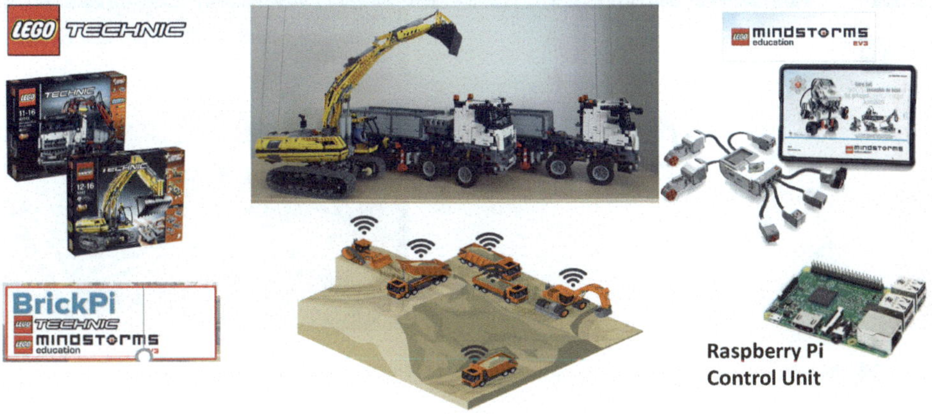

Fig. 3.38 Structure of the "real world" of the autonomous building area with LEGO components

The digital world realizes the monitoring of the system and the recording and analysis of data which the physical twins create and provide during their operation. On the one hand, the digital world contains a digital twin of the real system and, on the other, the possibility to carry out data analyses. The digital twin is based on a selection of necessary elements of the virtual image of a system which is managed in the SysLM backbone. According to the objectives of the demonstrator, the digital twin monitors the field device, has the possibility to trigger certain services, and collects all necessary data for system optimization and monitoring [27]. The real systems are monitored via its digital twin. The data of the field device is sent to its digital twin via the messaging protocol MQTT[96]. The operational states of the physical twins can be observed in the IOT monitoring tool[97] (Fig. 3.39) [27].

In summary, it can be determined that the presented process model consisting of:

- The ^{VPE}SystemDevelopment$_{Methodology}$ on the macro-methodology level and
- The mecPro^2 methodology on the micro-methodology level.

is in the position to clearly describe a complex Cybertronic System (CTS) in the development phase through a system model and to derive a simplified Digital Twin from this for the prototype. Using the example of this DT, we can recognize that the logical structure and the behavior description of the twin in this example are significant for the

---

[96]MQTT—message queuing telemetry transport is a public network protocol for M2M communication.

[97]CONTACT elements were used as an IOT monitoring tool, https://www.contact-software.com/en/news/2020/10/contact-elements-155/ (retrieved on 06.04.2020).

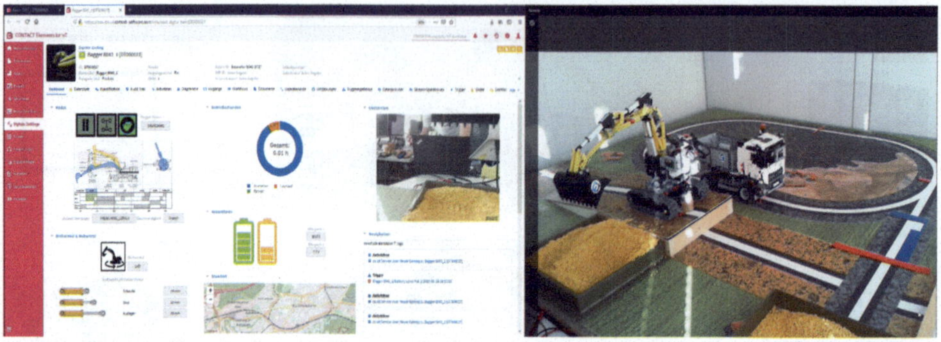

Fig. 3.39 Monitoring the operational statuses of the CTEs of the autonomous construction area [27]

simulation and monitoring of the data reported back. The SysLM backbone as the store for the system elements created in the development phase has also proven itself. Through the real reproduction of a subset of the autonomous building site from LEGO components, communication between the PT and DT could be proven and thereby graphic monitoring could be realized on an IOT monitoring tool[98].

References

1. Andreasen, M.M.: Machine design methods based on a systematic approach. Ph.D. Thesis, Lund University (1980)
2. Andreasen, M.M.: Vorgehensmodelle und Prozesse für die Entwicklung von Produkten und Dienstleistungen. In: Schäppi, B., Andreasen, M.M., Kirchgeorg, M., Radermacher, F.J. (eds.) Handbuch Produktentwicklung, pp. 247–263. Carl Hanser, Munich (2005)
3. Annunziata, M., Evans, P.: Industrie 4.0 – Pushing the Boundaries of Minds and Machines, GE General Electric, Fairfield. Connecticut, USA (2012)
4. Annunziata, M.: Four steps to succeed in industry 4.0: How Quebec forges ahead, Forbes, May 14, 2018
5. Ashton, K.: That 'Internet of things' thing. (Hrsg.). RFID Journal Online. https://www.rfidiournal.com/that-internet-of-things-thing (2009). (retrieved 2.15.2020)
6. Aurich, J., Meissner, H.: Entwicklung cybertronischer Produktionssysteme – Vorgehen für einen integrierten Entwicklungsprozess cybertronischer Produkte und Produktionssysteme. ZWF 109(1–2), pp. 70–73 (2014)
7. Aurich, J., Gülsüm, M.: Produkt-Service-Systeme für Werkzeugmaschinenhersteller. ZWF Zeit-schrift für wirtschaftlichen Fabrikbetrieb 110(4), pp. 177–181. Hanser, Munich (2015)
8. Beasley, R.; Partridge, R.: The three T's of systems engineering – Trading, tailoring, and thinking. Paper presented at the 21st Annual Symposium of the International Council on Systems Engineering (INCOSE). Denver, CO, USA. June 20–23, (2011)

[98]CONTACT elements have been used as an IOT monitoring tool.

9. Beck, K., Andres, C.: Extreme Programming Explained: Embrace Change, 2nd edn. Addison-Wesley Professional, Boston (2005)
10. Bender, K.: Embedded Systems – qualitätsorientierte Entwicklung. Springer, Heidelberg (2005)
11. Bernard, Y.: Requirements management within a full model-based engineering approach. INCOSE Syst. Eng. 15(2), pp. 119–139. Wiley Online Library. https://onlinelibrary.wiley.com/doi/epdf/10.1002/sys.20198. (retrieved 10.12.2020)
12. Bitzer, M., Eigner, M., Faißt, K.-G., Muggeo, C., Eickhoff, T.: Framework of the evolution in virtual product modeling and model management towards digitized engineering. In: Proceedings of the ICED17/21st International Conference on Engineering Design, Vancouver, Canada, August 21–25, 2017, 379–388, The Design Society, Glasgow (2017)
13. Bley, K., Leyh, C., Schaffer, T.: Digitization of German enterprises in the production sector—Do they know how "digitized "they are?", in surfing the IT innovation wave: 22nd Americas Conference on Information Systems (AMCIS 2016) San Diego, California, USA, 11–14 August 2016. Curran Associates lnc, Red Hock, 736–746
14. BMWi: Smart Service Welt II – neue Anwendungsbereiche für digitale Dienste und Plattformen, German Ministry for Economy. https://bmwi.de/Redaktion/DE/Publikationen/Digitale-Welt/smart-service-welt-ii-neue-anwendungsbereiche.html. (retrieved 12.17.2020)
15. Boehm, B.W.: Guidelines for verifying and validating software requirements and design specifications. In: Proceeding of the '79 Euro IFIP Conference, London, September 25–28. North-Holland Pub. Co. Amsterdam (1979)
16. Böhmann, T., Warg, M., Weiß, P.: Service-orientierte Geschäftsmodelle erfolgreich umsetzen. Springer, Berlin (2013)
17. Brandstätter, M., Eckl, C.: Einsatz von Model-Based Systems Engineering in der Automobil Industrie. Konferenzband Tage des Systems Engineering (TdSE) (2015)
18. Büchner, S., Muster, J.: Digitalisierung. Jetzt. Eine Entgegnung. ZOE (Zeitschrift für Organisationsentwicklung) 4, pp. 70–72 (2017)
19. Boudette, N. E.: G.M. to ldle plants and cut thousands of jobs as sales slow. New York Times, 26.11.2018
20. Bracht, U., Geckler, D., Wenzel, S.: Digitale Fabrik. Methoden und Praxisbeispiele (VDI book). Springer, Berlin (2011)
21. Broy, M.: Cyber-Physical Systems. Innovation durch software-intensive eingebettete Systeme. Springer, Berlin (2010)
22. Buede, D.M.: The Engineering Design of Systems – Models and Methods. Wiley, Hoboken (2009)
23. Cadet, M., Meissner, H.: Cybertronische Systeme. In: Eigner, M., Koch, W., Muggeo, C. (eds.) Modellbasierter Entwicklungsprozess cybertronischer Systeme – Der PLM-unterstützte Referenzentwicklungsprozess für Produkte und Produktionssysteme, pp. 19–22. Springer Vieweg, Berlin (2017)
24. Cadet, M., Sinnwell, C., Fischer, J., Stephan, N.: Kernelemente für die Zusammenarbeit von CTP-Entwicklung und CTPS-Planung. In: Eigner, M., Koch, W., Muggeo, C. (eds.) Modellbasierter Entwicklungsprozess cybertronischer Systeme – Der PLM-unterstützte Referenzentwicklungsprozess für Produkte und Produktionssysteme. Springer Vieweg, Berlin (2017)
25. Capgemini Consult: lndustry 4.0 – The Capgemini consulting view. https://www.capgemini.com/consulting-nl/wp-content/uploads/sites/33/2017/08/industrie_4.0_0.pdf (2017). (retrieved 4.12.2020)
26. Dahmen, U., Roßmann, J.: Simulation-based verification with experimentable digital twins in virtual testbeds. In: Schüppstuhl, T., Tracht, K., Franke, J. (eds.) Tagungsband des 3. Kongresses Montage Handhabung Industrieroboter, pp. 139–147. Springer, Berlin (2018)

27. Dickopf, T.: A holistic methodology for the development of cybertronic systems in the context of the internet of things, PHD thesis. Eigner, M. (ed.), University of Kaiserslautern, publication series VPE, Bd. 23, Kaiserslautern (2020)
28. Downes, L., Nunes, P.: Big Bang Disruption: Strategy in the Age of Devastating Innovation. Penguin, London (2014)
29. Ehrlenspiel, K., Meerkamm, H.: Integrierte Produktentwicklung. Carl Hanser, Munich (2013)
30. Eigner, M.: Digitaler Zwilling – Stand der Technik, ZWF Sonderheft Digitaler Zwilling, pp. 3–6 (2020)
31. Eigner, M.: Das Industrial Internet. In Sendler, U. (ed.): Industrie 4.0 grenzenlos. Springer Vieweg, Berlin. ISBN: 978-3-662-48277-3 (2016)
32. Eigner, M.: Industrie 4.0 – nur Produktionsautomatisierung oder doch mehr? Konstruktion – Zeitschrift für Produktentwicklung und Ingenieur-Werkstoffe 2015(6), 3. ISSN: 0720-5953 (2015)
33. Eigner, M., Detzner, A., Tharma, R., Schmidt, P.: Definition des Digital Twin im Produktlebenszyklus. ZWF 114(6), pp. 345–350 (2019)
34. Eigner, M., Dickopf, T., Apostolov, H.: Interdisziplinäre Konstruktionsmethoden und -prozesse zur Entwicklung cybertronischer Produkte – Teil 1. Konstruktion, 11–12–2018, VDI Fachmedien (2018)
35. Eigner, M., Dickopf, T., Apostolov, H.: Interdisziplinäre Konstruktionsmethoden und -prozesse zur Entwicklung cybertronischer Produkte – Teil 2. Konstruktion, 01–02–2019, VDI Fachmedien (2019)
36. Eigner, M., Dickopf, T., Apostolov, H.: The evolution of the V-model: From VDI 2206 to a system engineering based approach for developing cybertronic systems. In: Proceedings of the PLM17/IFIP 14th International Conference on Product Lifecycle Management, Seville, Spain, July 9–12, 2017
37. Eigner, M., Dickopf, T., Apostolov, H., Schäfer, P., Faißt, K.G., Keßler, A.: System lifecycle management- initial approach for a sustainable product development process based on methods of model based systems engineering. In: Fukuda et al. (eds.): 11th IFIP WG 5.1 International Conference—PLM 2014. Springer, Heidelberg (2014)
38. Eigner, M., Dickopf, T., Huwig, C.: An interdisciplinary model-based design approach for developing cybertronic systems. In: Proceedings of the Design 2016 14th International Design Conference, Dubrovnik, Croatia, May 16–19, 2016, S. 1647–1656. The Design Society, Glasgow (2016)
39. Eigner, M., Dickopf, T., Schneider, M., Schulte, T.: mecPro^2 -A holistic concept for the model-based development of cybertronic systems. In: Proceedings of the ICED17 121st International Conference on Engineering Design, Vancouver, Canada, August 21–25, 2017, pp. 379–388. The Design Society, Glasgow (2017)
40. Eigner, M., Dickopf, T., Schulte, T., Schneider, M.: mecPro^2 – Entwurf einer Beschreibungssystematik zur Entwicklung cybertronischer Systeme mit SysML. In: Schulze, S.-O., Muggeo, C. (eds.) Tag des Systems Engineering 2015. Hanser, Munich (2015)
41. Eigner, M., Gilz, T., Zafirov, R.: Proposal for functional product description as part of PLM solution in Interdisciplinary product development. In: Proceedings of the Design 2012112th International Design Conference, Dubrovnik, Croatia, May 21–24, 2012, pp. 1667–1676. The Design Society, Glasgow (2012)
42. Eigner, M., Koch, W., Muggeo, C. (eds.): Modellbasierter Entwicklungsprozess cybertronischer Systeme: Der PLM unterstützte Referenzentwicklungsprozess für Produkte und Produktionssysteme. Springer Vieweg, Berlin (2017)
43. Eigner, M., Muggeo, C., Dickopf, T., Faißt, K.-G.: An approach for a model based development process of cybertronic systems. In: Proceedings of the 58th Scientific Colloquium, Technical University Ilmenau, 8–12 September, 2014

44. Eigner, M., Roubanov, D., Zafirov, R.: Modellbasierte virtuelle Produktentwicklung. Springer, Berlin (2014)
45. Eigner, M., Schmich, M., August, U.: Smarte Produkte erfordern ein Umdenken bei Produktstrukturen und Prozessen – Digitalisierung, Integration, Interdisziplinarität und Föderation. Whitepaper der SIEMENS Industrie Software GmbH (2017)
46. Estefan, J.: Survey of Model-Based Systems Engineering (MBSE) Methodologies. INCOSE MBSE Initiative. http://www.omgsysml.org/MBSE_Methodology_Survey_RevB.pdf (2008). (retrieved 12.16.2020)
47. Forrester, J.W.: Industrial Dynamics, 9th edn. MIT Press, Cambridge (1977)
48. Frank, M.: Engineering Systems Thinking: Cognitive Competencies of Successful Systems Engineers, New Challenges in Systems Engineering and Architecting Conference on Systems Engineering Research (CSER), St. Louis, MO. www.sciencedirect.com (2012)
49. French, M.J.: Conceptual Design for Engineers. Springer, London (1999)
50. Friedenthal, S., Greigo, R., Mark S.: INCOSE MBSE Roadmap, in "INCOSE Model Based Systems Engineering (MBSE) Workshop Outbrief ' (Presentation Slides), Presented at INCOSE International Workshop 2008, Albuquerque, NM, S. 6, Jan. 26, 2008
51. Friedenthal, S., Moore, A., Steiner, R.: A Practical Guide to SysML: The Systems Modeling Language, 2nd edn. Morgan Kaufmann Pub, London (2012)
52. Gajski, D.D., Abdi, S., Gerstlauer, A., Schirner, G.: Embedded System Design. Springer, Boston (2009)
53. Gausemeier, J., Lückel, J.: Entwicklungsumgebungen Mechatronik: Methoden und Werkzeuge zur Entwicklung mechatronischer Systeme. Heinz-Nixdorf-Institut, Paderborn (2000)
54. Gill, H.: A continuing vision: cyber-physical systems, fourth annual Carnegie Mellon conference on the electricity industry, March 10–11, 2008
55. Gilz, T.: PLM integrated interdisciplinary system models in the conceptual design phase based on model-based systems engineer. PHD thesis. Eigner, M. (ed.): University of Kaiserslautern, publication series VPE, Bd. 14, Kaiserslautern (2014)
56. Grieves, M., Vickers, J.: Digital twin: Mitigating unpredictable, undesirable emergent behavior in complex systems. Transdisciplinary Perspectives on Complex Systems, pp. 85–113. Springer (2017)
57. Grieves, M.: Product lifecycle management: The new paradigm for enterprises. Int. J. Prod. Dev. 2 (2), pp. 71–84 (2005)
58. Grubenmann, M.: Der Mensch in der Digitalen Transformation. https://innoscope.com/der-mensch-in-der-digitalen-transformation/ (2019). (retrieved 4.12.2020)
59. Harari Y. N.: Homo Deus: Eine Geschichte von Morgen. München, Beck (2017)
60. Harashima, F., Tomizuka, M., Fukuda, T.: Mechatronics -What is it, why, and how? An editorial, IEEE/ASME Transactions on Mechatronics 1(1) (1996)
61. Hille, M., Janata, S., Michel, J.: Leitfaden Digitalisierung: Strategien, Technologien und Ökosysteme – Ein Kompendium für Entscheider im Mittelstand im Auftrag der QSC AG. Crisp Research by Cloudflight GmbH, Kassel. https://digitales-wirtschaftswunder.de/wp-content/uploads/2016/10/Leitfaden_Digitalisierung.pdf (2016). (retrieved 4.20.2020)
62. Humpert, M.: In 10 Schritten digital: Ein Praxisleitfaden für Mittelständer. https://www.bitkom.org/sites/defaultlfiles/filei/mport/170601-ln-10-Schritten-digital-Praxisleitfaden.pdf (2017)
63. Hybertson, D.W.: Model-Oriented Systems Engineering Science. A Unifying Framework for Traditional and Complex Systems. CRC Press, Boca Raton (2009)
64. INCOSE Systems Engineering Handbook: A Guide for System Life Cycle Processes and Activities. Wiley, New Jersey (2015)
65. lsermann, R. : Mechatronische Systeme Grundlagen. Springer, Heidelberg (2008)

66. ISO/IEC/IEEE 42010:2011 Systems and Software Engineering—Architecture Description (2011)
67. Kennebrew, L.: PDM/PLM Industry—Trends and best practice. Lecture ArcherGrey am 18.7.2018. https://archergrey.com/wp-content/uploads/2018/09/PLM-Ciurrent-Trends-and-Best-Practices.pdf. Zugegriffen: 24. Apr. 2020
68. Kruchten, P.: The rational Unified Process (Addison-Wesley object technology series). Addison-Wesley, Reading (1999)
69. Künzel, M., Schulz, J., Gabriel, P.: Engineering 4.0 – Grundzüge eines Zukunftsmode lls, Begleitforschung AUTONOMIK für Industrie 4.0 (Publisher). VDI/VDE Innovation+ Technik GmbH, Berlin (2016)
70. Lienig, J.: Layoutsynthese elektronischer Schaltungen – Grundlegende Algorithmen für die Entwurfsautomatisierung. Springer, Berlin (2006)
71. Martin, J.N.: Systems Engineering Guidebook—A process for Developing Products. CRC Press, Boca Raton (1996)
72. Malmqvist, J., Svensson, D.: A design theory based approach towards Including QFD data in product models. In: Proceedings of the 1999 ASME Design Engineering Technical Conferences. Las Vegas, Nevada, USA (1999)
73. Matt, C., Hess, T., Benlian, A.: Digital transformation strategies, business & information. System Engineering 57(5), pp. 339–343 (2015)
74. Mert, G., Herder, C.F., Menck, N., Aurich, J.C.: Innovative services for customized, availability-oriented business models for the capital goods Industry. In: Proceedings of the 8th CIRP IPSS Conference on Product-Service Systems across Life Cycle, pp. 501–506, Bergamo, Italy, June 20–21, 2016
75. Muro, M., Maxim, R.: What GM's layoffs reveal about the digitalization of the auto Industry. Harvard Business Review (2018)
76. Muro, M., Liu, S., Whiton J., Kulkarni, S.: Digitalization and the American workforce, Report. Brookings. https://www.brookinqs.edu/research/digitalization-and-the-american-workforce/ (2017). (retrieved 4.21.2020)
77. Nattermann, R., Anderl, R.: Approach for a data-management-system and a proceeding-model for the development of adaptronic systems. In: Proceedings of the IMECE2010, Vancouver, Canada, ASME, New York (2010)
78. Pahl, G., Beitz, W.: Konstruktionslehre. Methoden und Anwendung, 4th edn. Springer, Heidelberg (1997)
79. Pfenning, M.: Durchgängiges Engineering durch die Integration von PLM und MBSE, PHD Thesis. In: Eigner, M. (Hrsg.): University of Kaiserslautern, publication series VPE, Bd. 20, Kaiserslautern. ISBN: 978-3-95974-060-9 (2017)
80. Pfenning, P., Eigner, M: A novel procedure model for developing individualized digitalization strategies. Konferenzband Design 2020. Dubrovnik (2020)
81. Pomberger, G., Pree, W.: Software-Engineering. Hanser, Munich (2004)
82. Ponn, J., Lindemann, U.: Konzeptentwicklung und Gestaltung technischer Produkte – Systematisch von Anforderungen zu Konzepten und Gestaltlösungen. Springer, Berlin (2011)
83. Porter, M., Heppelmann, J.: Wie smarte Produkte den Wettbewerb verändern. Harvard-Business -Manager- das Wissen der Besten 36(12), pp. 34–60. ISSN: 0945-6570 (2014)
84. Porter, M.E., Heppelmann, J.E.: How smart, connected products are transforming companies. Harv. Bus. Rev. 93(10), pp. 96–111 (2015)
85. Precht, R.D.: Jäger, Hirten, Kritiker Eine Utopie für die digitale Gesellschaft. Wilhelm Goldmann, Munich (2018)
86. Rauscher, R.: Entwurfsmethodik hochintegrierter anwendungsspezifischer digitaler Systeme. Pro-Universitate-Verlag, Sinzheim (1996)

87. Roubanov, D.: Durchgehende IT gestützte Methode für eine effiziente Zusammenarbeit zwischen der interdisziplinären Produktentwicklung und Montageplanung, PHD Thesis. In: Eigner, M. (ed.): University Kaiserslautern, publication series VPE, Bd. 20, Kaiserslautern (2017)
88. Rurnbaugh, J., Jacobson, I., Booch, G.: The Unified Modeling Language Reference Manual. Pearson Education, Boston (2010)
89. Rupp, C.: Requirernents-Engineering und Management – Aus der Praxis von klassisch bis agil. Hanser, Munich (2014)
90. Schulte, T., Müller, P., Eigner, M.: mecPro^2 – Zusammenfassung und Ausblick. In: Eigner, M., Koch, W., Muggeo, C. (eds.) Modellbasierter Entwicklungsprozess cybertronischer Systeme – Der PLM-unterstützte Referenzentwicklungsprozess für Produkte und Produktionssystem. eSpringer Vieweg, Berlin (2017)
91. Senge, P.: The Fifth Discipline: The Art And Practice of the Learning Organization. Doubleday, New York (1994)
92. Sinnwell, C. F. D.: Methode zur Produktionssystemkonzipierung auf Basis früher Produktinformationen, in Produktionstechnische Berichte aus dem FBK, Publisher, Aurich, J., Volume 02/2020
93. Spur, G., Krause, F.-L.: Das virtuelle Produkt: Management der CAD-Technik. Fachbuchverlag, Leipzig (1997)
94. Stark, R., Kind, S., Neumeyer, S.: Innovations in digital modeling for next generation manufacturing system design. CIRP Ann. Manuf. Technol. 66(2017), pp. 169–172 (2017)
95. Stark, R.: Opportunities and obstacles in establishing and using digital twins: The mind of the machine—Digital twins and processing lntelligence. Chalmers University, Gothenburg, 03.26.2019
96. Ströer, F., Faißt, K.-G., Eickhoff, T., Apostolov, H., Sivasothy, P., Seewig, J., Eigner, M.: Big Data in verfügbarkeitsorientierten Produkt-Service-Systemen am Beispiel einer Landmaschine. In: Schulze, S.-O., Tschirner, C., Kaffenberger, R., Ackva, S. (eds.) Tag des Systems Engineering, Paderborn, pp. 285–294. Hanser, Munich (2017)
97. Tafvizi Zavareh, M.: Systematik zur Reifegradmodell-basierten Validierung der Digitalisierung des Engineerings, PhD Thesis. In: Eigner, M. (Hrsg.): University of Kaiserslautern, publication series VPE, Bd. 24, Kaiserslautern (2021) (submitted PhD)
98. Tafvizi Zavareh, M., Sadaune, S., Siedler, C., Aurich, J., Zink, K.J., Eigner, M.: A study on the socio-technical aspects of digitization technologies for future integrated engineering work systems. In Proceedings of NordDesign. Linköpig Sweden. ISBN: 978- 91-7685-185-2 (2018)
99. Tao, F., Cheng, J., Qi, Q., Zhang, M., Zhang, H., Sui, F.: Digital twin-driven product design, manufacturing and service with big data. Int. J. Advanced Manufacturing Technology 10, pp. 2233–2247 (2017)
100. VDI 2221 – Methodik zum Entwickeln und Konstruieren technischer Systeme und Produkte. Beuth, Berlin (2019)
101. Ulrich, K.T., Eppinger, S.D.: Product Design and Development. McGraw-Hill Education (ISE Editions), New York (1999)
102. VDI 2206: Entwicklungsmethodik für mechatronische Systeme – Design Methodology for Mechatronic Systems. Beuth (2020) (Gründruck)
103. VDI Nachrichten, No. 20/21, der Brückenbauer, 28/29 (2020)
104. Wagner, T., Herrmann, C., Thiede, S.: lndustry 4.0 impacts on lean production systems. Procedia CIRP 63, pp. 125–131 (2017)
105. Weilkiens, T.: Systems Engineering mit SysML/UML: Anforderungen, Analyse, Architektur, 3rd edn. dpunkt, Heidelberg (2014)

106. Eigner M.: WiGeP-Positionspapier Digitaler Zwilling. ZWF Sonderheft Digitaler Zwilling (2020)
107. Wollert, J.: Industrie 4.0 – Warten bis die Revolution vorbei ist? VDI – Wissensforum. https://www.vdi-wissensforum.de/news/industrie-40-warten-bis-die-revolution-vorbei-ist/ (2015). (retrieved 5.12.2020)
108. Zafirov, R.: Model-based systems engineering methods for integrated product design, process planning, and production system design, PHD. In: Eigner, M. (ed.): University Kaiserslautern, publication series VPE, Bd. 21, Kaiserslautern. ISBN: 978-3-95974-068-5 (2017)
109. Eigner M.; Detzner A.; Tharma R.; Schmidt P.: Holistic Definition of the Digital Twin, submitted to International Journal of Product Lifecycle Management 2021
110. Lee, J., Jin, C., Liu, Z.: Predictive big data analytics and cyber physical systems for TES systems. In: Lee, J., Jin, C., Liu, Z. (eds.) Advances in Through-life Engineering Services, pp. 97–112. Springer, Cham (2017)
111. Glaessgen, E.H., Stargel, D.S.: The digital twin paradigm for future NASA and US air force vehicles. In: 53rd AIAA/ASME/ASCE/AHS/ASC Structures, Structural Dynamics and Materials Conference. Special Session on the Digital Twin. Honolulu, HI, United States, pp. 1818 ff., 23–26 April 2012
112. Schuh, G., Blum, M.: Design of a data structure for the order processing as a basis for data analytics methods. In: Portland International Conference on Management of Engineering and Technology (PICMET), pp. 2164–2169 (2016)
113. Cadet, M.: Prozessrahmenwerk zur agilen und modellbasierten Produktentwicklung in der frühen Phase, PHD submitted, University of Kaiserslautern, Institute IMAD, Prof. Teutsch (2021)

4 Engineering 4.0—Implementation of the Digitalization of Engineering

If you don't know your destination, you can't find the way.

German proverb

Abstract

This chapter explains the implementation of the trends and methodologies of digitalization presented in Chap. 3, specifically in the application area of engineering. Digitalization of the product and the connected service developed in the framework of service-orientated business models are presented here. A further significant point of digitalization is the horizontal and vertical integration of the technical and administrative work processes along the product lifecycle. Vertical integration concerns the integration of the authoring systems along the lifecycle phases, requirements management, system architecture, CAD in mechanics, electronics and computer-aided software engineering (CASE), and simulation. Horizontal integration focuses on the administrative functions such as release, change, and configuration management across the entire product lifecycle (PLC) and the technical integration of the information created in the individual partial models of the PLC phases. The typical application functions along the product lifecycle are represented on the basis of a market leading SysLM system. The chapter is rounded off with a requirements analysis of SysLM and a presentation of a maturity model for the evaluation of the digitalization of engineering.

This chapter is structured according to Fig. 4.1. The framework conditions for the successful implementation of the digitalization regarding people, organization, and data from Sect. 3.1 are summarized once again. When the applications of the vertical and horizontal integration are presented, these are generally based on the SysLM system Aras Innovator.

M. Eigner, *System Lifecycle Management*, https://doi.org/10.1007/978-3-658-33874-9_4

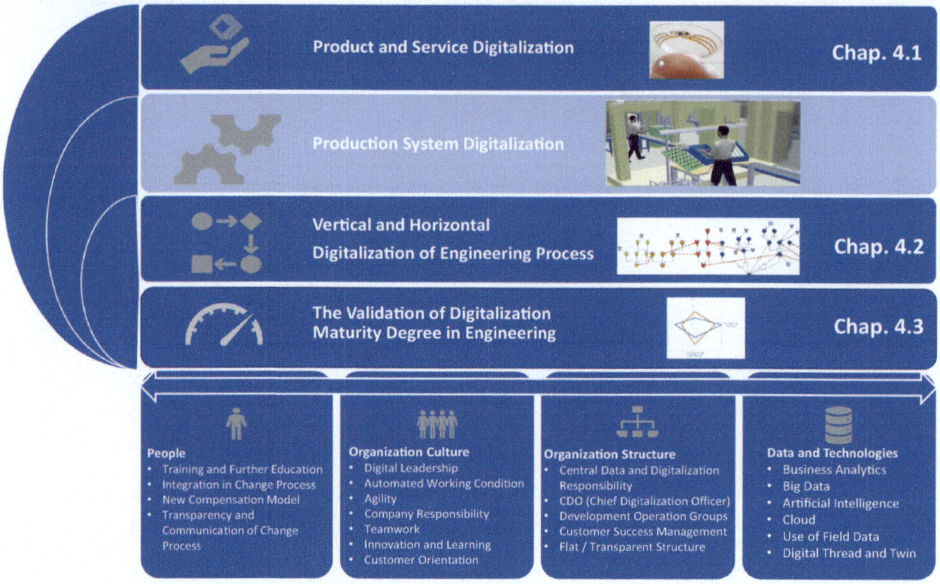

Fig. 4.1 Digitalization of engineering/engineering 4.0

The digitalization of production systems is not dealt with in this book due to the focus on the area of engineering.

4.1 The Digitalization of Products and Services

As described in Chap. 3, IOT assumes networked and inter-communicating systems and services built on these (→ IOS[1]). The functional scope of current mechatronic systems is significantly extended through the mutual networking and influencing. When these communicate with one another—generally over the Internet—we refer to *Cyberphysical Systems (CPS)* or *Cybertronic Systems (CTS)*. Both represent a further development of mechatronic systems in the direction of intelligence and communication capability. The complexity of system and process information will continue to increase due to the increasing networking of technical systems across the virtual world of the Internet. The term Cyberphysical System (CPS) correlates strongly with the area of embedded systems and, according to Lee, describes a network of highly connected and interacting devices. Or more precisely, Lee defines CPS as the integration "of the calculation with physical processes." Embedded computers and networks monitor and control the physical

[1]The combination of IOT and IOS results in the "new" designation IOTS.

processes, generally with feedback loops, in which physical process calculations are influenced and vice versa [9, 42, 61, 62, 83]. Broy states: "Cyberphysical Systems realize the networking of embedded systems for monitoring and controlling physical processes through sensors and actuators across communication devices with global digital networks (→ Cyberspace)" [19, 42].

The term *Cybertronic System (CTS)* was defined in the German research project mecPro[2] (model-based engineering of products and production systems) [35]. Cybertronic Systems and Cyberphysical Systems are extremely similar. However, unlike CPS, which focuses more on electronics and IT, CTS is always based on a mechatronic system (MTS) with a basic mechanical system as a significant component, in addition to electronics and software which control the basic system[2] [20]. The entire system is capable of communicating internally and externally with the Internet through networking. The connection between the elements of a CTS is based on bidirectional information exchange. CPS and CTS can network further and form systems of systems (cf. Sect. 3.3, Fig. 3.8).

Smart products are significant components of CTS and CPS and offer new functions which are based on intelligence and connectivity. Smart products contain sensors which deliver information about their environment, their current use, and their current status (→ *aware*). The data sent by the sensor are analyzed and actuators are controlled on this basis. Connectivity in smart products offers the possibility to communicate within the product and between products. Integrated interfaces enable interaction with human users (→ *connect*). This also forms the foundation for *Cybertronic Production Systems (CTPS)* and ultimately for that of the smart factory. Further smart product capabilities are the adaptive function and adjustment possibilities to better align its environment and tasks to one another (→ *responsive and intelligent*). In the long-term, smart products will also be in a position to receive configuration and functionality themselves during their entire lifecycle and stay in connection with manufacturers in order to provide them with a wealth of information for product and production optimization and innovation.

A simple example which demonstrates the possibilities of a smart product is the well-known contact lens from Google (Fig. 4.2). This digital contact lens aims to simplify diabetes management by measuring the blood sugar levels in tear fluid. Sensors are embedded between the two layers of lens material, and a hole in the lens allows tear fluid to seep into the sensor to measure the blood sugar level. An antenna, thinner than a human hair, is used to communicate information to a wireless device. The data is then sent to an external device, for example a smartphone. After four years of research, Google's mother company, Alphabet, decided to stop the project as proving blood sugar in tears is a complicated—and potentially insurmountable—technical and scientific

[2] A simply differentiation results as follows: A smartphone is a CPS which plays a subordinate role to mechanics. An autonomous driving vehicle is a CTS. The basic mechanical system is the car.

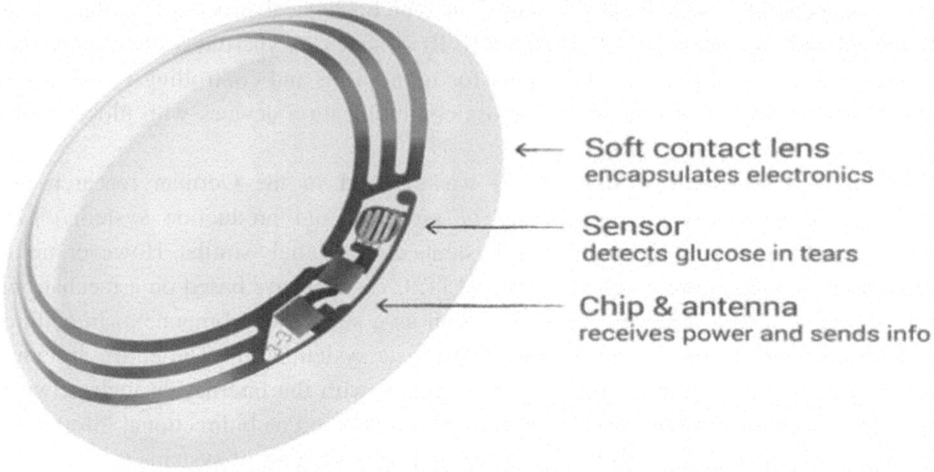

Fig. 4.2 Intelligent contact lens from Google to measure blood sugar. *Source* Google

undertaking,[3,4]. Nonetheless, this simple example shows the basic features of a smart product extremely well.

Currently, methods, processes, and integrated IT tools are being created for the development and management of information from CTS and smart products and services. Building on MBSE, research is currently underway into the process for the cross-discipline and integrated development of CTS, which includes both products, production systems, and also service-orientated business models, so-called *Product Service Systems* (PSS) [10, 35, 98]. New, often disruptive service-orientated business models for various applications are being developed on the basis of communicating "things" (→ product/systems/services), for example smart products, smart factory, smart energy, smart mobility, smart farming, and smart buildings. Services within the new overall systems will be a central success factor of the Internet of Service (IOS). The value proportion of electronics and software will continually increase with these kinds of products and embedded services. The functional scope of current mechatronic systems will be significantly extended to Cybertronic Systems (CTS) through mutual networking and influencing. They can communicate and cooperate with other products in open networks and thereby network themselves to intelligent, partly autonomous, self-adapting full-service systems. Figure 4.3 shows a research concept of an optimization of the driving route of a trash lorry through the level indication of the trash cans.

[3]https://www.cnbc.com/2018/11/16/alphabet-verily-stops-smart-lens-glucose-measuring-contact-lens.html, (retrieved on 6.20.2020).

[4]https://www.labiotech.eu/in-depth/contact-lens-glucose-diabetes/, (retrieved on 6.20.2020).

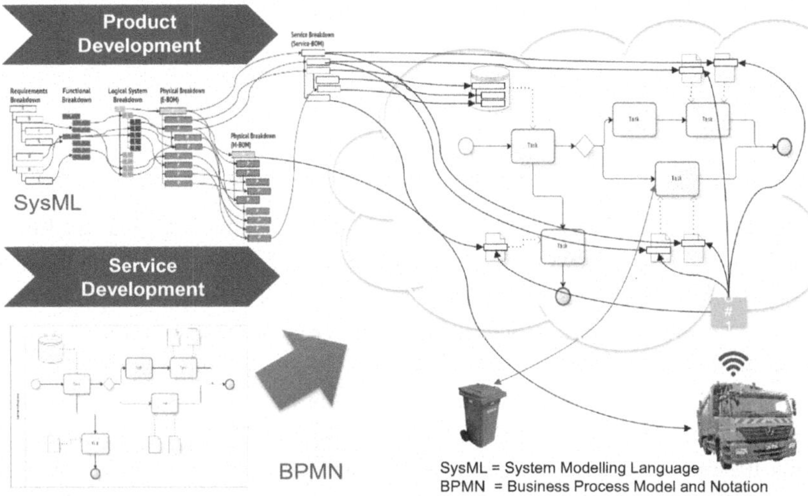

Fig. 4.3 Full-service system with the example of Trash 4.0. *Source* M. Menges

The basic strategy of a service-orientated business model is the diversification of the market performance through the sale of products or systems with a subsequent service option. The objective is to increase competitiveness through an additional customer benefit. An extended strategy in the future could consist of the extension of the service business according to the principle of build-own-operate (BOO). This means that a manufacturer not only offers service after the sale of their product, but also goes beyond this to offer the functionality fulfilled with the product as a service. Thus, the product itself becomes the service. For example, manufacturers of airplane turbines no longer sell the engine to an airline as such, rather they offer the engine as a service charged by operating hours and conclude maintenance service contracts for it. A trend which is also associated with the fact that the increasing product complexity of the necessary know-how for the engine maintenance continuously grows with it and sometimes can be better developed where direct access to development is possible. A further strategic possibility which digitalization makes possible is a data-based service, where the new or extended value creation can be generated from customer data. These PSS concepts are also becoming more and more strongly implemented with software solution providers. Further strategic option which result from digitalization are operating data-based services, where the new or extended offer can be generated from customer data. For reasons of efficiency, a manufacturer has a great interest in recording the operating data of their system in the field and, building on this, firstly, be able to carry out optimization, for example fuel consumption, secondly, collect feedback for product development, and thirdly optimize maintenance uses with the assistance of condition monitoring. Three approaches should be differentiated between here [77, 104]:

- Preventive maintenance (PM),
- Condition-based maintenance (CBM), and
- Predictive maintenance (PdM).

Preventive Maintenance is regularly planned and designed in such a way that machines, components, and other equipment work well. In an ideal case, preventative maintenance extends the lifespan of the machines and reduces the number of breakdowns. *Condition-Based Maintenance* is maintenance which is based on the continual monitoring of the state of the system for the purpose of optimizing timely maintenance deployment. If there is a system breakdown, immediate maintenance can be introduced and precisely planned on the basis of the information collected. Realizing *Predictive Maintenance* requires not only the monitoring of the state of the system with suitable sensors as close to real time as possible, rather it also primarily means the analysis of correlations in large amounts of data. PdM extends CBM with a suitable predictive model. With the assistance of data analytics approaches, the system failure should be predicted and a total standstill completely prevented [77]. Through the use and intelligent evaluation of field data, including the derivation and use of defined status information, product systems capable of communication can be optimized in operation and the risk of breakdowns reduced. Furthermore, new possibilities for the individualization of service products in capital goods result. Thus, the creation and provision of customer and machine-specific full-service concepts to ensure the highest availability to enable the functional extension of the product—provided it is software-based—deployment planning, more precise determination of the spare part potential, and the automated initiation of service processes. Service has become an essential part of positioning against the competition, improving customer retention and the basis for successful business models, in particular in the scope of digital transformation for companies in all sectors. A significant component to guarantee the function and extension capability as well as the fault-free status of the software installed on the product is *update over the air (OTA)*. Over the air is the possibility to make software updates available to a product via technologies such as Wi-Fi and mobile broadband. This means devices do not need to be directly connected to one another in order to install updates. Such OTA updates are already standard in smartphones. There are differences in the transfer of general software (→ SOTA) and firmware (→ FOTA).

However, it is becoming more and more difficult for managers to correctly estimate the opportunities and challenges of service-oriented business models for their companies. Missed opportunities, excessive expectations, misinterpreted customer requirements, and underestimated transformation risks are the results [30]. A four-stage process model developed in the scope of the research project[5] GENIAL![6] together with the consultancy company UNITY is shown in Fig. 4.4.

After the presentation of the positive results—in particular of the validation of the attractiveness of the customer service through active customer retention—the implementation on the internal business processes takes place. The determination of the digital infrastructure, the processes, and an interdisciplinary design methodology are relevant for the engineering processes concerned here (cf. Sect. 3.6). In summary, smart products, systems, and service-oriented business models open up a new era of product innovation as well as business model changes with greater social, societal, economic, and ecological importance. They are based on increasingly intelligent, modern, communication-capable components through Internet-capable and cost-effective sensor technology and actuators and new connectivity, enabled by the Internet. Key technologies are sensor/actuator technology, intelligent hardware, communication technology, and embedded systems. An Internet-capable system and service platform is developed on this that enables the aggregation of the products and services, fusion of customer and manufacturer benefits, intelligent evaluation of data, optimized management and regulation of the processes, and a graphic visualization of the statuses and interfaces. Innovative software technology forms the foundation of these levels (→ business analytics, cloud computing, big data, context-sensitive systems, etc.), artificial intelligence, machine learning, intelligent, interdisciplinary, and holistic engineering backbone systems (→ System Lifecycle Management).

4.2 The Digitalization of the Engineering Processes

In addition to the digitalization of products, systems, and services, there is also the digitalization of the engineering process in the roadmap of a digitalization strategy. The further development of the established technology of PLM to SysLM and the new methods of model-based systems engineering (MBSE) and systems thinking (ST) was recognized in its relevance for the implementation of Industry 4.0 [21]. Thus, "Central topics here are PLM-supported engineering that connects the product and production design and enables continuous support along the entire value creation. This addresses specialist points such as the integrated consideration of systems engineering, modeling, and simulation"[7]. Below, the digitalization of the engineering process should not only be understood as the existence of lifecycle-specific IT solutions, but as the integration,

[5]https://www.unity.de/

[6]GENIAL!: Joint electronic roadmap for innovations in the automobile value chain!, https://www.edacentrum.de/genial/.

[7]Plattform Industrie 4.0, https://www.plattform-i40.de/PI40/Navigation/DE/Home/home.html, (retrieved on 5.7.2020).

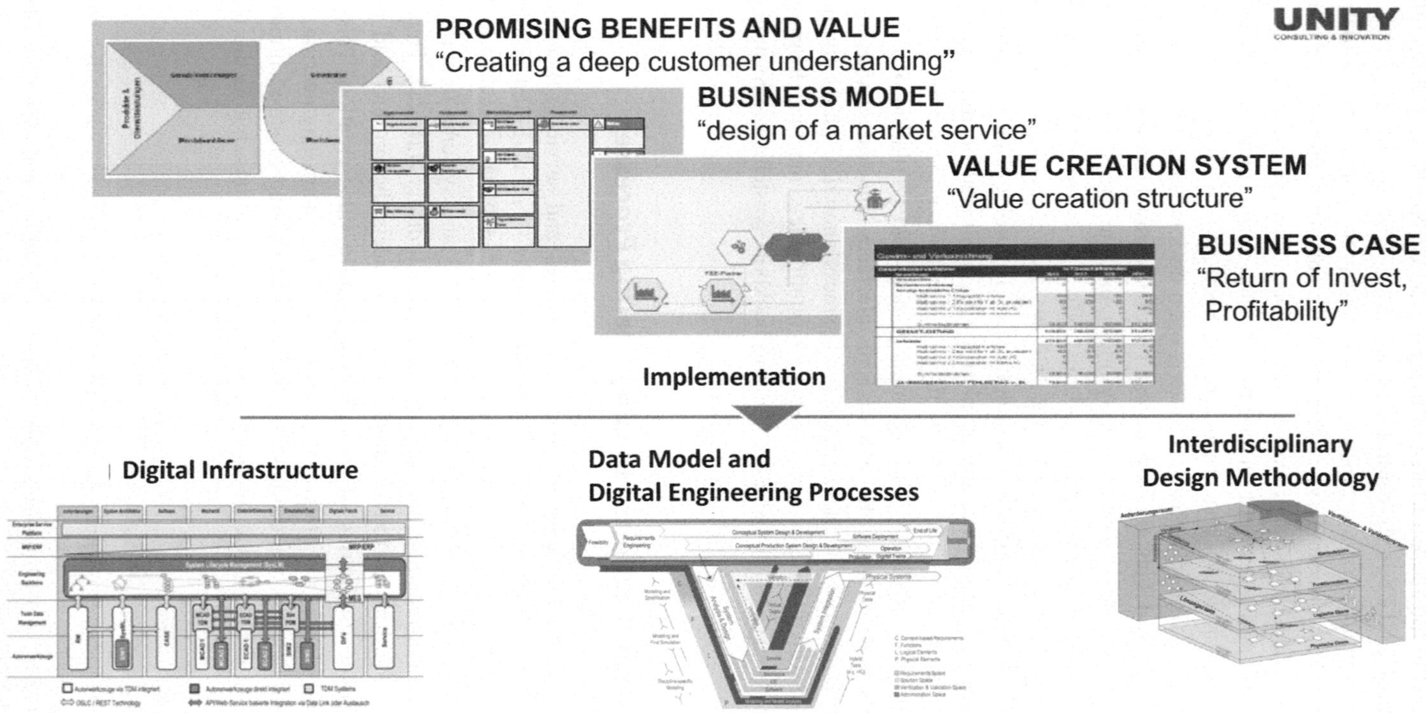

Fig. 4.4 Structured process for business model development in the context of IOT/IOS. *Source* Unity

i.e., model-based transfer, of information beyond IT systems. The use of models instead of documents is particularly advantageous in development scenarios in which several disciplines are involved. This sub-chapter presents digital partial models, their vertical integration from creation to administration (→ authoring systems, TDM, SysLM), and their horizontal integration along the product lifecycle. The terms introduced in Sect. 3.4 Digital Model, Digital Thread, and Digital Twin are shown in connection with the horizontal integration in particular, with concrete examples [2, 3, 13, 92, 93, 102].

4.2.1 Vertical Integration and Digitalization

Under vertical integration, here the integration of authoring systems directly or via the TDM solutions to an existing SysLM backbone and vice versa, should be presented according to Fig. 4.5. The "up stream" potential, especially, that is system modeling and the requirements modeling integrated in it, including the connection to the administrative levels (→ SysLM), are only little used in industry. The processes are not yet fully matured. However, the projects in this area increase continuously. The relevance of this integration in the early phases of the product lifecycle is generally recognized by experts and will initially increasingly penetrate the areas of high tech, automotive, medical, aerospace, and defense in the coming years. In contrast, the integration of M-CAD and E-CAD systems in TDM and PLM/SysLM has been standard for many years. Other less used areas of integration today are software development and

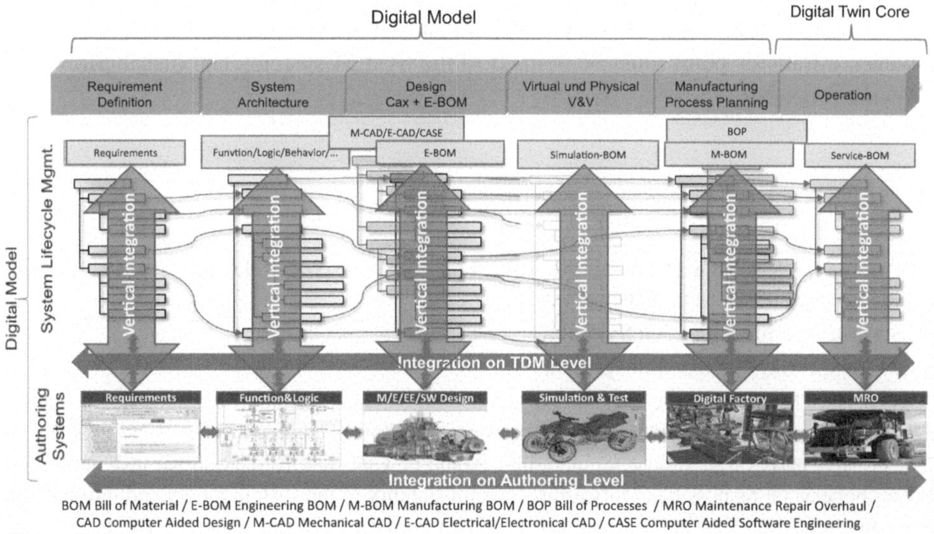

Fig. 4.5 Vertical integration and digitalization of authoring systems in TDM and SysLM along the product lifecycle

simulation. Integration in production planning and management (MRP[8]/MES[9]) is less of a technical challenge and more of a political one, as the applications of SysLM, MRP and MES overlap considerably. This subject area will be dealt with in Sect. 4.2.2, as MES and MRP are not seen as authoring systems but as central application systems in production planning, procurement, and production.

A general question which concerns the entire vertical integration strategy along the product lifecycle is the use of Team Data Management solutions (TDM). The advantage of this level is – through the close connection to the authoring system – a better acceptance by the user, as they generally use the same user interface. A further advantage is the possibility not to have to keep authoring system-related data and local processes in the SysLM backbone. This includes:

- Native data formats, for example original CAD files, software code, complex calculation results, and architecture diagrams,
- Different specific BOM formats, for example, in electronics,
- Other structuring formats, for example trunks in software development,
- Simulation-specific workflows,
- Various discipline-oriented ECM[10] processes, and
- Local and temporary ECM processes without effects on the entire system.

Of course, if required, this information should also be visible for the user on the backbone level via linked data[11]. The process level may also be sensible in certain use cases to execute configuration-relevant changes on the backbone level and temporary changes relevant for the entire process on the TDM level. This is particularly appropriate during system modeling[12] and software development[13] in the very early development phase. However, a transfer point must be defined in the SysLM system. The distribution of the following functions

[8]MRP Material Resource Planning and ERP Enterprise Resource Planning (=MRP+finance, human resource and assert management).

[9]MES Manufacturing Execution System.

[10]ECM Engineering Change Management.

[11]Linked Data here should be understood as on modern, Internet-based technologies (→REST/RDF) that can connect information from other systems without redundancy.

[12]The creative early phase of system modeling should be kept free from administrative activities in the backbone as far as possible. Only after a certain stability in the system model and internal company communication should the Requirement, Function, Logic (RFL) elements by filed over these in the backbone.

[13]In software development, changes which are not relevant for the overall system are carried frequently and quickly for local testing.

- Company-wide ECM,
- Revision management as a part of ECM,
- Version or iteration/generation management[14],
- Role and rights management, and
- Configuration management

must be defined phase and authoring system specifically. The company size and many of the applications on the authoring system level also play a role. In the past years, a trend has developed where authoring systems take over a part of the administrative intelligence, for example, which make the M-CAD system Onshape based on a database and cloud, and thereby the TDM level, unnecessary. In general, it is a permanent challenge to optimally orchestrate the distribution of functions, data, and processes across all phases of the entire lifecycle and to define the depth of the integration parallel to this. This can go from the simple saving of metadata and file (→ native and/or lightweight) from the authoring system in the SysLM backbone simply based on "drag and drop," to a 1:1 transfer of the elements and structures in the SysLM backbone supported by the process and the system. There is no "patent recipe," rather the internal company framework conditions and cultures lead to various IT architecture designs.

4.2.1.1 Requirements Management and Requirements Development (→ Requirements Engineering)

In order to develop products successfully, it is necessary to know the customer and user requirements from the start and to let these flow into the development. This generally happens before the start of a development through market analyses and user behavior analyses. Building on this, the technical and numerous behaviors, quality, and integration requirements are determined and managed in the requirements development, in order to ensure the product quality and customer satisfaction early on. Despite increased effort from the use of requirements engineering methods and tools, the benefit is clear to grasp and indispensable with the increasing complexity of systems and products. Lack of quality or failed projects can often be traced back to problems in requirements management [12, 47]. Requirements are the foundation of every product or service development. If they are unclear, incomplete, contradictory, or even erroneous, there is a risk that the development project will deliver the "false" result at the end, even if the project is carried out correctly. Furthermore, requirements define features for ensuring the quality of a product (cf. EN ISO 9000:2005—Quality Management [55]).

Requirements can be found in all phases of the product lifecycle. This could be requirements of customers in contract manufacturing or from the market in market-oriented production on the product and its entire environment, i.e., on the system

[14]This is a second or third level of change management which arises in some applications. For example, many CAD systems create a so-called iteration or generation when checking out a file.

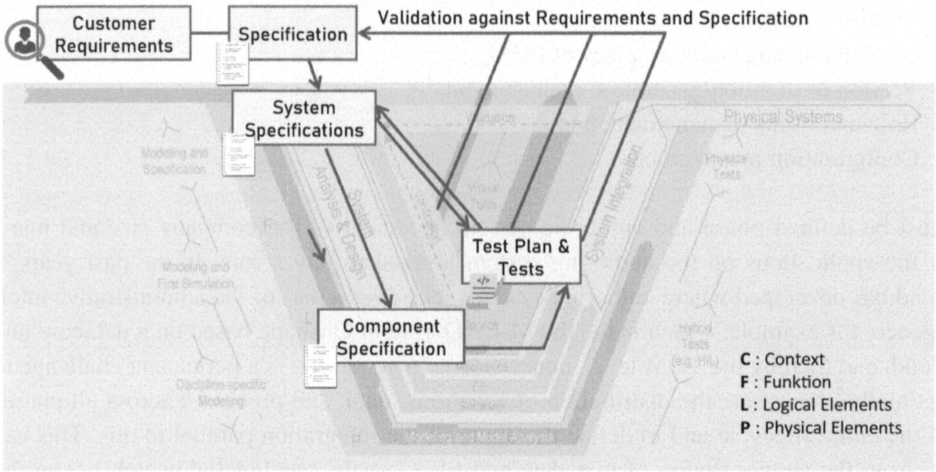

Fig. 4.6 Requirements and technical specifications in different levels of detail along the V-model [43]

architecture (→ functional/logical breakdown, behavior, use cases, stakeholder), specific requirements from disciplines, test requirements, and requirements on production and the operational use. The definition of natural language requirements and the requirements engineering built on this form the foundation for the following phases of the product lifecycle. Requirements engineering refers to all activities at the start of a development to specify the problem and tasks. In some cases, it is also seen as a partial application in addition to the system model. Requirements engineering should ensure that all relevant requirements are known and understood to the required level of detail, the project participants reach sufficient agreement about the known requirements, and the requirements are documented and specified in compliance with the regulations [82, 86]. Only explicitly formulated requirements can be taken into consideration in the development process. Figure 4.6 shows the documentation produced in requirements engineering (requirements specifications[15]) on different levels in connection with the V-model and the relations to downstream phases. The documented requirements form the starting point of a development and are closely related with the system models subsequently created in the product lifecycle [43, 105].

Requirements today are entered and managed in natural language formulations in text and hierarchy-oriented requirements management systems. It is generally possible to integrate a graphic. The requirements are structured with the help of so-called chapters. In order to enable traceability, revision of all configuration-relevant requirement elements is assumed.

[15]The requirements specification generally represents the view of the customer, the technical specification derived from this is the view of the supplier.

A more modern alternative—but currently only used in industrial research projects—would be to describe the requirements with the help of a modeling language. The use of a modeling language has the advantage that the effort for the modeler can be reduced, but the disadvantage that this modeling language must be known and learnt and should be understood by all those involved in the requirements documentation. A further advantage is that the visual formulation enables faster cognitive recording of relations. This is a big advantage in the requirements documentation of desired software behavior, use cases, and scenarios, among other things. SysML has established itself for system development in industrial research (see the following section "System Architecture Model and System Model"). Here, there should only be a short overview of requirements models and their connection to the natural language documentation [36]. SysML makes the requirements diagram available for modeling requirements. Natural language requirements can be formulated with the help of this diagram and connected via relationships to functional and logical elements. Furthermore, it is possible to connect the requirements with further system architecture elements in order to define clear validation and verification relationships (→ simulation model and test case). The reference to the term *Context* introduced in Sect. 3.6.1 – superior to the requirements – is also easily transferrable as both the stakeholder and the use case can be modeled in the SysML language.

In practice, a hybrid model may be implemented in the future. The typical circle of people of the first requirements definition (→ sales) prefers the typical textual requirements management solutions due to the simplicity of natural language entry. Implementation in SysML is recommended for the more detailed requirements engineering, in particular for complex interdisciplinary products and systems. This has the big advantage that the connection in the system model can simply take place and the requirements can be enriched by further model elements. Requirements including associated metadata between software tools of various manufacturers can be exchanged with the standard ReqIF[16] – an exchange format for requirements in XML file format. Today, this allows almost every standard market requirement management system to transfer into a system model described in SysML via ReqIF. A technical merger of requirements engineering with the system model may be sensible. A further extension in the sense of the integration of the requirements management would be the identification and adoption of parameters and formulas from the requirements for further use in the subsequent applications, for example simulation and test.

There is an extremely diverse implementation for requirements in industry. Today, handling requirements is largely fragmented and isolated. The range of application reaches from

- Office tools (→ Excel and Word) or solutions built on these,
- Dedicated requirements management solutions (→ Jama, visure, ORCANOS, Accompa, osseno, ReQtest, etc.),

[16]ReqIF Requirements Interchange Format.

- Partially integrated requirements management solutions from ALM[17] systems (→ IBM Rational, Doors, Integrity, Polarion, codeBeamer),
- Solutions from other integration concepts (→ SystemWeaver),
- Solutions integrated in PLM/SysLM (→Aras Innovator, Enovia, Windchill, and Teamcenter) and
- Full integration in a SysLM solution, possibly already a with a transfer and full approach to system modeling based on SysML.

A TDM solution is generally not required as requirements are simple and generally hierarchically structured. However, if a standalone solution is integrated into a SysLM backbone, this takes on the function of a local TDM system. The ALM integrated solutions also work with an ALM-specific TDM system (→ integrity lifecycle manager, rational engineering lifecycle manager). Due to the high significance of requirements for project success, their integration in the system model and thereby into the entire engineering process is of the utmost importance.

4.2.1.2 System Architecture Model (SAM) and System Model (SM)

The digitalization of engineering demands an interdisciplinary and holistic consideration of the product, system, and service across the complete product lifecycle. MBSE (Sect. 3.5.4) and ST (Sect. 3.5.3) support this approach through phase-typical integrated partial digital model formation. MBSE represents the bridge between the various engineering disciplines and across the various phases of the product lifecycle. System modeling languages such as SysML are used as a start in the early phases. These represent a tool for an interdisciplinary description of the system model [51]. The system model (SM) (→ requirements and system architecture model = system model) can be, at the upper levels, an interdisciplinary metamodel can, however, at the lower levels be a discipline-specific sub-model. There is a broad range of commercial and non-commercial modeling tools on the market (Sect. 3.6.2).

The topic of the integration of the SM in SysLM, and previously already in PLM, has been discussed for a few years. There are efforts in OMG[18] and INCOSE[19]/GfSE[20] as well as ProStep[21] to drive forward the integration of both system worlds and standardize it in the mid-term. In the meantime, the first PLM/SysLM system providers are offering interfaces to certain modeling tools. It was indicated in Sect. 2.4 that the support of the early development phase and the integration of an MBSE process along the

---

[17] ALM Application Lifecycle Management.

[18] OMG Object Management Group, Homepage: https://www.omg.org/.

[19] INCOSE International Council of Systems Engineering, Homepage: https://www.incose.org/.

[20] GfSE Gesellschaft für Systems Engineering, German group of INCOSE, https://gfse.de/.

[21] ProStep iViP is understood as a globally active, independent network from industry, IT, and research. The focus of its work is on the digital transformation of engineering and production.

entire lifecycle is a significant feature of System Lifecycle Management. Therefore, only the term SysLM instead of PLM will be used below. The SysLM approach is based on the management of product information administrated in multiple partial digital models along the PLC supporting process models[22]. System models have an extremely high significance in the holistic digitalization of engineering, and therefore, the integration of the SM into SysLM is highly important. As today system modeling solutions today cannot offer mature process support. Especially in the concept phase, the influence on the later overall solution is particularly high, and therefore, the use of system models and the integration with SysLM represent a lot of added value. Gilz [44] developed three basic concepts which are fundamentally needed for SM/SysLM integration in addition to the IT tool for modeling (e.g., Cameo, Enterprise Architect, etc.) (Sect. 3.6.2):

- A functional product description (SysML profile),
- A process methodology (SE[23]-VPE method/process), and
- A data model which is capable of taking on and integrating artifacts from the SM in SysLM and applying the typical processes ERM, ECM, and CM to it (SE-VPE data model).

The focus of the concept clearly lies in the concept phase and deals with the question of SM/SysLM integration in the context of the early lifecycle phase. An interaction of a TDM solution close to the authoring system for system modeling and an enterprise SysLM system for the further transfer of the generated artifacts for later lifecycle phases (primarily CAx and simulation processes) is planned for support with a process model. All SM/SysLM solutions currently presented by research or system providers still have a prototype character, as many questions can only be specified in detail if the concepts are used in industrial practice. It is therefore anticipated that profiles, methods/processes, and data models will be standardized with the progressive use of system modeling nationally, internationally, and/or industry specifically. This would be a significant prerequisite for the exchange of SMs along the supply chain. As both the profile and the process methodology were presented in Sect. 3.6.2, the description below is limited to the VPE data model, which was also the foundation for the Aras data model for system modeling.

The SE-VPE Data Model

A central question for the integration of a SysLM system and MBSE tool is the question of which elements of the system model have their representation in the SysLM data model and in which granularity they are depicted there. Seven packages are defined from which the data model is composed (Fig. 4.7).

---

[22]The release (ERM), change (ECM), and configuration (CM) processes in particular are understood as SysLM relevant process models.

[23]SE Systems Engineering.

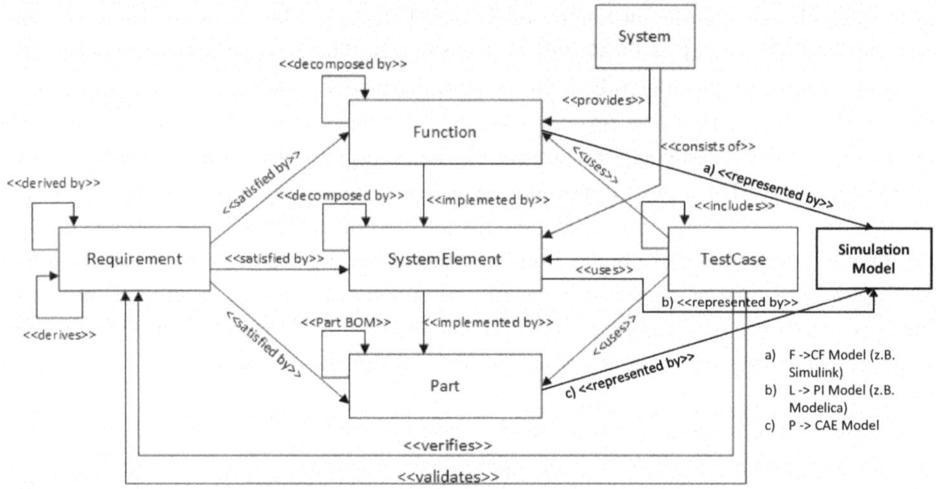

Fig. 4.7 SysLM data model for system modeling. *Source* Aras

This incorporates the core SysLM functions (→ part, workflow, lifecycle), which describe the existing SysLM system and six further ones for the various use cases in the SE-VPE method:

- Requirements inc. structure (R),
- Functions inc. structure (F) (→ functional breakdown),
- Logical elements inc. structure (L) (→ logical breakdown and system element),
- Virtual simulation (→ simulation model),
- Test cases (T),
- Discipline-specific design/part inc. Structure (P) (→ E-BOM[24]).

Here, the physical abstraction level part (P) is depicted by the item structure (part structure/BOM) of the target system (Aras Innovator®) contained in the SysLM. The data model defined in SimPDM[25](Sect. 4.2.1.4) is referenced with the simulation model. Hierarchical structures are depicted in SysLM on the levels R and F. In addition, port type and flow port, as well as the relationships satisfy and verify contained in the system model, are defined and depicted in SysLM on the logical and functional levels, so that the function and logic network can also be represented.

Two alternatives for a SysLM integration are possible on the basis of the presented functionality [44]:

[24]E-BOM Engineering BOM/Engineering Bill of Material.

[25]SimPDM Simulations PDM (the TDM level of the simulation).

- Functional product description (FPD) on the TDM level:
  Data for the early system modeling phase of product development can be found on this layer in particular. FPD[26] can use all data created in the authoring systems and is not restricted to "released" data. This provides the flexibility to create, use, and interpret data in the early phases of the design process, which could support engineers in their work, but does not have company-wide relevance. A connection to the SysLM layer is necessary in order to tackle strategic processes.
- Functional product description on the level of the SysLM backbone:
  It is possible to use all data from different disciplines which is made available by the backbone on this level. Data with the status "released" has a greater relevance for the support of company-wide processes. It may be sufficient to only keep "released" data on this level, depending on the variety of functionalities which are executed on the SysLM layer.

Both alternatives describe the extreme solutions of an integration. A sensible and pragmatic solution is a co-existence from both solutions, the use of a TDM to manage the system model in the early phase, and the publication of model information after release or a defined maturity in the SysLM system. TDM systems provide functionalities for team collaboration and SysLM provides the lifecycle and workflow management for interdisciplinary and company-wide processes. Figure 4.8 shows this kind of hybrid structure [44].

The integration of the Cameo system modeler with Aras Innovator can be taken from Fig. 4.9 (Cameo side).

Arrow 1 shows the SysLM menu embedded in Cameo, and Arrow 2 shows the model structure from Cameo shown in SysML. As already mentioned at the beginning of the chapter, the orchestration of the vertical integration can be a challenge. This applies in particular for this new application of system modeling, which is still largely in the industrial research and prototype phase.

Topics which are to be discussed in the implementation of an SM/SysLM integration are the level of detail of the integration in regard to the system element and the definition and distribution of the lifecycle and workflow process [44]. Figure 4.10 shows various levels of detail of the system model integration in SysLM. Of course, it is also possible with the SM tools to manage the diagrams directly in TDM-like solutions as files. This solution is not pursued in the sense of vertical and horizontal integration.

The following detail stages result in detail:

- Management of SM diagrams only (alternative A),
- Management of model elements with a hierarchical structure (alternative B), and
- Management of model elements with structure and topology to create the prerequisite for network structure (alternative C).

---

[26] FPD Functional Product Description.

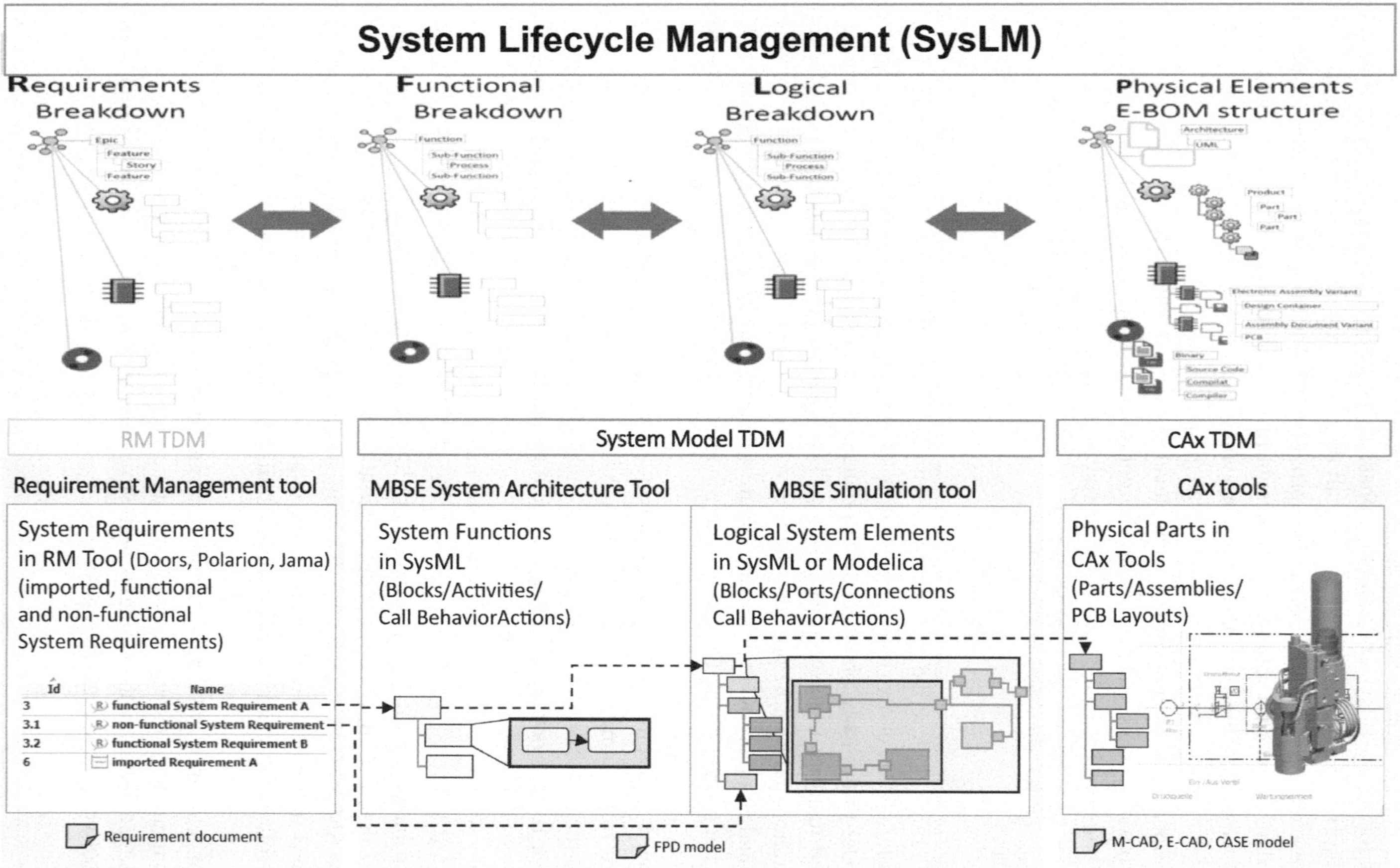

Fig. 4.8 Infrastructure of a hybrid TDM/SysML system for the phases of system modeling [44]

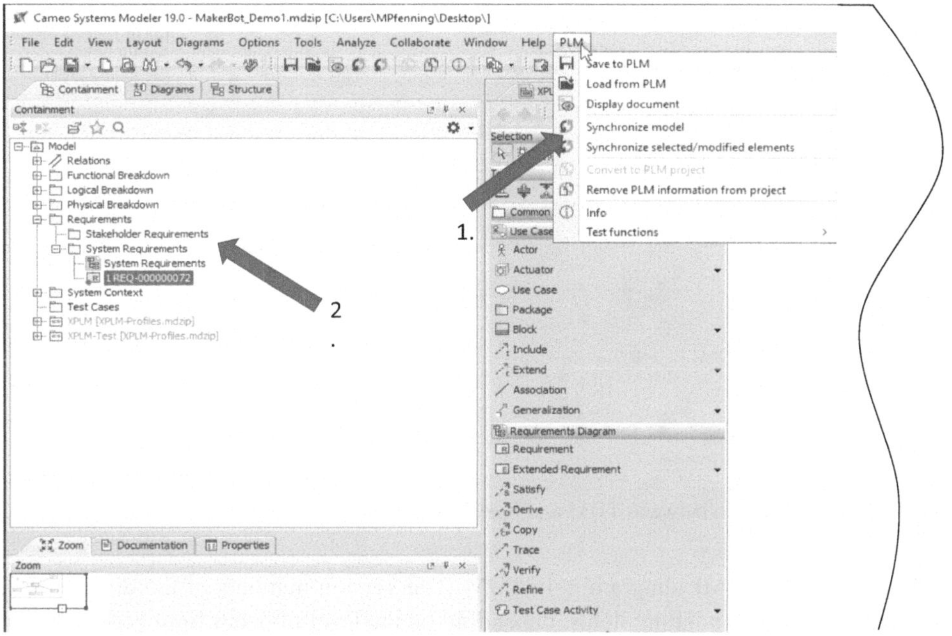

Fig. 4.9 Integration of the Cameo system modeler in Aras Innovator I. *Source* Aras and XPLM

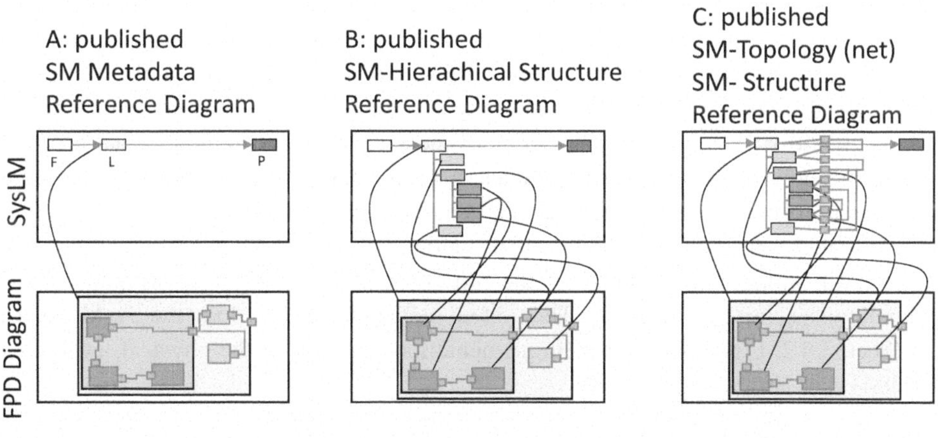

Fig. 4.10 Various levels of detail of the system model integration in SysLM [44]

Figure 4.11 shows the possibility if a collaboration between SysLM and TDM during change management on the basis of a workflow. In practice, here too, the level of detail of the change management and the time from which the information is transferred in SysLM must be determined in the scope of the application strategy.

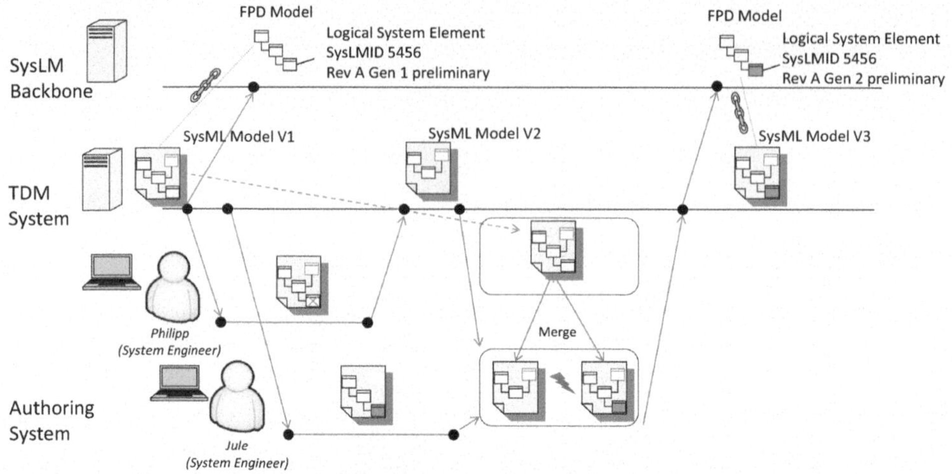

Fig. 4.11 Collaboration between TDM and SysLM levels in change management [44]

The complete SysML diagram is in TDM. The version numbers of the diagrams are assigned to the corresponding items, depending on the level of detail from Fig. 4.10. The items contained in the SysLM backbone receive metadata, for example item number, revision, and generation. The development of the system model between Jule and Philipp can take place in parallel with the help of the TDM system. A lock-modify-unlock or copy-modify-merge strategy can be used for this (Fig. 4.11 uses the latter strategy). If a new status of the model is reached, the system model on the SysLM backbone is updated. A suggestion for a pragmatic process would be to only work on the TDM level in the very early development phase and only execute the SysLM transfer from a certain level of maturity [44]. As the connections of the system model in the original language SysML are only transparent for specialists, a simple form of the visualization must be found for users on the management level and in the subsequent areas.

A visualization client should visualize hierarchies, networks, and, of course, links between all information elements of the SM. Here, the path via a networked browser technology is selected (Fig. 4.12). The system elements are linked via the SysML matrices (Sect. 3.6.2, Fig. 3.35). In this implementation, the link to the physical level (→ P, E-BOM) was carried out manually.

4.2.1.3 CAx Integration

CAx data is a central component of a PLC. The transition from the abstract modeling to the concrete is carried out here. This phase results in the

- Mechanics: 2D, 3D models, and BOM,
- Electric/Electronics: schematic and layout drawings, manufacturing, and assembly data and BOM, and
- Software: source code, flash code, and links (→ baselines) to software management tools.

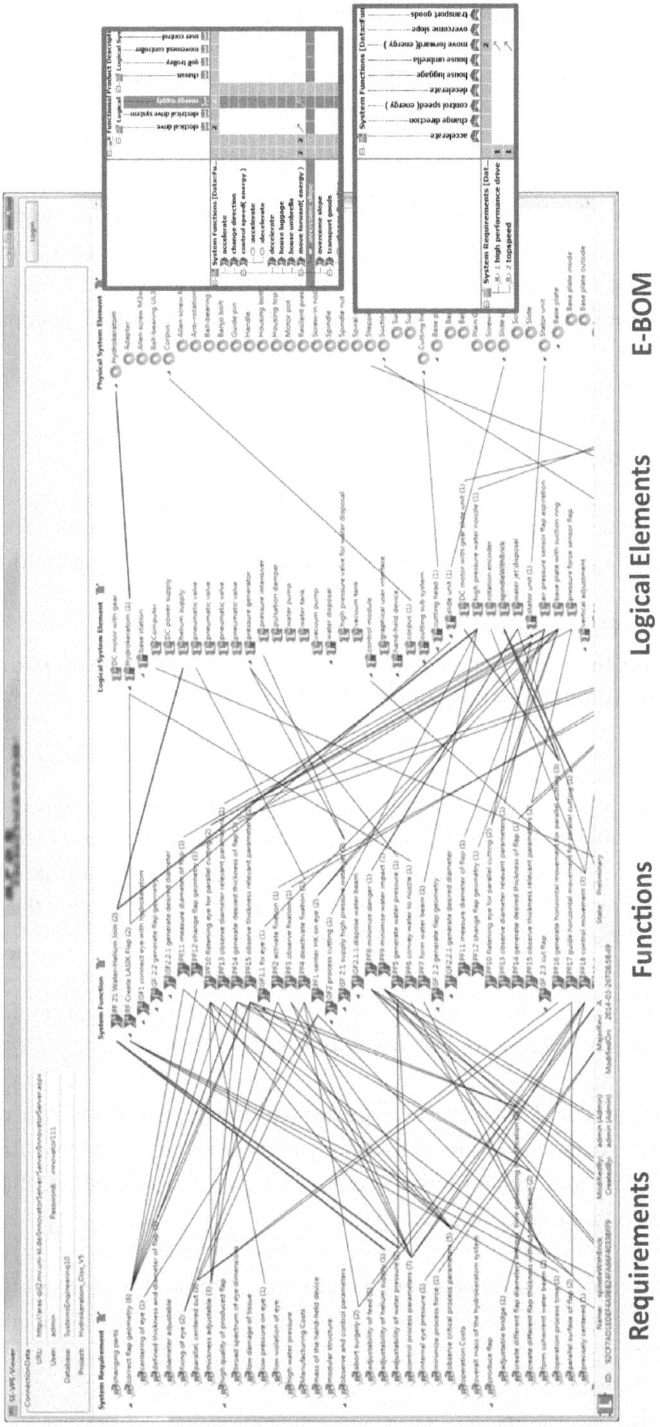

Fig. 4.12 Representation of the system model in Aras browser technology [44]

This information supports a broad range of downstream processes, including virtual prototyping, simulation, process planning, and technical publications. The management of CAx data by the engineering processes running on the backbone provides an enormous advantage for the company and often has the highest priority of SysLM initiatives.

M-CAD Integration

M-CAD integration is the longest used integration and was already a component of the market offering with the first PDM systems in the 80s. It is described in detail in [37]. Here, again, is a brief representation of the foundations and indication of the new trends. Various consideration levels are assumed in the coupling of a 3D CAD system to SysLM:

- The work level of the designer (→ CAD system),
- The internal representation of the CAD system, and
- The internal representation of the SysLM system.

Figure 4.13 shows the representation of a 3D model in a CAD system. The document structure in the CAD system is represented in a browser in module construction.

The M-CAD/SysLM Integration Format

This is important for the further consideration of the individual elements in the CAD and SysLM system. We differentiate between:

- CAD assembly:
  This is the CAD model of an assembly (→ document structure). An assembly can contain sub-assemblies. The term "assembly" may be misunderstood here, because it is actually only a 3D document structure. However, this assumption is often used in document-centric working processes. The document structure contains the topological information used for all types of visualization in SysLM systems.
- CAD part:
  The smallest unit in the CAD model is the part[27]. Again, the term part is not quite correct. It corresponds to a part, but is actually only a 3D CAD model.
- CAD drawing:
  While SysLM parts and assemblies correspond directly to 3D CAD models, the drawings derived from these are stored separately and are then also linked to the corresponding parts and assemblies. Of course, with a 2D working technique, these drawings can also be directly linked to parts.

In today's CAD systems, these fundamental elements are stored in separate files. Figure 4.14 shows the CAD parts, assemblies, and drawings with the associated internal

---

[27]The term part is used here in the true sense as an individual part. Multiple parts can form an assembly in a structure (BOM). PLM/SysLM use the term part partially differently. Part is the generic term for components (individual parts) and assemblies.

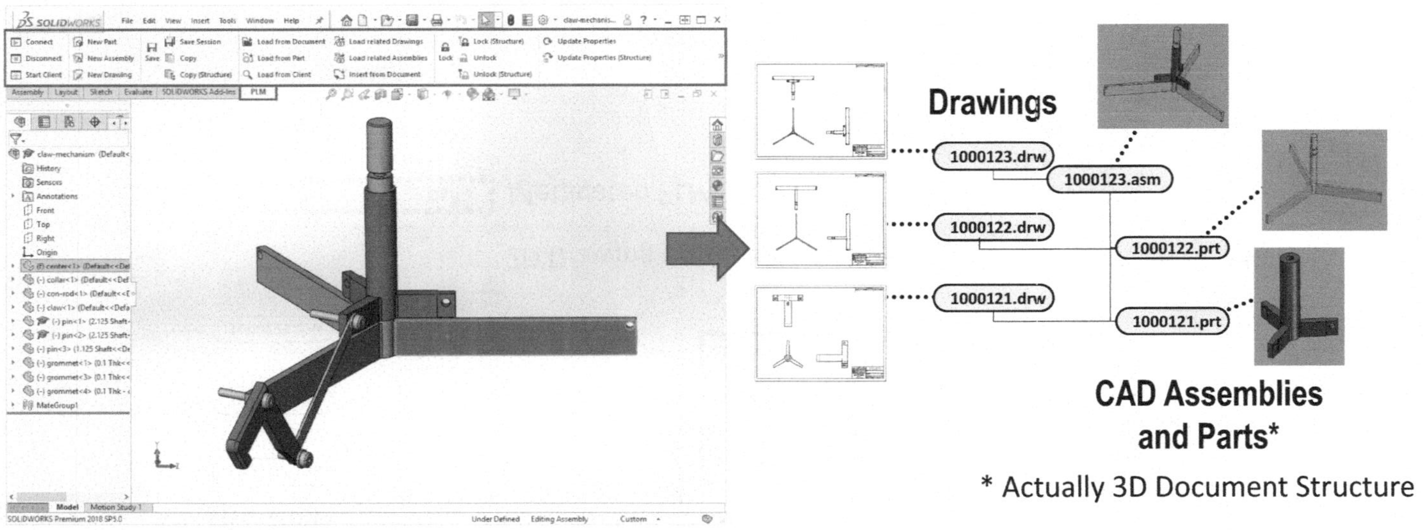

Fig. 4.13 Representation of a 3D CAD model with CAD data structure in a CAD system (solidworks)

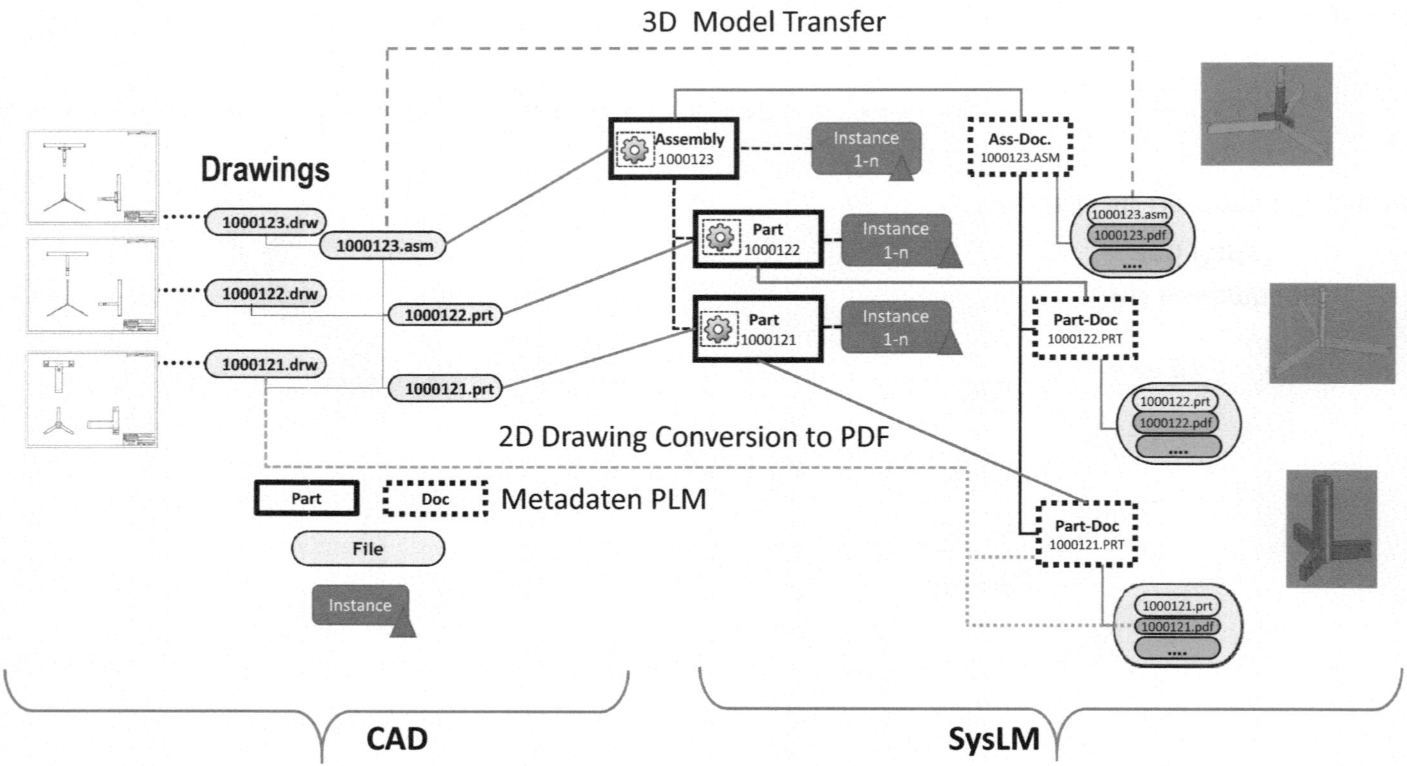

Fig. 4.14 Transfer of the CAD file structure in SysLM and derivation of the parts and BOM

file structure which are created and managed by the CAD system[28] as well as the transformation into SysLM.

The separate storage in various files in the CAD system is very important, as it is the prerequisite for the simple and efficient preparation of the model structures for transfer according to SysLM (Fig. 4.14). We recognize that the transformation takes part in two steps:

- Transfer of the CAD file structure and formation of the SysLM document structure
  - CAD documents get metadata (→ document number, revision and name) and are generated 1:1 from the CAD files,
  - CAD assemblies get also metadata and are derived directly from the CAD document structure (→ parts, assemblies[29]),
  - Native files are linked to the metadata of CAD documents,
  - Lightweight[30] (→ viewable) formats for 2D and 3D models are optionally generated in batch mode and are also linked to the metadata of the CAD documents, and
  - The numbering, revisioning, and naming should be automated as far as possible.
- Derivation of the part/BOM structure from this[31] (→ parts and BOM metadata) in SysLM
  - The part assignment takes place on the basis of a part per 3D model (1:1)[32],
  - All CAD documents are automatically assigned to parts and assemblies in SysLM,
  - BOM generation uses the CAD document structure as a template to initially create them. However, this is done only in the procedure that the document is created first and a BOM is derived from it. Another common approach is based on the existence of the BOM before the documents are created, and
  - Simple access to native and lightweight 3D and 2D documents by clicking the part or the assembly.

Many CAD applications today, particularly in small- and mid-sized companies, do not extend beyond the management of the CAD file structures typically in the PDM[33] Systems and work on the philosophy that a CAD part document is similar to a part

[28]The CAD system Onshape shows it can also be done differently. It runs in the cloud, and there is no file system; rather, there is a database for CAD model storage. This could be a trend.

[29]The designations part and assembly are sometimes used differently in PLM/SysLM systems; for example., a part is the generic term for individual parts and assemblies.

[30]Lightweight formats are, for example, TIFF, PDF, JT, and 3D PDF, which only require a fraction of the storage capacity of native formats (the original CAD formats).

[31]There are two alternatives for the BOM: Derivation from CAD or the BOM exists initially, and CAD models are later hung on the nodes.

[32]there are cases where the 1:1 assignment is unassigned, for example, for color-neutral CAD models that reference to multiple color-specific parts.

[33]These companies are far from a SysLM deployment. Therefore, the term PDM is used (see Sect. 2.2).

(→ 1:1 relationship) and a CAD assembly document corresponds 1:1 to a BOM. This method of working is described by the term "document-centric." The metadata of the part and assemblies are often not stored in the PDM system at all but passed directly to the MRP system. There, part master data and BOM's are then derived. Consequently, the document number often corresponds directly to the part number of the MRP system. This assumption is in many cased correct, but there are repeatedly exceptions:

- documents which were not created in a CAD system but are necessary for the description of a product, for example text documents, which leads to a 1:n relationship,
- parts that have a m:1 relationship to the documents, for example a "colorless" 3D model, that references multiple-colored parts, or for the documents themselves the good old "table drawing" with references on many parts, and
- parts which were not created in the CAD system, for example glue or filling material, mirrored parts (n:0).

The real reason why you can rely on a SysLM structure with an independent and autonomous management of documents and parts are however the need for n:m relationships between them. In addition, there is the potential that the part master is the absolute link to all other relevant information elements of the product lifecycle phases (cf. Chapter Attachment Fig. A1) and the formation of traceable mechatronic structures (cf. Fig. 4.43). This approach is called part centric and is a prerequisite for model-based engineering. The knowledge that software which was flashed onto a chip which is placed in a PCB[34] board which in turn belongs to a housing can only be generated via network discipline-specific parts lists, and that is the essential prerequisite for the construction of a Digital Thread. This process is difficult to convey to the users in the individual disciplines, in particular because the handling of two structures (→ document and part level) which are also still very similar for a discipline is complex and prone to error. This is why many document/part processes of the CAD/SysLM integration should run extensively automated, for example, automatic SysLM parts metadate creation from the CAD part and automatic SysLM BOM derivation from the CAD assembly structure. CAD documents as well, for example 2D sections derived from 3D models, should automatically be assigned to the associated parts and modules.

Some more applications for the optimization of the M-CAD/SysLM integration are presented below.

3D Dimension and 3D Master

Unfortunately, today derived 2D documents in native and lightweight formats are still required. This is due in particular to the required geometry-annotated dimensions, tolerances and surface specifications. But the trend is strongly towards a native and lightweight

[34]PCB Printed Circuit Board.

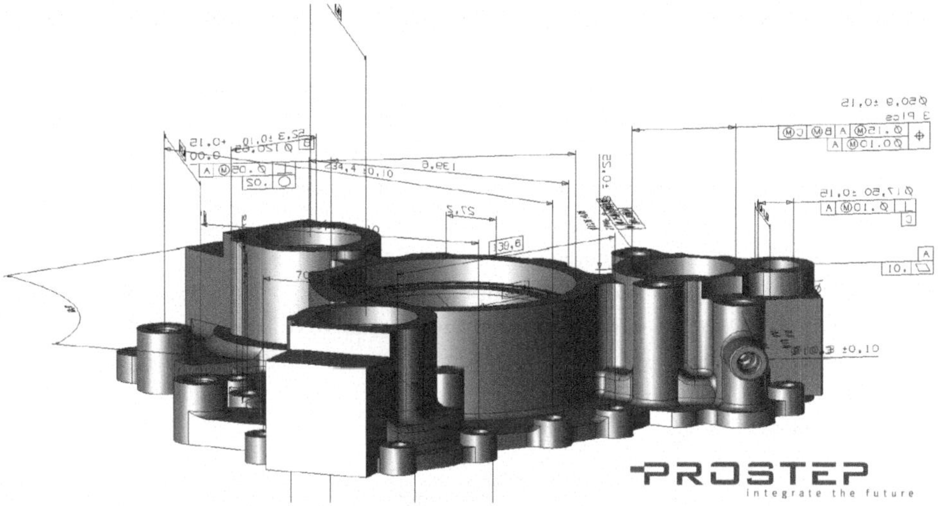

Fig. 4.15 3D dimensioning as part of the 3D master concept. *Source* ProSTEP

visualization of three-dimensional dimensioning of the 3D model along with master data derived from SysLM and MRP (→ often called "3D master" in industrial projects), which enable a drawing less internal and external information processes (→ in German automotive industry defined by VDA 4953 -1 and -2). With the application of the VDA concept, it would be possible to work in the information processes around the SysLM system, without media breaks and additional conversions. Why "would"? Unfortunately, lightweight formats are rarely accepted by customers in external data exchange, and many IT solutions used in subsequent processes do not yet accept these formats. In addition, none of the typical 3D lightweight formats (→ JT, STEP, 3D PDF) is standardized as the real standard for the 3D master. Internally, however, it was recommended to implement such a 3D master concept with high priority and to make the information available to MES, MRP, and the SysLM users (Fig. 4.15). If this is still connected with intelligent viewing technology (→ cuts and explosions), the acceptance of the users is considered assured.

Part Families (→ Family Tables)

Many CAD systems support parametric parts families. This means that there are different parts with different parameter-determined CAD instances. However, these instances are not physically saved, rather if required they are generated from a table, for example. SysLM should support this process.

Module and Component Libraries

Products are created for a part from components from suppliers. The variety of purchased parts is almost incalculable and extends from profiles of simple parts and modules (→ screws, washers, bearings) to complex components (→ couplings, transmissions, drives). The manufacturers of these purchased parts and components have a commercial interest in making these parts available to the engineers in CAD formats and, of course, connecting the manufacturer-specific order number with the components. Portals act as mediators between the manufacturers, their data, and the CAD users. There are two predominant providers for mechanical components: Traceparts[35] with headquarters in France with the CAD library of the same name and Cadenas[36] in Germany with the Part Community. Both portals have similar basic functions[37]:

- Search functions for component and parts selection,
- Visualization of the parts and components,
- Loading the CAD model in the desired format (→ native or lightweight),
- Infrastructure for the construction of a CAD catalog for the manufacturer, and
- Provision of the accompanying data such as images and documentation, suppliers, order number.

CAD libraries provide their data in lots of formats. Both traceparts and the part community support several, provider-neutral formats such as Step[38] and STL[39] and also various CAD-specific file formats. The download of the CAD model in its own format saves the import and potential adaptation work. While the basic functions of the providers are similar, there are differences in the detail. Thus, the CAD file satisfies the engineer, but other departments such as purchasing and enterprise resource planning also need data. Formal information such as correct designations, order numbers or keywords for the classification make it easier to enter the supplier product in the ERP and SysLM systems. This makes orders from BOM directly possible, for example. Data sheets and operating instructions should be included in the delivery of somewhat more complex products, such as motors or sensors. This supplementary information is often not available.

Intelligent Design (Topology Optimization, Generative Design, and Application of AI)
The generative design software, for example from nTopology[40] and Dassault[41] (→ Function-Driven Generative Design), combines synthesized geometry and simulation results with optimized manufacturing models, especially compatible for 3D printing. 3D printing technology increases the design freedom of the engineer on one hand, and on

[35]https://www.traceparts.com/de/, (retrieved on 6.10.2020).

[36]https://cadenas.partcommunity.com/, (retrieved on 6.10.2020).

[37]WeSt Blog, Die besten CAD-Bibliotheken für die mechanische Konstruktion, http://www.west-gmbh.de/blog/cad-bibliotheken (retrieved on 6.10.2020).

[38]STEP Standard for the Exchange of Product model data.

[39]STL Standard Triangulation/Tessellation Language.

[40]https://ntopology.com/ (retrieved on 6.10.2020).

[41]https://www.3ds.com/events/single-eseminar/catia-function-driven-generative-design-webinar/, (retrieved on 2.14.2021).

the other hand, highly complex after a strength optimization geometry can manufacture products more quickly and therefore more efficiently than with traditional processes. In many cases, conventional manufacturing processes cannot produce these geometries at all. Further generative engineering software was developed by Frustrum[42], which enables the creation of optimum geometry for certain requirements on the basis of loads and framework conditions. Optimum depiction on the 3D printing technology is also a focus for the engineering software from Frustrum. The difference between *topology optimization* and *generative design* is relevant in connection with the two technologies presented[43]:

- Topology Optimization
  This technology enables CAD developers with defined objectives to change a construction in such a way that it is optimized for a subsequent application, for example, manufacturing. It is more of a foundational technology from the 90s and primarily serves for material optimization and load distribution. It is largely used by simulation experts [18].
- Generative Design
  This technology is more of a tool for designers. Several designs which are orientated on functional and design-orientated objectives are calculated by the software in parallel [18]. Artificial intelligence (AI) is increasingly used in the selection of the optimal solution.

The third generative approach designated as *target-driven design* or *intention-driven design* can be explained with the example of reflection lines or geometric character lines. The product or design features desired by the designer are stipulated as a design and control parameter for free-form geometry – for example in the form of reflection lines or characteristic geometry contours. The lines or contours can be changed directly; the fundamental surfaces follow this modification[44] [27].

Figure 4.16 shows the difference between generative design and target-driven design.

Static and Dynamic Visualization

The static approach assumes that access is made on a monolithic view file chained to the CAD assembly to be displayed. The topology is "hidden" in the view file. The block is loaded in the viewer with all the sub-assemblies. This approach has advantages in regard to the performance, as several files do not have to be loaded in the viewer and the various

---

[42]Frustrum was taken over by PTC in November 2018.

[43]https://www.engineeringspot.de/2018/11/ptc-frustum-ki-generative-design/ (retrieved on 6.10.2020).

[44]F I O R E S Formalization and Integration of an Optimized Reverse Engineering Styling Workflow, Brite Euram-Project BE 96—3579, project homepage: http://www.fiores.com/FIORES.html.

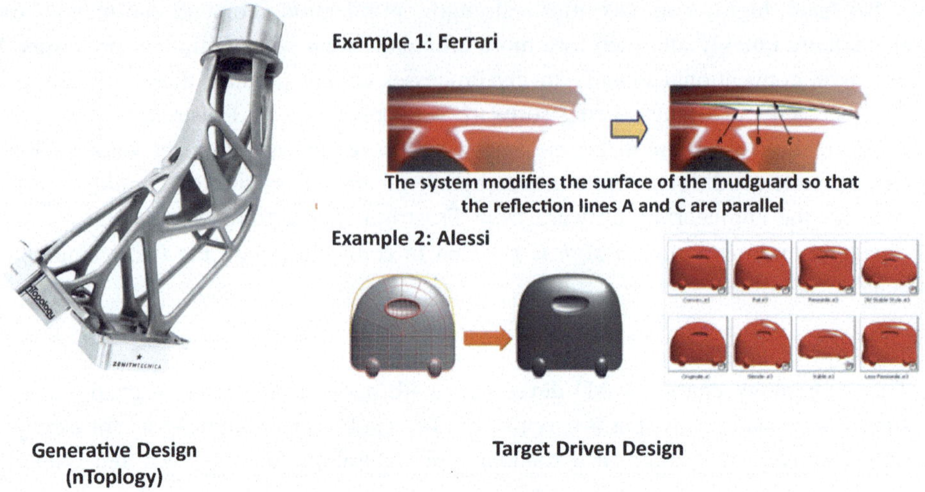

Fig. 4.16 New design approaches are based on generative- and target-driven design

topologies analyzed. It also supports more complex functions at a module level. Changes to sub-modules and parts which occurred after the generation of the monolithic display are not reflected. The dynamic approach expands the entire document structure and uses the transformation matrix which is chained with every instance of a CAD part. It makes it possible to manually set the representation mode, for example, of the highest approved version or on the actual version in the development or a configuration that was manufactured on April 2, 2018. As the individual parts are loaded in the viewer, every version of a part can be represented. Figure 4.17 shows the CAD model. The static visualization accesses (1), i.e., the view file of the assembly and the dynamic visualization can, in principle, load all revisions of the CAD parts and position them via the transformation matrix (2). The indication of this is that the transformation matrix is openly accessible in the SysLM system and can also be used for other applications, for example variant visualization in native as well as lightweight[45] formats. Another possible representation form for dynamic visualization is to graphically document the exact current state of an instantiated Digital Twin with a special serial number (cf. Fig. 4.84).

Functional Scope of the M-CAD/SYSLM Integration

Figure 4.18 summarizes the basic functions of a good M-CAD SysLM coupling and should be a component of the investigations[46] in a PoC[47]. Advanced features in 3D

[45]'lightweight' means formats which can be displayed in a viewer, such as JT, PDF, 3DPDF, PRC, et.

[46]PoC Proof of Concept.

[47]XPLM Homepage: https://www.xplm.com/de/.

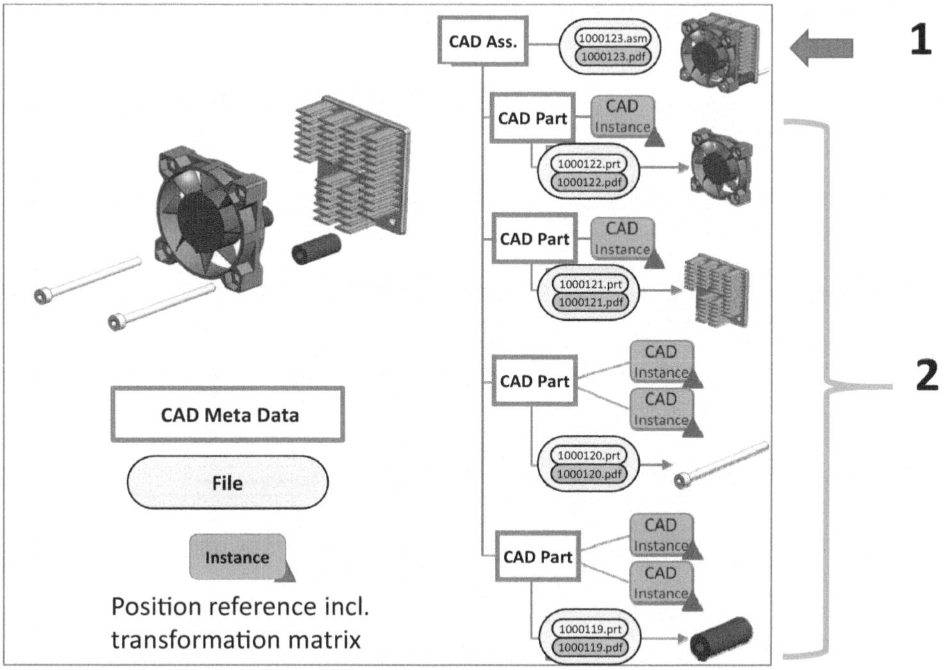

Fig. 4.17 CAD data structure with topology instances assigned to the CAD parts (according to Aras)

modeling associated with SysLM include component comparison, partial or mixed loading (→ performance by loading only components to be machined or a mix of native and lightweight formats) and replacing new modified components in an existing assembly.

E-CAD Integration[48]

The application areas of the computer-supported electrics and electronics (E/E) were initially demonstrated using the example of a mechatronic system—the injection module of a combustion engine (Fig. 4.19).

The motor control device, the sensors and actuators, and their energy supply are on the highest level of the electronic system. The sensors and actuators connected with it, as well as their power supply, are documented in a schematic diagram. The control unit itself is usually realized on a printed circuit board. As a further development of the free electric wiring, circuit boards consist mainly of a rigid or rigid-flexible board on which the electronic components are fixed (soldered). The compact and robust construction

[48]In this chapter, the author was supported by Paweł Chądzyński (Aras) and Robert Huxel (XPLM).

Function
Running the Integration
Connect
Disconnect
Start Client
New on
New Part
New Assembly
New Drawing
Save to PLM
Save
Save Session
Copy
Copy (Structure)
Load from PLM
Load (from Document)
Load from Part
Load related Drawings
Load related Assemblies
Load from Client (Search)

Function
Insert
Access Control Management
Lock
Lock (Structure)
Unlock
Unlock (Structure)
Property Exchange
Update Properties
Update Properties (Structure)
Update Title Block
Material and BOM Management
Create BOM
Create Part
Workspace Management
Select Workspace
Workspace Manager
Remove Local Files
Tools
Display Document
Assign Document
Synchronize Part Library
Info/About

Fig. 4.18 Basic functions of an M-CAD SysLM integration. *Source* XPLM

of active and passive electronic components is characteristic of printed circuit boards (→ PCB[49]). Due to the specific characterization as a board of electronic components, circuit boards represent the second level for the classification of E-CAD tools. The tools to design integrated circuits (→ logic chips, user-specific circuits (→ ASIC[50]) and programmable digital circuits (→ FPGA[51])) are one level lower. Similarly, to circuit boards, integrated circuit component elements (mainly transistors, diodes, capacitors, resistors)

[49]PCP Printed Circuit Board.

[50]ASIC Application Specific Integrated Circuit.

[51]FGPA Field Programmable Gate Array.

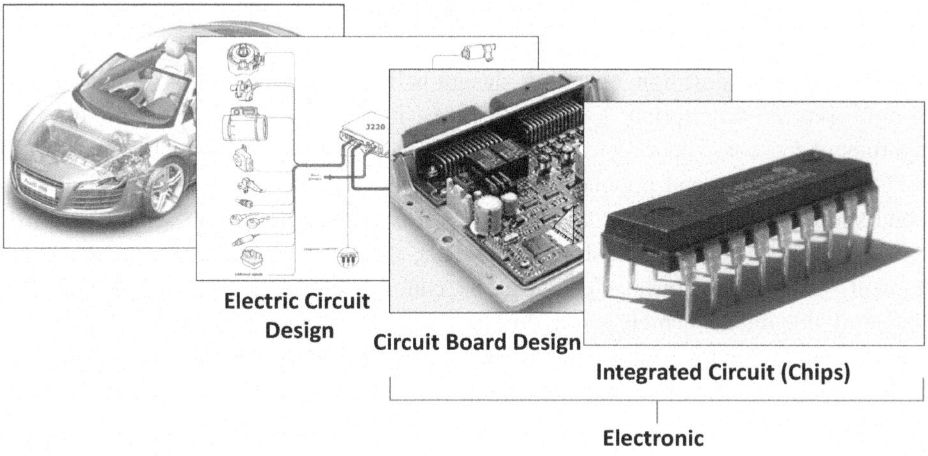

Fig. 4.19 Application areas of E/E development

are connected to one another in the finest structures. This concerns a much denser group of purely digital functionalities. These are realized with their connections to one another directly on a substrate. External connections are also realized (wire bonds[52]), in order to create contact to the circuit board and thereby to further discrete component elements. These three levels of consideration represent three possibilities to realize the electro-technical circuits: through electro-construction, circuit board design, and chip design. The development and production processes on these levels are different and are supported by various classes of process models, methods, descriptive formalisms, and authoring tools for E/E development. In this section, only the topic of circuit board design or PCB design is dealt with in the scope of the E-CAD coupling [107].

Definition E-CAD:
E-CAD stands for computer-aided design in electrical and electronic design. It includes all IT-based assistance for design, production preparation, and protection on the three levels of electrical design, circuit board design, and chip design. Today, the term E-CAD not only includes the creation of the schematic and layout, but also relates to all activities in connection with electrical and electronic design. The term electronic design automation (EDA) is also used for the pure electronic design application.

SysLM and PCB design are closely connected with one another. Companies which continually develop innovative solutions must permanently improve their designs as soon as new technologies are available. Due to technological improvements or shifts, electronic components have short lifecycles that require changes or even completely new

[52]Wire bonding is the method of making interconnections between an integrated circuit.

architecture. SysLM solves this problem by providing supply chain transparency, traceability, and component discontinuation.

Unlike mechanical design, electronic design is based on a much more comprehensive and multi-layered description of various aspects. It includes, for example, the electrical properties of the construction elements, but also information for production preparation (→ PCB manufacture and population) as well as the delivery status (→ end-of-life and second source information). As this information is typically saved in separate environments (e.g., CAD, libraries, SysLM, and MRP), there is a risk that development, procurement, and manufacturing specifications can no longer be kept up to date. The number of documents which result during electronic design is so large that E-CAD systems often summarize them into so-called project folders. In order to achieve fine-grained traceability, the individual documents (→ schematic and layout) should be stored in SysLM.

Approach in the Model-Based Electronic Design (→ PCB Design)

A detailed view of the PCB design and manufacturing process can be taken from Fig. 4.20.

The specification of the requirements and functions form the prerequisite for the creation of the circuit diagram (→ schematic) in the development of circuit boards. Circuit diagrams contain information about the selected construction elements and their logical relationships to one another. An initial test of the electronic logic can already take place on this basis. The third step in the development process incorporates the composition of additional information, for example the parts lists of already determined components,

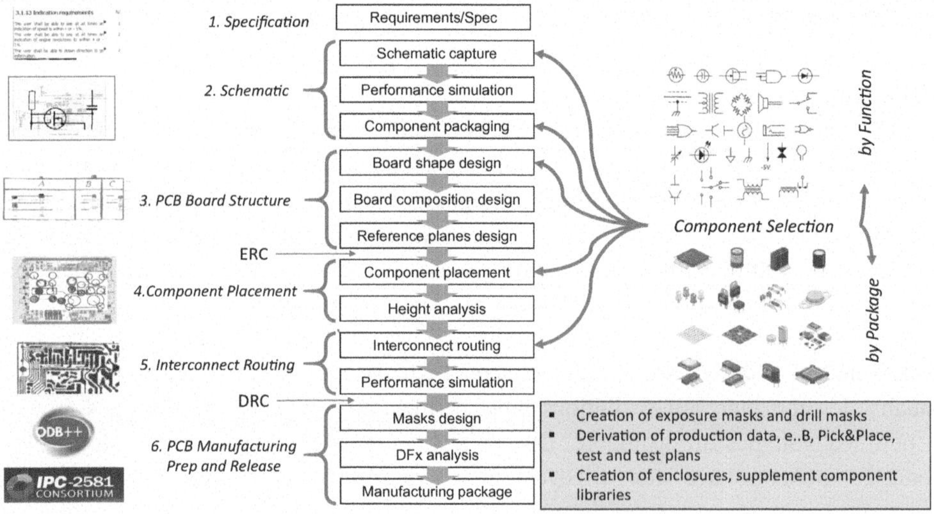

Fig. 4.20 Detailed consideration of the electronic design and production process [22]

which are specifically important for the wiring and placement on the board. There is a comparison with the component in the library (→ library) throughout the entire development process. Available components in the library can be used or new components are entered into the library. The so-called electrical rule check (ERC) ensures the electrical operability of the designed scheme. For this, the ERC verifies the electrical consistency of the schematic. The type of assembly technology plays an important role from the fourth step. We differentiate between through-hole technology (THT) and the modern and now standard surface-mounting technology (SMT). Installation-relevant information about the component connections (connecting wires and connecting surfaces) according to the manufacturer's specification are drawn on in this phase of the development process, depending on the type of component. The recording of this information is com pany-specific. This information is normally managed in electronic component libraries. The physical layout of the circuit board is created in the fifth phase of the circuit board design, based on the circuit diagram and the compiled, installation-relevant component information. In comparison to the circuit diagram, the layout not only represents the logical but also the geometric relationships between the components in regard to the circuit board. Here, component-specific framework conditions identified in the third phase are taken into consideration for positioning. The circuit board unbundling is a significant work step in the layout design. Here, the designed electrical circuit diagram is implemented according to the manual or automatic positioning of the construction elements on the circuit board in a conducting path network. Today, it is carried almost always out manually on the computer or automatically with the help of a so-called auto-router[53]. The so-called design rule check (DRC) is carried out on this basis. The design rule check (DRC) is a step in the layout verification of the circuit board design and of integrated circuits. It should ensure the technological manufacturability of the designed layout. The technical documentation includes drawings for the manufacture of the board (master drawing, artwork master, production master) as well as drawings and data for assemble and then soldering the components on the board (printed board assembly drawing, pick, and place) (→ DIN EN 60194[54]). Here, the board can be populated single-sided, double-sided, or even with embedded components (→ DIN IEC/TS 62326-16[55]) [107].

A summarizing view of the significant design tools of the electronic design is shown in Fig. 4.21. Often, significant process steps are carried out by various internal or external (→ supplier) teams.

The original format for the transfer of production data and probably still the most widely used file format for the PCB manufacturer is called Gerber. They are ASCII files

[53]https://de.wikipedia.org/wiki/Leiterplattenentflechtung, (retrieved on 5.20.2020).

[54]DIN EN 60,194: Konstruktion, Herstellung und Bestückung von Leiterplatten—Begriffe und Definitionen (2007.

[55]DIN IEC/TS 62,326–16: Leiterplatten—Teil 16: Trägermaterial mit eingebetteten Bauteilen—Allgemeines (2013).

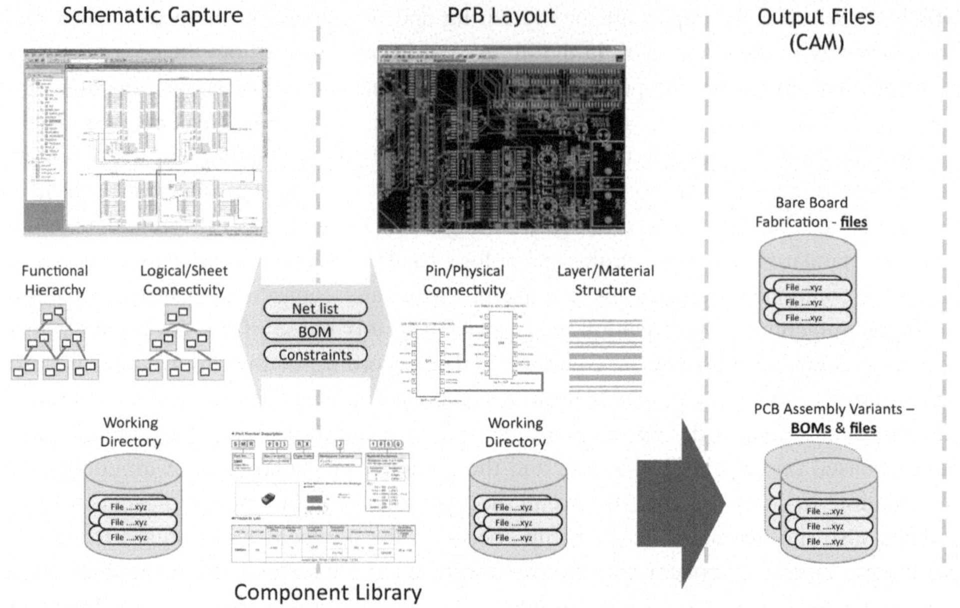

Fig. 4.21 PCB design tools [22]

which contain Gerber-formatted data. The state-of-the-art IPC-2581 for CAM data transfer uses XML structures. IPC-2581 is supported by a consortium of manufacturers and suppliers to the electronic industry; however, it has not yet caught on with many developers and manufacturers[56]. A requirement in the forward transfer of data to SysLM is the free selection of the CAM format. Thus, the model-based approach enables a seamless development process from the creation of the circuit diagram to the production. IT tools in the circuit board design are designated as electronic design automation tools, or EDA tools, due to the high degree of further use of the created models and automation of work process in the development process [107]. Overall, process models and IT tools in electronic development have achieved a very high level of intelligence and have become a significant element of model-based engineering.

We can also refer to an MBSE approach in chip development, a further component of electronic design which will not be looked at in greater detail here. Hardware and embedded system architecture are designed and specified in SysML and also depicted as embedded software in UML. The hardware description language (→ HDL), which represents the foundation of the chip design, can then be derived via the ICÒNIX method. Special

[56]https://www.multi-circuit-boards.eu/support/leiterplatten-daten/ipc-2581.html, (retrieved on 6.16.2020).

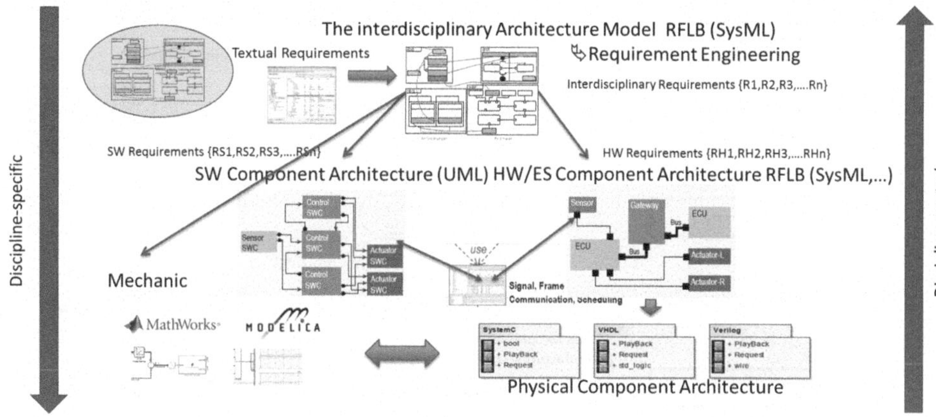

Fig. 4.22 MBSE approach in CHIP development inc. embedded software in order to generate HDL code from SysML (basis: Enterprise Architect und Iconix)

SystemC can depict both the hardware and the embedded software[57]. A simulation of the entire system can take place through capsules of the SystemC code and connection to Simulink or Modelica (Fig. 4.22). However, it is doubtful whether, in industrial use, code changes at the SystemC level can be returned back into the system architecture.

The E-CAD/SysLM Integration Format

An example of an E-CAD/SysLM data model is represented in Fig. 4.23.

The parts and part structure (→ BOM) are constructed identically to M-CAD, except for the following differences:

- The parts list is generally only two level. The module is the populated board, the electronic components lie beneath this on one level. The PCB module bundles the assembled circuit board. The unpopulated circuit board is in position 1 (→ PCB bare).
- An instantiation through position indicators is usually in the electronics, as the position (→ x-, y-coordinates) of the respective components is relevant for the assembly process (→ pick and place). Thus, it is usual to additionally identify an identical component through position indicators in both the schematic and the layout drawing as well as the parts list (e.g., a capacitor with the same capacity has a part number, a revision/version, a name and the positions C1, C2, C3, C4, etc.). A capacitor appears in the parts list either four times with the same part number and revision, or it is attributed in a line with all four position indicators. In the aggregated, mechatronic overall parts list in SysLM, the capacitor appears exactly once in a position line with the quantity 4.

[57]http://www.sparxsystems.com.au/downloads/ebooks/Embedded_Systems_Development_using_SysML.pdf, (retrieved on 6.16.2020).

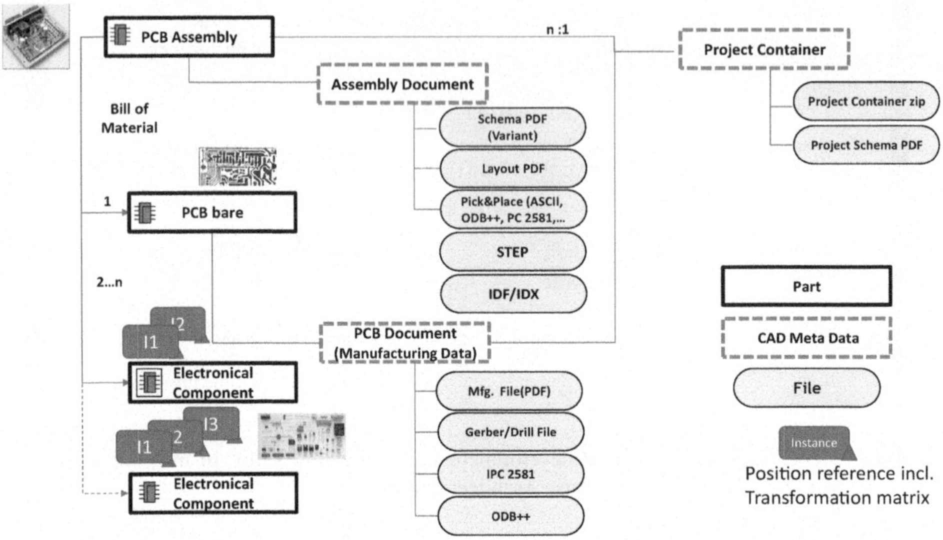

Fig. 4.23 E-CAD/SysLM data model [50]

The documents and files are structured with much more complexity:

- Assembly document
  defines a data model object in SysLM which bundles the population information of the circuit board. The following files hang under this:
  - The schematic
    defines an output file with the variant-specific circuit diagram as a PDF file,
  - The layout
    defines the output file with the variant-specific layout as a PDF file,
  - The pick and place list (→ population list in various formats, for example in ASCII or PDF, alternatively, it can also be a component of the ODB++ or the IPC 2581 file.),
    Both the E-CAD/M-CAD exchange formats STEP[58] and IDF[59] and the proprietary EDX[60] define an output file with the proprietary design and manufacturing information of the circuit board. These are the PCB external contour, the position of the parts, positions and sizes of holes, regions to be kept clear and outlines of the heights of the components.

[58]STEP Standard for the Exchange of Product model data.

[59]IDF Intermediate Data Format.

[60]EDX Enterprise Data eXchange (Mentor).

- PCB document
  defines the data model object in TDM/SysLM that bundles the manufacturing information of the circuit board. The following files hang under this alternatively depending on the specific manufacturing and delivery situation:
  - The manufacturing file contains the manufacturing drawing (drill plan) as a PDF file,
  - The Gerber file is an open ASCII format for circuit board designs. It was a de-facto standard in the PCB industry for a long time in order to describe circuit board images with copper layers, soldering masks, prints, drill data, and keep-out areas, and
  - ODB++ and ICP 2581 contain the defined manufacturing information of the circuit board according to the respective industry standard.
- The project container defines a data model object in SysLM that primarily bundles the native files of the circuit board design. In the case of a variant project, the relationship to the PCB module is n:1; otherwise, of course, it is 1:1. The variants are typical in consumer electronics in particular. The following files hang below these:
  - The zip container contains the native design data of the authoring system, and
  - The project schematic contains the complete population, while the schematic below the module document can contain a reduced or alternative population (→ variants).

The information mentioned overlaps heavily and must be flexibly generated by the E-CAD system. This is often dependent on the supplier and/or the customer. For example, the circuit board manufacturer wants to have a Gerber file while an internal thermal analysis works with ODB++. Presented below are some more applications for the optimization of the coupling integration.

Variants for PCB's

It is typical for circuit boards in the consumer goods industry in particular that a product on the foundation of a basis design requires variants through customer requirements that can all be selected via various options or functions. Similar to the methodology of mechanical variants, the PCB design also uses the possibility to implement a variant series in such a way that customer-specific variants can be derived from a fundamental design through reduced or alternative population. In practice, a variant uses the same basis design; however, the circuit board module is loaded with the set of components which is defined by the variation and the corresponding options. A variant can then be named in the creation of the assembly and manufacture of the circuit board (→ BOM, pick and place, assembly drawings, etc.), which in turn determines the assembly of the (cf. Fig. 4.23). The possibility to derive variations from the same basis design in the E-CAD system significantly increases the efficiency of circuit board projects. Numerous variants of the PCB design can be defined with variants, whereby each component can be configured according to criteria [7]:

- Assembled on the board,
- Not assembled on the board,
- Assembled on the board but modified with component parameters, and
- An alternative component.

If it is in any way possible, of course one tries to transfer the variation point from the hardware into the software.

Libraries

Many types of information are compiled in a library for electronic design (→ component library) (Fig. 4.24).

The responsibility for the construction and maintenance of the library varies from company to company and is a cost factor which should not be underestimated. In larger companies, there are often more librarians; in smaller company, the developer is often responsible for everything. The trend is toward external databases. The provider with the largest portfolio is far and away SiliconExpert, which also makes APIs available in order to integrate information with other systems. Octopart also warrants a mention—this platform also provides CAD models (partially cost neutral, partially for a fee). The sense of integration in other systems is the transfer of logistical information such as price, availability on the market, delivery time, obsolescence, etc., to the hands of the developers. The entire thing can also be placed in connection with the term "design to cost." The developer should be able to make an "intelligent" component selection in a very early stage, in order to keep the costs for a module low and, if possible, without leaving his development environment. The topic is really complex and includes endless variations.

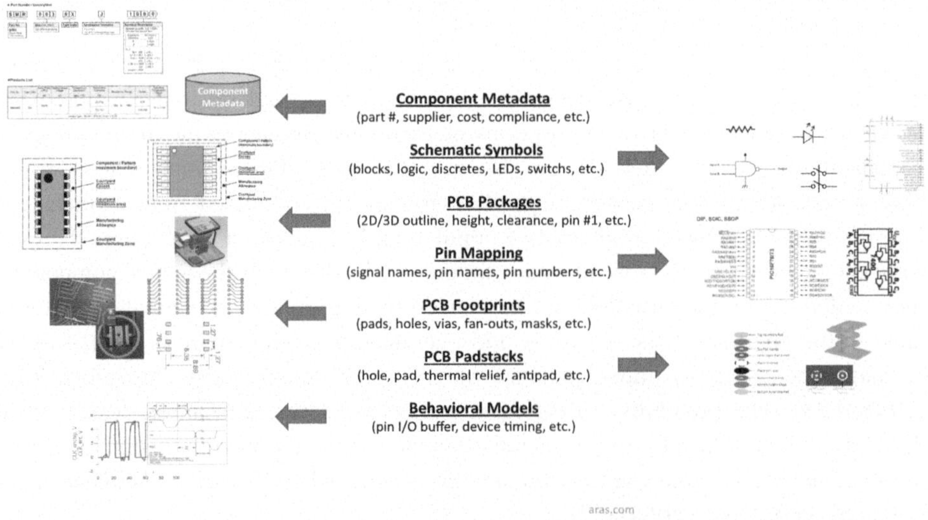

Fig. 4.24 Component library (according to [22])

Some providers concentrate on support across the E-CAD systems, others are conspicuous with analysis tools for the calculation of risks in the supply chain, while another group advertise direct integration in an E-CAD system, including order options with a supplier [50].

M-CAD/E-CAD Coupling

Most electronic designs begin with the schematic and the layout of the circuit board. The product design, for example of the housing or the installation space, is developed in the mechanical area. The difficulty is often in the lack of synchronization of both disciplines. How can the mechanical construction process be connected with the electronic design process and thereby guarantee that the designers from various disciplines work together? In all M-CAD/E-CAD integrations, the necessary information from the mechanical design is distributed over the coupling with the electronic design, in order to support the fast and precise exchange of data between the M-CAD and E-CAD tools. The data elements that are used for E-CAD/M-CAD coupling include the board contour, component positioning, and connections. In most cases, the circuit board contains dependencies from the construction space available in the housing. Examples of such dependencies or specifications include I/O interfaces and plug connectors, internal structures and assembly features, thermal requirements, and sometimes also aesthetic features [58]. This information is exchanged via proprietary, or better, via standardized interfaces on the tool level. The two older exchange formats DXF[61] and IDF[62] are extended by the XML schematic IDX[63]. Data objects which were previously missed have been added and objects for collaborative exchange were supplemented. This IDX standard offers users a reliable basis for data exchange. Support for iterative changes is also enabled[64].

The way in which the exchange process is carried out depends heavily on the use case. Sometimes, the mechanics dominate, for example, with tight construction spaces and aesthetic requirements; sometimes, it is the electronics. Let us assume the mechanical engineer starts, then the information to be exchanged, which is stipulated by the circuit board outline, highly restricted areas and stipulated component positions (→ plug connections), which the electrical engineer uses as a starting point. If the E-CAD/M-CAD collaboration is intensified, design process information is created with tracking of the changes which arise during the development cycle. This can be very useful if a design requires a change, in order to reset to an earlier status, or if data is required for the conformity documentation. [58]. Of course, these releases and change processes must be synchronized on the TDM/SysLM level. Figure 4.25 shows the result of such M-CAD/E-CAD coupling.

[61]DXF Drawing Interchange Format.

[62]IDF Intermediate Data Format.

[63]IDX Interdomain Design Exchange.

[64]ProStep iViP PSI ECAD/MCAD Collaboration, https://ecad-mcad.org/, (retrieved on 6.18.2020).

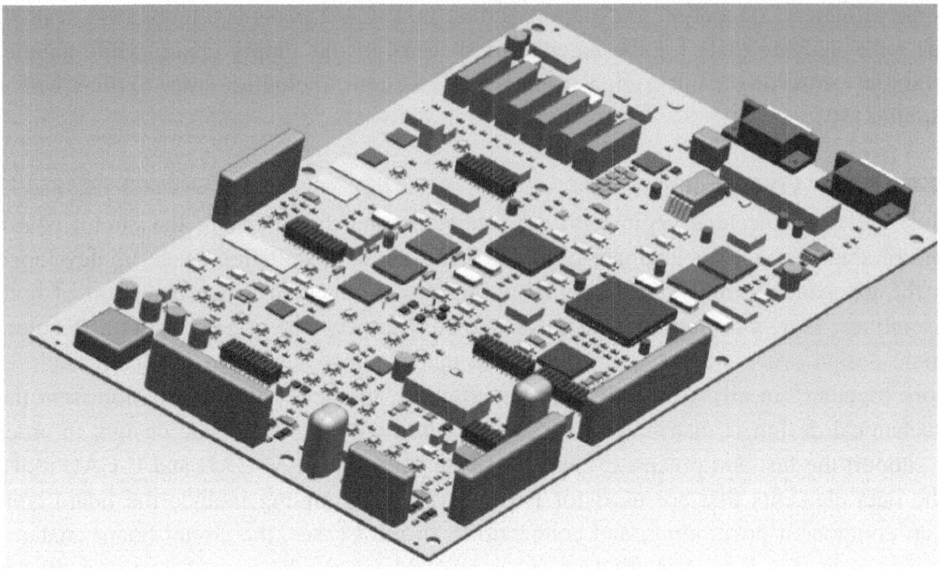

Fig. 4.25 M-CAD view after IDX data transfer from E-CAD [95]

Interdisciplinary cooperation between the design disciplines is significant for the design process in order to adhere to tight schedules. The visualization and use of design content which was created by other colleagues and disciplines in the overall project team enables faster supply and renders decision making better and faster [58].

CASE[65] Integration

Computer-aided software engineering encompasses a wide field of tools and methods. In order to correspond to the requirements of modern software development, the path for pure manual programming has been in the direction of model-based software development for several years. This takes place largely without programming. Processes and algorithms as well as structures and processes are modeled with graphic tools in various modeling languages and then translated by compilers/linkers automatically into program code (→ build). Different software tools are used to model the software. This is why we also refer to computer-aided software engineering (CASE). Building on the general process model (Sect. 3.5.1), we should now present a selection of the most important case tools and case techniques [7]. While the language SysML is used in general system modeling, in software engineering, the language subset Unified Modeling Language (UML) is used. UML is a graphic modeling language for the specification, design, documentation,

[65]case Computer Aided Software Engineering.

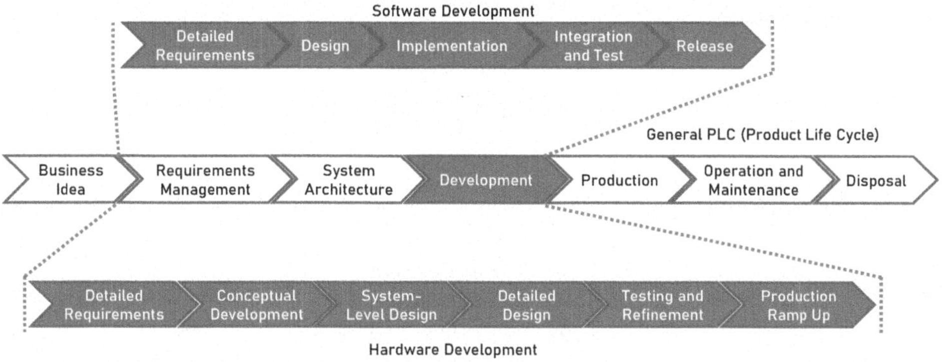

Fig. 4.26 Differences in software and hardware development [25]

and visualization of software parts and other systems. It is developed by the Object Management Group (OMG) and is standardized both by this group and by the ISO (ISO/IEC 19505 for version 2.4.1)[66]. The focus in this chapter does not lie on the authoring systems as in the previous chapter, rather on development and management tools. The term authoring system is also unusual in software engineering. It is most likely that tool kits for software engineering could be compared with the term authoring system.

Case Functions in the Software Development Process

Software and hardware development processes are different (Fig. 4.26).

In order to describe the most important steps of pure hardware development, a generic development process by Ulrich and Eppinger can be presumed [103] (Sect. 3.5.1). The software development process consists of similar phases. The process incorporates the following phases: detailed requirements, analysis, design (overall and detailed), implementation, integration, and test and release [25, 54]. In contrast to the original image from Crnkovic et al., great importance is attached to running a super-ordinate requirement management and super-ordinate system architecture in the product lifecycle before distribution in the disciplines.

Figure 4.27 represents various structure views of the software development process.

The software development process here can be separated into various phases. The best known is the subdivision into upper and lower case tools. Upper case tool incorporates the planning, definition, and design of the software analog to the hardware (cf. Sect. 4.2.1.2). The late phase—designated with lower case tools—incorporates the implementation and integration as well as the associated tests down to operation and continued maintenance. In integrated mechatronic and cybertronic product development, the description processes of hardware and software should of course run in synchronization. The upper case tools permit the description of the requirements and the complete

[66]https://de.wikipedia.org/wiki/Unified_Modeling_Language, (retrieved on 7.26.2020).

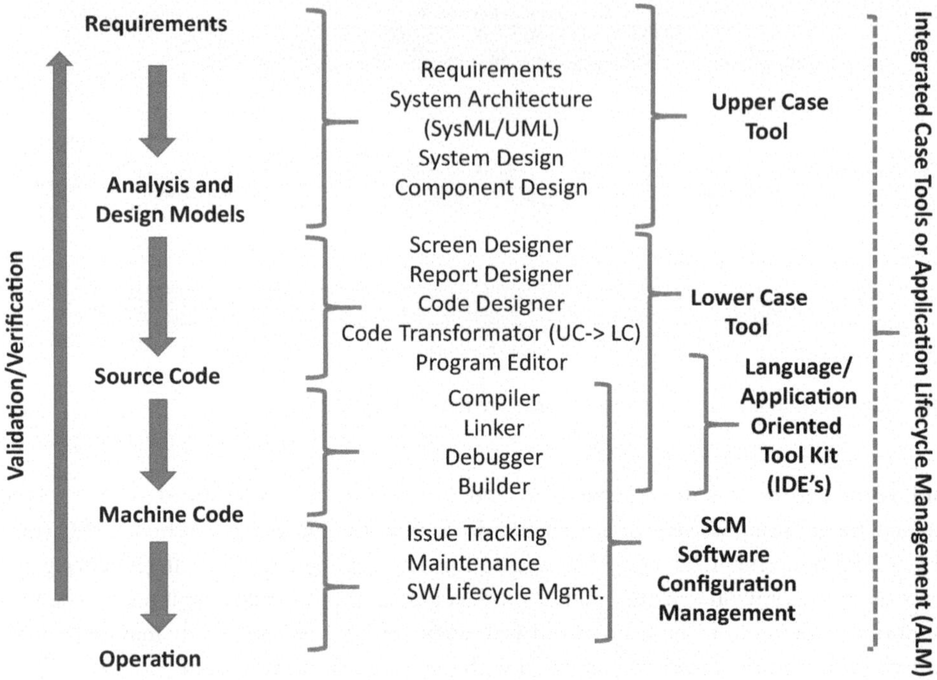

Fig. 4.27 Various structure possibilities for the software development process

architecture of a software solution on an abstract level and thereby make it possible to consider the requirements of a software solution, independently of concrete target platforms, program languages, and technical details. This enables developers to concentrate on the problem solution and removes the complexity from the development process. In addition, a problem, architecture, and solution description with graphic elements on an abstract level leads to more transparency in large and complex systems. In the software development process, the manual and semi-automatic transfer of the system model into program code for any target platform in any programming language takes place on the interface between the upper case tools and the lower case tools [57]. The scope of automated code generation depends on the level of detail of the system model. Often, software systems are not modeled down the last detail, as the effort for the creation of the models compared to the benefits would be too large. It is usually restricted to the software system being described with models in as much detail as necessary to make the complexity it contains understandable and manageable. This is why a programmer phase is normally connected to the automatic code in the transfer between upper case and lower case tools in the development process, in which the automatically created program skeleton is completed. The lower case tools used for this support the programmers with highly developed text editors with automatic program code formatting, Intellisense[67] functions, integrated assistance functions, code reuse, refactoring, code highlighting, and much more. In addition to pure programmer support for various languages, many lower case tools also offer even more integrated debuggers to find errors easily, fully integrated and very easy to use compilers, and semi-automatic software generation depending on the modeling depth. Due to the integration of many functions for the lower case in a development environment, we also refer here to so-called Integrated Development Environment (IDEs). Examples of widely used IDEs are probably the two most popular Visual Studio from Microsoft and Eclipse from the Eclipse Foundation (founded by the Eclipse consortium led by IBM). In order to connect the individual phases of the software development processes and all development artifacts such as requirements, models, code pieces, tests, documentation, etc., across all upper case and lower case tools and design a development with as few gaps as possible, the functions described above are integrated with so-called Application Lifecycle Management (ALM) solutions. ALM is a combination of the development and maintenance of applications (application software) across their entire lifecycle. This also includes comprehensive user support and the continued development of the software[68]. ALM is therefore a very broad term which demonstrates a changed perception of software development and is mirrored in the DevOps term. DevOps designates the mixture of tasks which are realized in a company by development and IT operations teams[69]. The dotted parenthesis for ALM (Fig. 4.27) is intended to indicate that the ALM tools are far from complete and also need to be supplemented by other systems.

The following ALM providers can be found in a relatively unclear market, who more or less cover parts of the software development process and of the typical ALM function scope[70,71]:

- Aligned elements,
- Atlassian with JIRA, Confluence, Stash, Bamboo,
- CA Technology Agile Central (previously Rally),
- Digital.ai with VersionOne, TeamForge, Continuum,
- Digité Softwareift ALM,
- Inflectra SpiraTeam,
- IBM Rational Collaborative Lifecycle Management,

[67]Intellisense is an assistant provided by Microsoft for automatically completing program code.

[68]https://de.wikipedia.org/wiki/Application-Management, (retrieved on 8.18.2020).

[69]https://www.computerweekly.com/de/definition/Application-Lifecycle-Management-ALM, (retrieved on 8.18.2020).

[70]https://www.softwaretestinghelp.com/best-alm-tools/, (retrieved on 8.18.2020).

[71]https://www.sam-solutions.com/blog/top-7-application-lifecycle-management-alm-tools/, (retrieved on 8.18.2020).

- Intland codeBeamer,
- Jama Software,
- MICRO FOCUS Quality Center (previously HP), StarTeam, AccuRev,
- Microsoft Team Foundation Server (TFS),
- PTC Integrity Lifecycle Manager,
- Rommana Software,
- Seapine Software,
- SIEMENS Polarion,
- TechExcel DevSuite,
- ThoughtWorks,
- Tuleap,
- VisionFlow,
- and others.

To increase the confusion even further, other terms are used—partly with ambiguous abbreviation:

- Software Configuration Management (SCM)[72]
  SCM is a subset of ALM and is a specialization of configuration management in the area of software engineering. It has several objectives:
  - Definition and tracking of processes,
  - Documentation of all processes,
  - Versioning and conflict handling,
  - Management of prerequisites,
  - Increasing efficiency in automated application creation,
  - Integration of existing tools, and
  - Access control.
- Version control software VCS including source code management (SCM) or revision control system (RCS).
  VSC is a subset of ALM and SCM. The best-known systems are GIT, GitLab, Bitbucket, and Subversion. Version control systems area category of software tools which help a software team in the management of changes in the source code across the lifecycle. Version control software traces every change to the code in a database.

In reality, two approaches have established themselves for the implementation of the software development environment[73]:

[72]https://de.wikipedia.org/wiki/Software-Configuration-Management, retrieved on 8.18.29020.

[73]https://issuu.com/kovair/docs/3_approaches_to_integrated_alm__a_case_for_alm_pla, (retrieved on 7.30.20209).

- Individual provider solutions possibly with functional extensions
  Two of the largest providers have developed relatively complete solutions for integrated ALM. Although their solutions vary in the actual implementation, they can be divided into the category "individual provider solutions." The IBM solution consists of Jazz integration technology, and various tools which were developed on the Jazz platform. This includes Rational Team Concert, Rational Quality Manager, Rational Requirements Composer, and Rational Project Conductor. Jazz has an open API which enables third-party providers to integrate their tools in Jazz. The heart of the integrated solution from Microsoft is the Team Foundation Server (TFS). Like Team Concert from IBM, TFS incorporates project management, source code management, continuous, build, and work task management. Similarly, to Eclipse IDE, Microsoft has its Visual Studio Team System as well as the tools for modeling, coding, and testing. However, for the vast majority of the companies with extensive investment in existing tools of various providers and development in Java, .NET, and other mixed technologies, neither the IBM nor the Microsoft solutions can be complete. The other ALM solutions mentioned above do not cover the entire scope and must be extended by various additional solutions.
- Best in Breed Open Source Solutions
  This solution has established itself very strongly in the industrial reality. The software developers who have worked under the radar of central IT for many years created their own biotope with open source and commercial part solutions. The very frequently used solutions can be taken from Fig. 4.28. In this solution, SysLM is able to take over the administrative portion of ALM (Fig. 4.29).

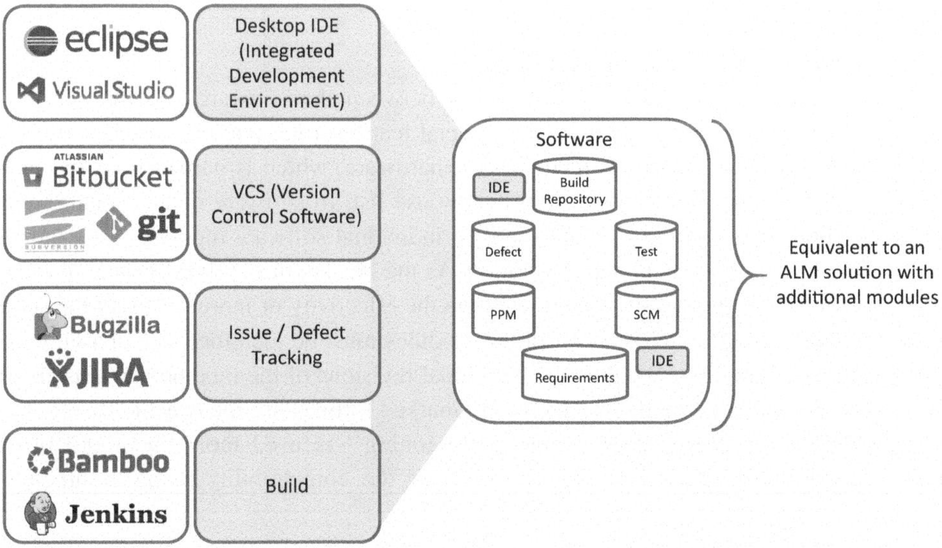

Fig. 4.28 Typical industrial software development environments, largely based on open source

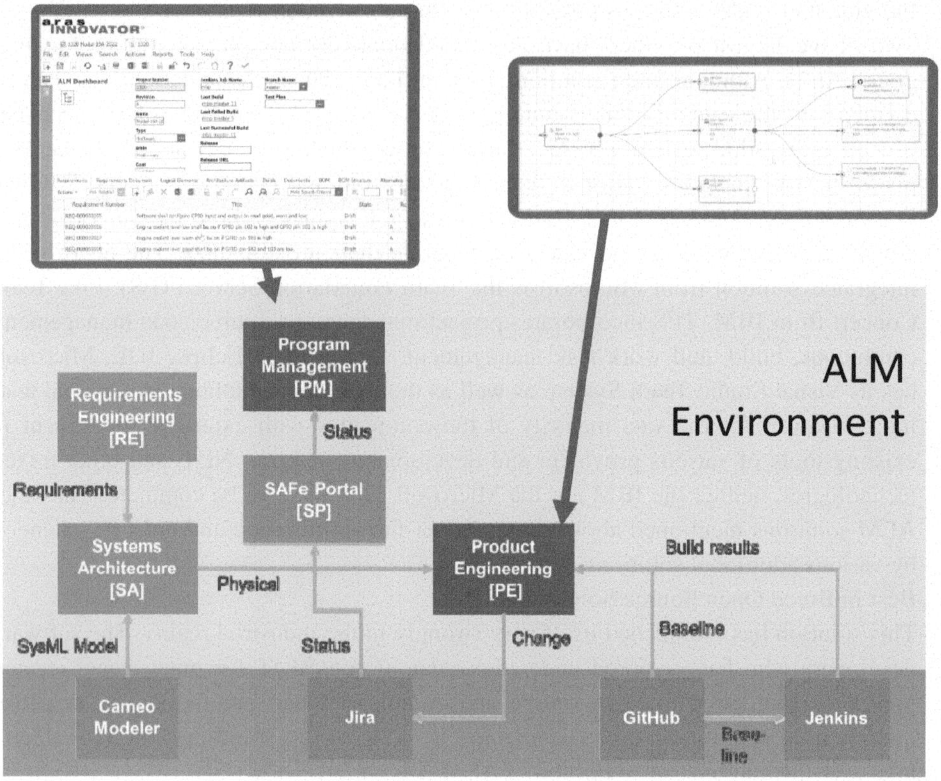

Fig. 4.29 Typical case solutions with open source products connected to SysLM. *Source* Aras

Special Features of the case Application

Figure 4.26 already indicated the differences between the software and the hardware development process. There are two more special features relevant with case. There is no hierarchical product structure as there is with hardware, which is named from the view from above parts list and the view from below use list. In software management, parallel trunks, branches, and mergers are used for individual software modules, for example application, operating system, and calibration. As the bracket of the BOM header, through which a current configuration is determined via the effectivity or more precisely the serial number, is missing, relevant versions of the modules must be identified via a baseline for the product configuration. For this, the individual revisions of the modules on a trunk are given tags. The software modules which were marked form a baseline (Fig. 4.30).

In addition, a more detailed semantic versioning[74] is used more and more often. This is a formal regulation for specification of the compatibility using a three-part

[74]Semantic versioning https://semver.org, (retrieved on 8.18.2020).

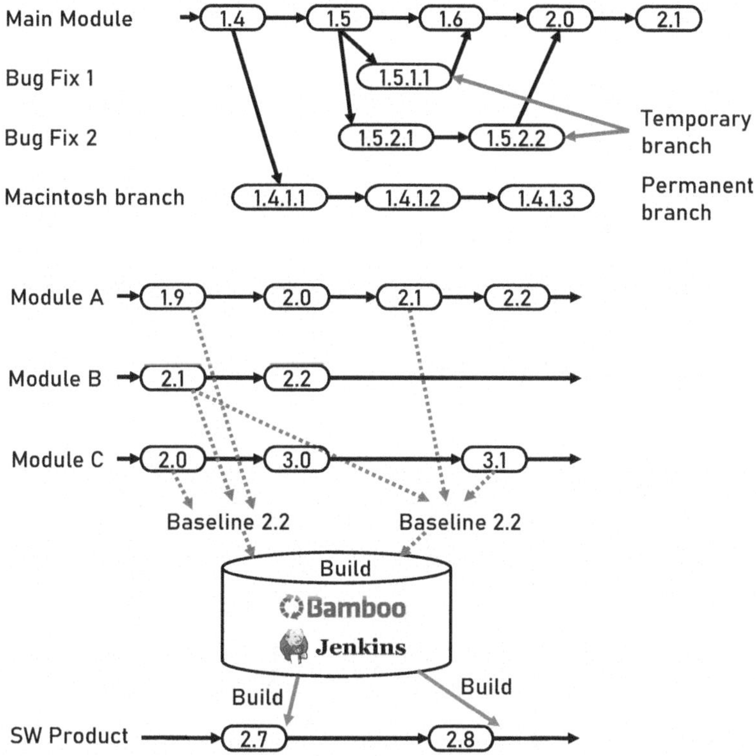

Fig. 4.30 Concepts of software management [25]

version number: major version, minor version, and patch version. The patch version is increased for smaller changes and error resolution which do not change the Application Programming Interface (API) of the software. The minor version is increased for versions that add new, but compatible features and the major version are increased for changes that are not compatible. Software that is based on version 2.1.5 of an API is compatible with version 2.2.3, for example, but not necessarily with 3.2.4[75].

CASE/SysLM Integration

Case integration is structurally simpler than a M-CAD or E-CAD integration. However, it is more complex due the amount and variety of case tools and the fact that there is less experience with it compared to the other CAx integrations, which leads to completely different requirements for the customer regarding SysLM integration. Figure 4.31 uses the example of the automobile supply industry to make this clear.

[75]https://en.wikipedia.org/wiki/software_versioning#Degree_of_compatibility, (retrieved on 8.18.2020).

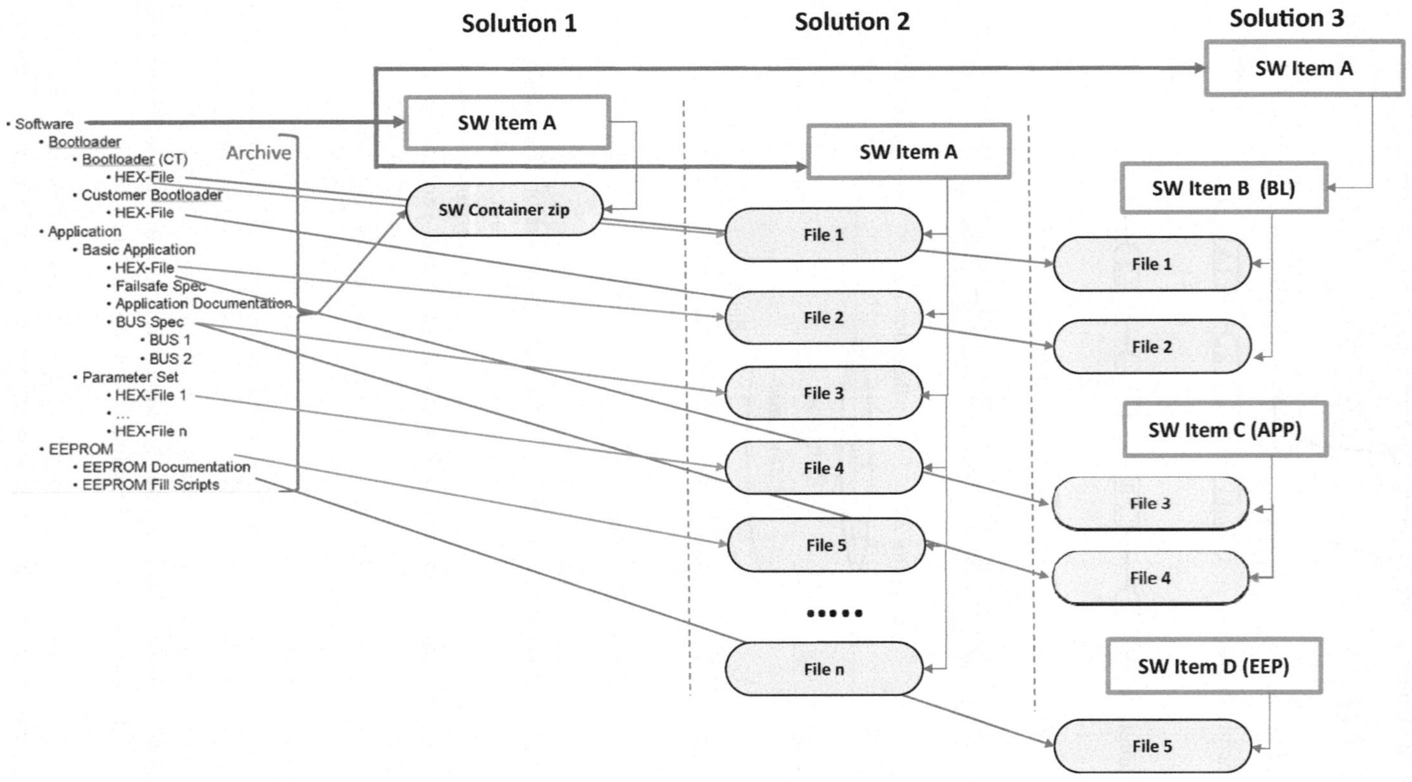

Fig. 4.31 Various software structuring possibilities in connection with SysLM

From simple filing of all relevant software modules in a ZIP file to a hierarchical depiction of functional software blocks, all possibilities were discussed and considered meaningful from the respective perspective. Here, it depends how the customer uses the SysLM integration. Does he want to control the flash process from SysLM/PPS or what partitioning of the software does the customer require. For some customers, an injection system would like to have the electronics application and the operating system pre-flashed by the supplier and download the calibration themselves in the final assembly process to protect the intellectual property. In turn, other customers want to use the option of the "where used" function from the SysLM system in order to determine which minor versions were delivered for which customer projects in order to optimize the update to customers (Fig. 4.32). For example, the following customer specific usecase:

- for customer A with an interface for the devices d and r and
- for customer B with an interface for the devices d and s.

In the event of a change to interface s (→ change to the minor version), customer A must not be updated to a new software status. Using the "trick" of essentially depicting a software structure below the software item, the "where used" function of the SysLM can be used as well. The delivery statuses can also be very easily traced when using a Digital Twin.

In general, a software item is a metadata artifact similar to the part that administrates the software content in the SysLM in connection with the software document (→ can be compared to the CAD document). The content can consist of many different types (→ firmware, tools, build, documentation, and descriptions) and different formats (→ Hex,

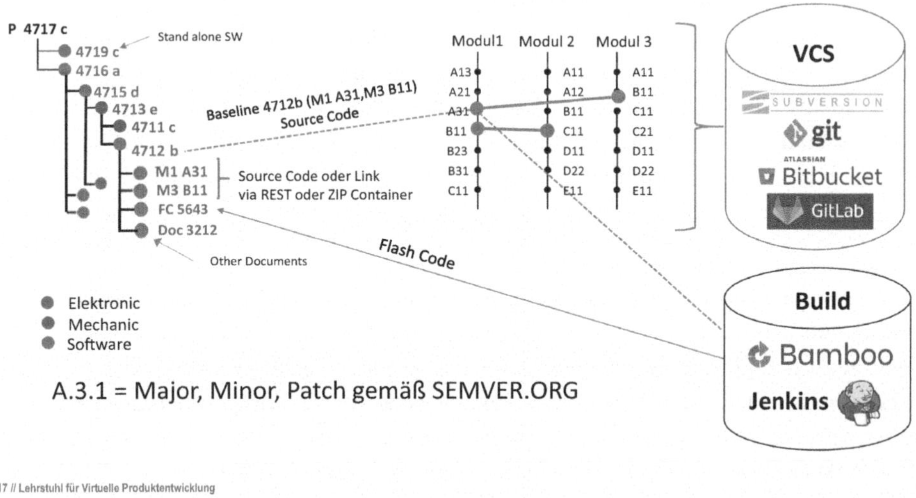

Fig. 4.32 Link of software modules in a software BOM in SysLM

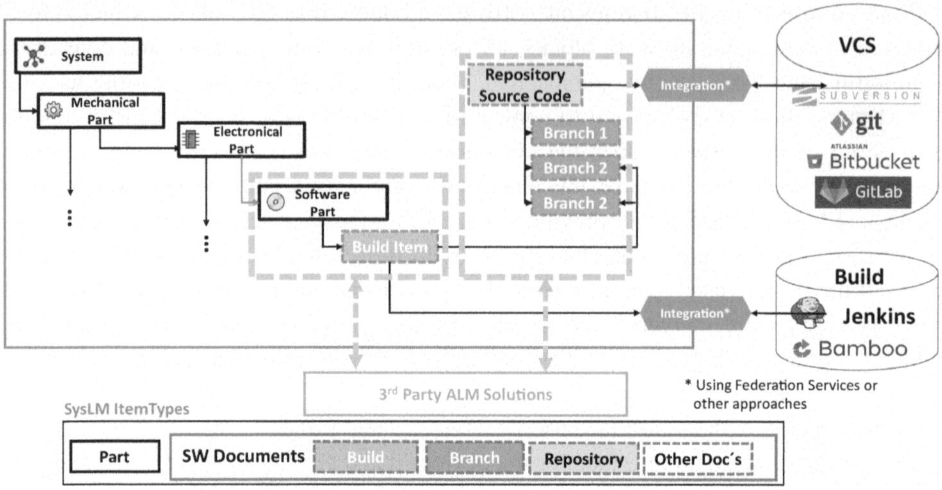

Fig. 4.33 Data model of a Case SysLM integration

BIN, EXE, ISO, DLL, ISO, MSI, ZIP, JAR, and PDF). Of course, a software item can be versioned. However, the three-stage revisioning on the revisioning present in the SysLM must be mapped. Figure 4.33 highlights the complex data model of a depiction of case in SysLM. It is important that the depiction concept is as flexible as possible, as the most varied customer situations must be implemented. Figure 4.34 demonstrates the corresponding user interface of a software item. Optionally, running software code (build) (1) or source code (2) can be stored in the SysLM.

4.2.1.4 Simulation Integration into SysLM

Many companies still use today PDM or PLM systems to control processes and manage CAD data (Sect. 4.2.1.3). Simulation and calculation data, however, is given insufficient consideration. The result of this is that, often, the current development status for the product and the determined calculation results no longer match. Here, CAE[76] providers[77], independent SimPDM[78] providers[79], and SysLM

[76]CAE Computer-aided engineering, umbrella term for all simulation and analysis tools.
[77]MSC Software, Simulation Process and Data Management,
https://www.mscsoftware.com/application/simulation-process-and-data-management
Altair, SimData Manager by PDTec, https://altairhyperworks.com/partner/simdatamanager
Ansys, Minerva, suMRPorted by Aras, https://www.ansys.com/de-de/products/platform/ansysMinerva
[78]SimPDM or SPDM is a local management of simulation data at the TDM level.
[79]PDTec, SimData Manager, https://www.pdtec.de/se-dm/simdata-manager/,
Comet, Sim AMRP, http://cometsolutions.com/, acquired by Aras.

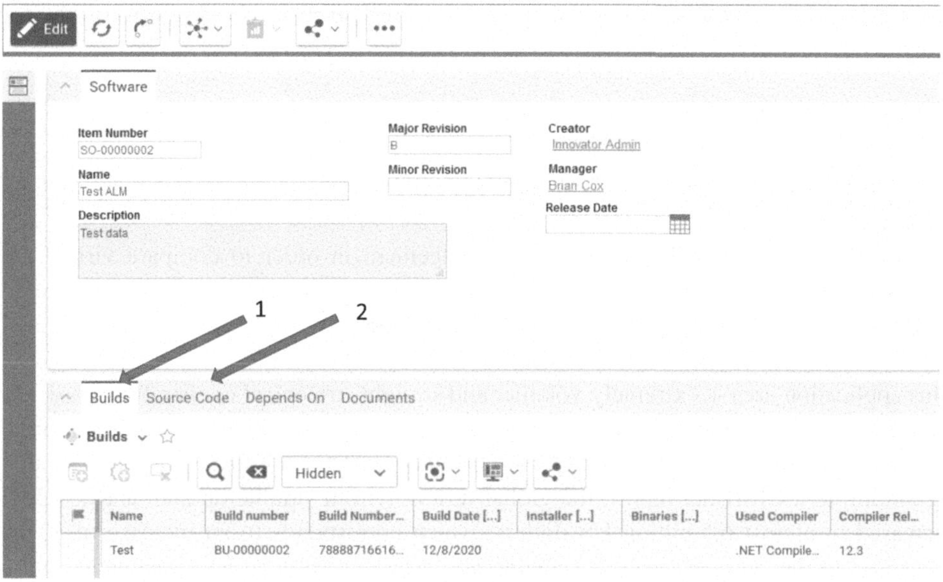

Fig. 4.34 User interface of a software item in SysLM. *Source* Aras

providers[80] have now started offering solutions. The CAE providers focus heavily on their own products. As a result of the increasing virtual product security through simulation and calculation, it is becoming more and more significant to incorporate simulation data in the entire development process on backbone levels, in order to support the traceability of functions (→ Digital Thread and Digital Twin). A further reason is that system architectures models are not executable in SysML, but it becomes possible to execute them in connection with simulation solution. Decisions and process results (→ design, prototype, production versions, and approvals) and the person who made the decisions are archived in SysLM. If SimPDM and SysLM are integrated, we can trace how and why decisions were made, and it is possible to improve and learn from process results.

These are a significant function of integrated SysLM and SimPDM systems. Use of a TDM level as opposed to a direct integration is highly recommended due to the

[80]SIEMENS TC UA, Simulation Process and Data Management,

https://www.plm.automation.siemens.com/global/en/products/collaboration/simulation-process-management.html (all retrieved on 6.15.2020)

PTC, Creo Simulate, https://ptc-solutions.de/produkte/ptc-creo-simulate

(all retrieved on 6.15.2020).

Dassault Systems, Simulia, https://www.3ds.com/de/produkte-und-services/simulia/

complexity of the simulation process. Here, this SimPDM platform should fulfill the following requirements:

- Open for all CAE applications, solvers, tool sets, optimization tools, and viewer,
- Configurable and extendable,
- Independent, intelligent workflow management in order to depict the complex CAE processes,
- Possible extension with physical test management in order to compare virtual and physical validation, and
- Clear functional splitting between SysLM and SimPDM.

This application area is extremely volatile, and straight strategical cooperation's are constructed. Altair distributes PDTec's SimData Manager, Ansys Minerva SPDM is based on Aras, and Aras itself has bought up the independent solution provider Comet[81]. According to a CIMData report, the global turnover from simulation and analysis 2018 increased to almost 6.5 billion US dollars, which corresponds to an increase of 13.1% compared to the previous year. It remains an "extremely fast growing segment" and is expected to reach 11 billion US dollars in 2023[82].

Simulation Management Processes

Simulation data is not merely the collection of files; rather, it demonstrates important differences from other design information. While a CAD file the descriptive object fully defined, a file of simulation results can only be interpreted if it is stored within the context of the data and software used to create it. This includes requirements, specifications, system architectures, especially functional structures, behavioral descriptions, and test cases. These are supplemented by the lifecycle-specific product models for the very early phase of simulation with reduced geometry, which is then concretized in the further development phases until the exact geometric shape, the electronic circuit, and the software code is available. This means that all decisions and assumptions that the simulation engineer have made at every step in the process which have led to the creation of the file must be traceable. The simulation can contribute towards accelerating design cycles, optimizing performance, reducing reworking, and controlling the costs. However, the lack of consistency between different tools and their specialists has heavily restricted the value of the simulation to date. The management of the simulation data and results in the Digital Thread connects inputs, processes, and results with a precise, traceable recording of the product configuration [23]. This improves access, repeatability, and cooperation and enables automated simulations or strategic initiatives such as additive manufacturing, generative design, Digital Twins, system design, machine learning, and artificial

[81]https://www.engineering.com/PLMERP/ArticleID/19895/Why-ANSYS-Bet-on-Aras-PLM-Technology-and-Why-Aras-Tends-to-Become-a-Silent-Partner.aspx, (retrieved on 6.15.2020).

[82]CimDATA, https://www.cimdata.com/, (retrieved on 8.14.2020).

intelligence. The focus of the following consideration is on SimPDM functionality and not on the authoring systems. These are so numerous and multi-faceted that this would exceed the scope of this book.

As an example, according to Fig. 4.35, a simple, two-stage, chained simulation of a bump over of a car should be used in order to show an application of a mixed-fidelity simulation with traceability. There are a number of requirements and parameters to be reviewed as well as other boundary conditions:

- The operational status for the product,
- A Modelica system model that describes the product behavior, and
- A detailed 3D representation of the lower control arm (LCA), which is used to check the suspension forces in comparison to the requirements.

First of all, a system analysis model that was created with Modelica and as an FMU[83] is defined. This is used to determine the forces and torque on the fixture points of the front left LCA. The parameters of the suspension geometry within SimPDM are varied in order to investigate different construction alternatives. Then, after detailing in CAD a second simulation of the lower control arm is carried out for a detailed tension analysis, whereby the loads from the Modelica model are used. Of course, a parametric CAD system is used so that the variation of the SimPDM system can be synchronized with the variation of the CAD system. With this, an FE analysis of the LCA is now carried out and optimized. Finally, the maximum LCA tension is checked based on the requirements. All simulation data is administrated in SimPDM or SysLM and made available to the Digital Thread [23].

Figure 4.35 demonstrates further relevant functions of a SimPDM system:

- The importance of the requirements and their possibility to identify significant system parameters and apply these across all applications of the PLC (green),
- SimPDM must allow multi-physics, multi-domain, and mixed fidelity (red), and
- A high-performance workflow system is used for more complex calculations (blue).

Depiction in SysLM

The full complexity of the simulation is made clear in Fig. 4.36. As already indicated, simulation results are only applicable and capable of being interpreted in connection with their context.

A context could be requirements, system architectures, or product models from the development. These are summarized with the overall order of the simulation into a simulation study. These are then broken down into the individual tasks (→ simulation task). Prepared simulation models flow into a simulation task resulting in so-called simulation artifacts. The simulations can be compared and traced via the graphic navigator [4].

[83] FMU Functional MockUp Unit.

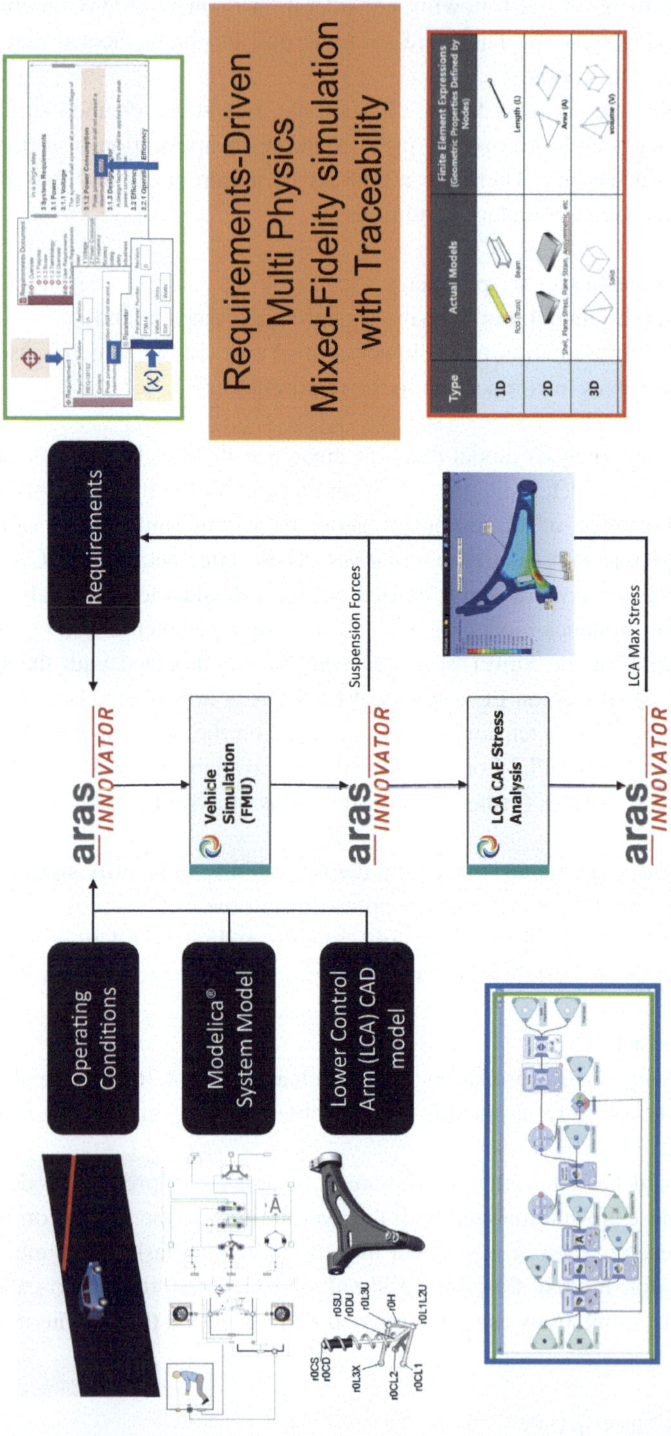

Fig. 4.35 Two-stage chained simulation of a car bump over [23]

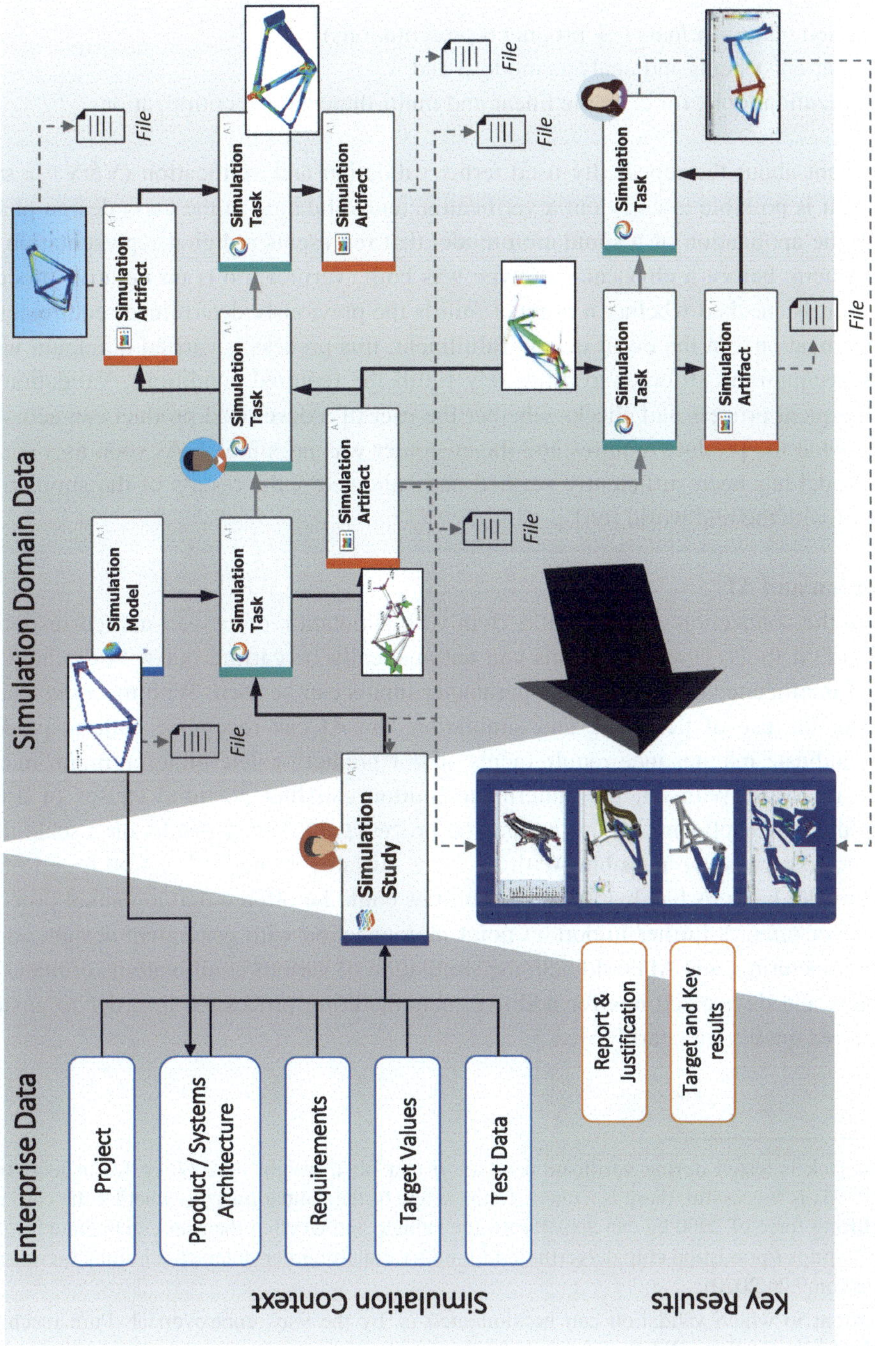

Fig. 4.36 SimPDM/SysLM data model. *Source* Aras

The variety of systems to couple in SimPDM is shown in Fig. 4.37. One distinguishes between:

- Input and output formats (→ product representations),
- Calculation systems and analysis models, and
- Optimization tools, for example linear and multi-dimensional optimizations.

A comment about the repeatedly used terms validation and verification (V&V) is still allowed. It is possible to carry out a verification and validation in the early design phase through the application of a simulation model that represents a digital representation of a real system, before a physical prototype was built. Verification is an iterative process in which it is checked whether a product fulfills the previously determined requirements and specifications. In the event of non-fulfillment, this process is carried out again with altered assumptions, in order to ultimately fulfill the required conditions. Validation is the subsequent process and checks whether the user of a developed product can actually achieve what the product requires and the customer was promised[84]. As soon as a simulation model has been sufficiently verified and validated[85], the results of the simulation should fit with the real world [66].

Simulation and AI

Through the connection of the Digital Twin with simulation processes, and corresponding SimPDM tools, new simulations can automatically be carried out with the help of artificial intelligence (AI) and varied parameter inputs can be used. With machine learning under the use of field data and simulation, the AI can recognize pattern, predict failure, indicate maintenance requirements and/or predict or determine optimum maintenance intervals. With AI, the pattern recognition can find potential causes of error and simulate their effects and use automated processes to deliver results on a scale that would not otherwise be possible. With AI more "what if scenarios" can be tested on a much broader basis. A bandwidth of possibilities could be offered that a manual process could never offer. A further important point in connection with generative design, additive manufacturing, and AI could help the simulation of various combinations of material properties and device settings for additive manufacturing processes, in order to ensure the required quality of a part [66].

---

[84]"A car jack is tested during verification to see if can lift a weight of 2000 kg for at least two hours. If this is successful, there is a check in the scope of the validation as to whether the car jack with a lifting force of 2000 kg can actually lift the vehicle and whether the user can operate it intuitively." https://praxistipp.chip.de/verifizierung-und-validierung-der-unterschied-einfach-erklaert, (retrieved on 9.26.2020).

[85]The extent to which validation can be simulated is, by the way, controversial. Pure teaching would basically say that validation can only be done with the customer.

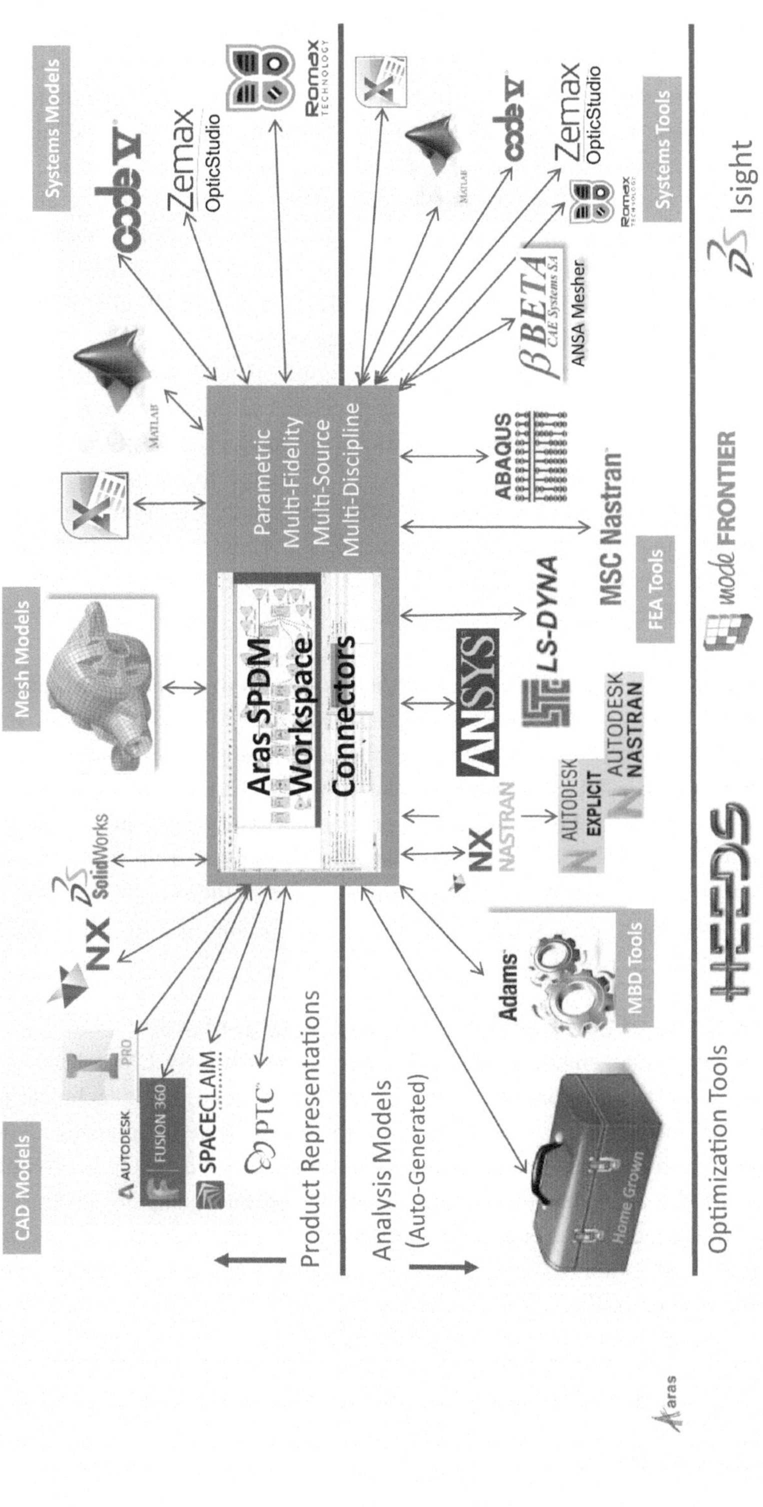

Fig. 4.37 Overview of a typical tool landscape to be integrated in SimPDM. *Source* Aras

4.2.2 Horizontal Integration and Digitalization

According to an investigation by Tech-Clarity, cross-departmental integration of the technical development represents a large problem for almost half of those surveyed. A little over half of the companies have separate teams for mechanical, electronic, and software design. Thus, the central development departments do work with one another, but not nearly enough to achieve optimum innovative development results. Without consistent cooperation and interdisciplinary development methodology, it is hardly possible to manage and master changes to the product, the production, and in all other related organizational units. The same applies for tracing damage events.

The digitalization concepts Digital Model, Digital Thread, and Digital Twin are implemented in an SysLM system as an engineering backbone concept. Figure 4.38 documents the target of consistent and company-wide integration of the product lifecycle phase with a simple example. In the event of a change to an individual engineering object, for example a CAD document, all other potentially Configuration Items must automatically be found by the Digital Thread (Sect. 3.4) in the Digital Model across all phases of the product lifecycle [52]. The more phases are integrated, the clearer and more complete the statement resulting from this.

In a traditional approach, engineers still use their discipline-orientated tools. Of course, there are regular file exchanges; nonetheless, the view of one department among the others is very restricted. The result of this: a separate, detailed construction, in which missing transitions in various disciplines or the competing constructions and failure are only discovered in a late development phase, when changes are much too expensive and time-consuming[86]. Presented below are the necessary prerequisites and framework conditions for a horizontal integration based on an engineering backbone concept (Fig. 4.39). The vertical integration of requirements down to the simulation is the prerequisite for this. The integration in the production, logistics, and use of information from the operative area are dealt with in this chapter (→ MRP and MES).

4.2.2.1 Standardized Company-Wide Technical Organization

Through positioning as an engineering backbone (cf. Fig. 3.30), SysLM solutions, unlike CAx, and simulation systems are much more deeply embedded in the technical process organization. For this reason, the implementation of these systems requires methodical and organizational preparation through analysis, optimization and, as far as possible, also harmonization of the engineering processes at a division and company level. Not only the industrial mid-sized sector, but also much larger companies are rather unsystematic in regard to the organization of the PDP[87] tasks, processes, and their documentation. The reasons for this are manifold:

---

[86] https://net-professionalservices.de/mcad-ecad-integration-was-wir-noch-bewaeltigen-muessen/, (retreived on 8.15.2020).

[87] PDP Product Development Process.

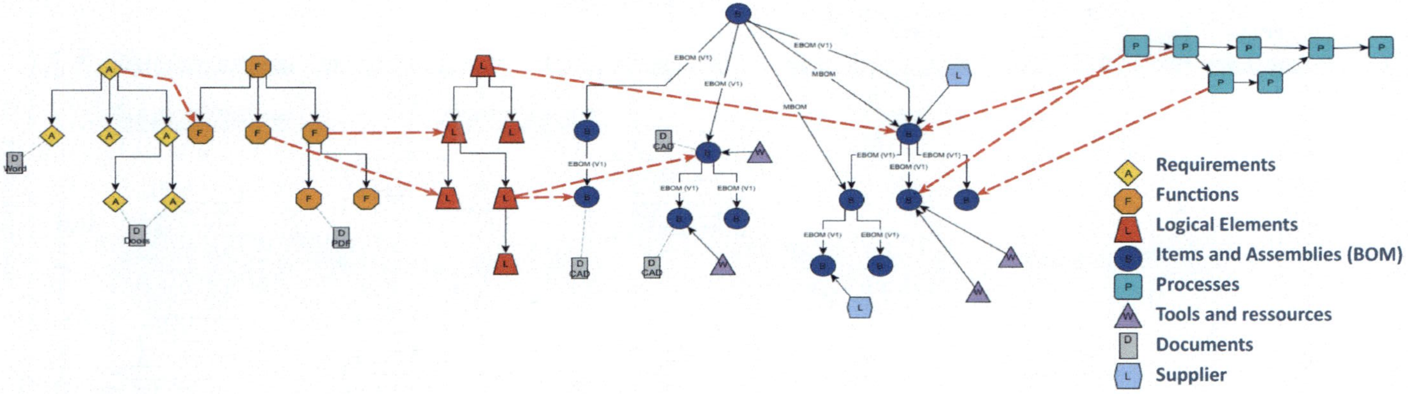

Fig. 4.38 Networking of all Configuration Items in the event of a change on the basis of a graph-based analysis across all object classes filed in SysLM

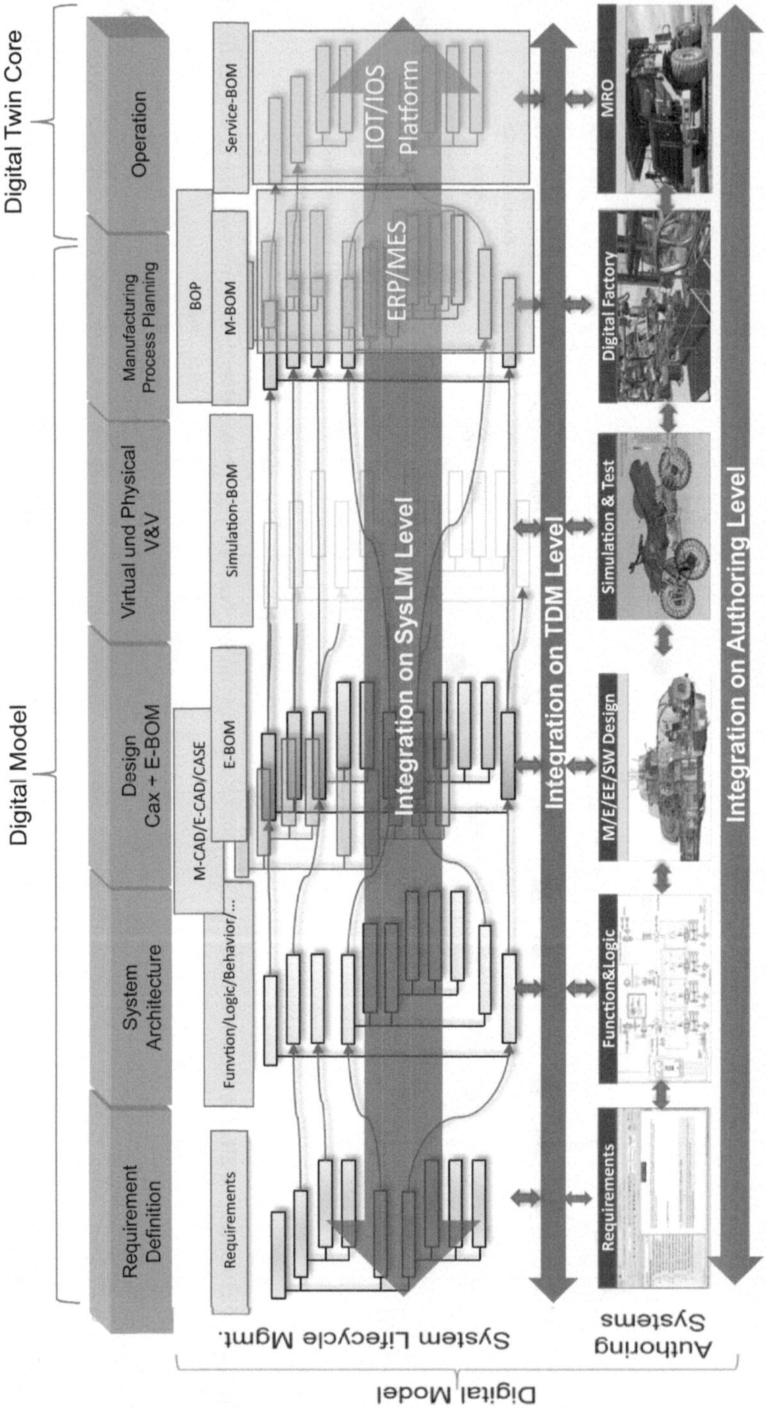

BOM Bill of Material / E-BOM Engineering BOM / M-BOM Manufacturing BOM / BOP Bill of Processes / MRO Maintenance Repair Overhaul / CAD Computer Aided Design / M-CAD Mechanical CAD / E-CAD Electrical/Electronical CAD / CASE Computer Aided Software Engineering

Fig. 4.39 Horizontal integration along the product lifecycle

- Management disinterest,
- Lack of company-wide function "Technical Organization" (has typically been rationalized away in the 80s and 90s),
- Lack of an overall view of the product lifecycle and across the various divisions and site locations,
- Persisting with the traditional paper- and document-orientated work style, and
- The repeated belief that an IT system will automatically resolve the organization deficit.

The significant topics of the organizational application preparation—partly already with SysLM examples—are presented below. In addition to the foundations of the numbering, naming, and classification, these also include the various types of product structuring and the technical processes release, change, and configuration management. As the topic of technical organization is extensively dealt with in two books by the author [36, 37], only a brief summary of the problem will be presented here. The focus in this book should be on the extended functionality in regard to interdisciplinarity and the breadth of application of SysLM. Technical organization is an area which is already built on eternal foundations and also which should be dusted off, simplified, and optimized in the scope of digital transformation. It would be a shame not to use this opportunity.

Foundations of Technical Organization

An overview of the elementary foundations of technical organization in regard to numbering systems, relations of the significant object classes of the SysLM system (→ part and document), change and configuration management, revision/version as well as classification can be taken from Table 4.1 (summary from [36, 37].

These "unpopular" topics in companies must not only be determined for the foundational object classes part and document, but also for all object classes planted to be filed in SysLM along the product lifecycle, for example requirements, functions, process plans, tools, equipment, etc. In general, we can state that there is often a theoretical optimum and a practical and often better accepted solution for an organizational problem. This should be made clear with the example of the document number:

- Solution a): the document number is completely independent of the part number
  This makes the depiction of an n:m relation between parts and documents optimally implementable. The designer who may have just transitioned from a document-orientated solution is confused, as he must now maintain two master data sets and moreover with two numbers.
- Solution b): then forms the document numbers from the part number (→ 47110815 plus a document type, for example CT5 (→ Catia V5 native). This results in the document number 74110815_CT5. Drawings, which now belong to several part masters, for example lubricating regulations and installation instructions, must then be specially numbered.

Table 4.1 Foundations of the technical organization

Problem	Solution alternatives	Recommendation
Parts number	a) Pull number from MRP b) SysLM and MRP no different	SysLM and MRP number should be identical
Clear understanding of ECM (Engineering Change Management)	ECM is globally defined in the standards and recommendations. It is a prerequisite for configuration management	Implement clear rules about the compatible and incompatible company-wide changes Show changes in lower BOM levels to the customer on the product level
Clear understanding of CM (Configuration Management)	CM is globally defined in the standards and recommendations. Important: A prerequisite is ECM, and variants	Implement clear rules and processes across the company
Configuration Items (→ CI)	CIs are information classes that are potentially affected by a change. These would then be affected items (→ AIs)	Define clearly and implement as a company-wide foundation of ECM and CM
Part: **Revision/change index**	a) Letters b) Numbers c) Alphanumeric	Any, for example, up to release b), afterward a) or c)
Relation part document	a) 1:n b) n:m	b) n:m
Relation document file to document	a) 1:1 b) 1:n	Both possible More document types in a)
Document number	a) Independent of document and part numbers b) Document numbers = Parts no. + document type	Both possible. Advantage a) n:m relation to part b) better acceptance
Documents: **Revision/change index**	a) Letters b) Numbers c) Alphanumeric	Any see part revision (+ iteration when checking in/out CAD)
Relationship part revision - Revision to documents	a) Identical b) Independent	Generally independent, but defining rules for document changes that influence part revisions or not
Classification	a) Thesaurus (semantic classification) b) Classification (property) lists c) Naming catalog	for standard parts b) otherwise generally introduce a multi-lingual naming catalog (c) based on a hierarchy like a)

Solution a is the theoretically "cleaner" solution, and b is better accepted by users. In principle, the old adage applies: rather an imperfect organization than none at all. Organizational wild growth is the worst prerequisite for a successful SysLM introduction.

BOM Management

This sub-chapter is dealt with in a little more detail, as the question about the BOM comes up again and again and a whole series of current weak points must certainly be discussed:

- The organizational, intellectual, and work-technical separation of development and production planning today leads to enormous transfer services with a high error rate (→ E-BOM/M-BOM[88]/BOB[89] transformation),
- Extremely high redundancy of product structures along the product lifecycle (Fig. 4.40) and
- Up to now, a silo-orientated mindset restricted the development of traceable interdisciplinary parts lists.

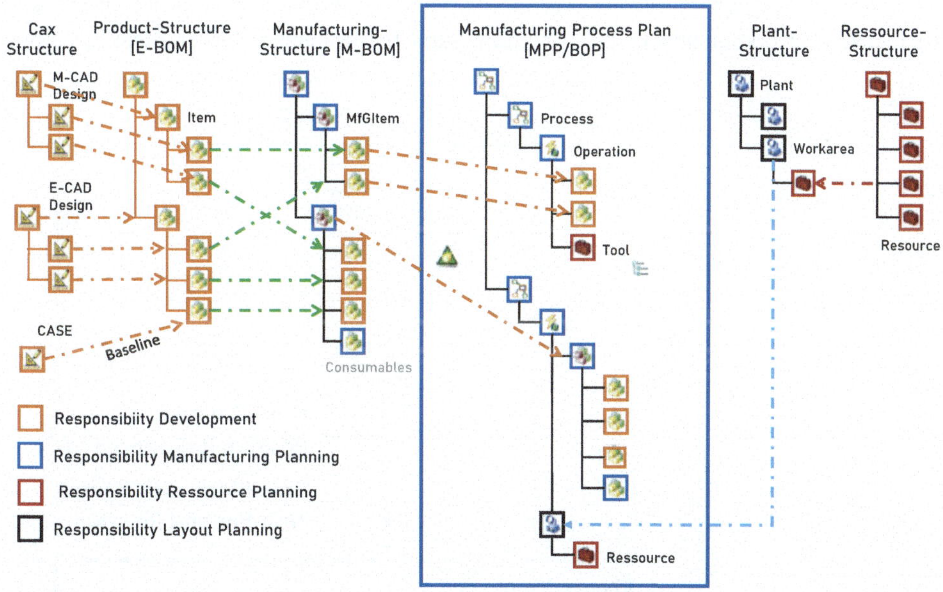

Fig. 4.40 Extreme redundancy of the product structures along the product lifecycle. *Source* based on Siemens

[88]M-BOM Manufacturing BOM.

[89]BOP Bill of Processes.

The interdisciplinary E-BOM is derivated from the design structures (→ CAx structures) for mechanics, electrics/electronics, and software filed in the CAx system. The manufacturing structure (→ M-BOM) results from this in turn from the adaptation to the assembly structure. The manufacturing process plan (→ MPP or BOP) is created from this with almost complete retention of the M-BOM structure by adding resources (→ machines, assembly lines, equipment, tools, operating personnel). M-BOM and BOP are in turn adapted in parallel manufacturing at locations by different supplier and assembly situations. This often leads to six or even more partially redundant structures.

In general, today the same methods are still applied in regard to describing the product structure as since the introduction of the BOM. In order to better understand the changes between the product structure and the technical documentation over the last decades, a typical example of an assembly drawing from the 70/80s is presented below (Fig. 4.41). It is typical for this phase that all technical and organization relevant information is contained on the drawing. Hence, the expression "the drawing is the language of engineers." It is identified on one hand by a largely flat structure of the individual parts and assemblies. The assignment of the parts to assemblies and products takes place visually via the parallel numbering in the parts list lines and the graphical representation. This method has the advantage that the assignment of parts is largely free from interpretations like function or assembly process-orientated. While product descriptions were still focused on documents in the 80s, now parts list-orientated hierarchical structures

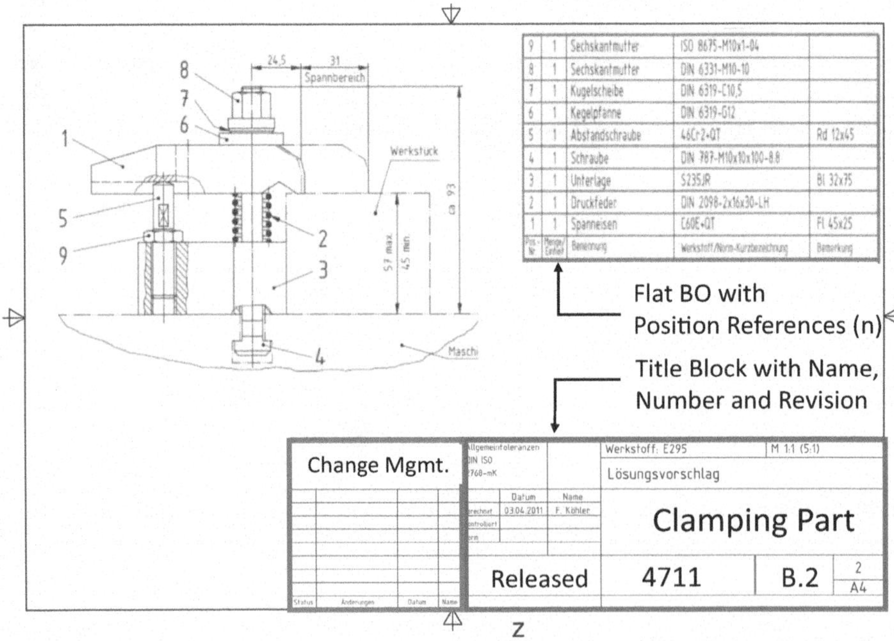

Fig. 4.41 Example of a module drawing from the 70/80s [38]

parallel to geometric description have prevailed—also due to increased CAD introduction in mechanics and electronics, as well as to support the scheduling process for emerging MRP systems. This process was supported in particular by 3D M-CAD and E-CAD systems that automatically created a topology-orientated structure for positioning the individual parts related to assemblies.

Hierarchical structures still form the core of today's product structures; however, due to the increasing significance of mechatronic and cybertronic in SysLM systems through linear and branched (→ software) and network-like structuring methods (→ System Model), they must be supplemented. If the individual structures from Fig. 4.40 are considered from the perspective of relevance, the following image results:

- Design structures from CAx systems are system-relevant for many functions, as they support significant processes through their additional topological information, for example DMU[90] and pick and place.
- In an interdisciplinary product world, the E-BOM has the task of merging the discipline-orientated BOMs to a traceable mechatronic BOM (Fig. 4.43). In a complex product environment (→ airplane, ship, car), the E-BOM has the task of the more functional view the product environment. For more simple products and in engineering to order configuration, it loses importance and should be identical with the M-BOM.
- The M-BOM represents the assembly process-orientated view of the product structure. It can vary many times in production at different site locations. The M-BOM is required for quantitative and qualitative manufacturing and procurement planning.
- At its core, the BOP consists of the stages of M-BOM supplemented by resources and time management and is ultimately the input information for production and assembly.
- Another interesting discussion is presented in the blog Beyond PLM of Oleg Shilovitsky. What is the optimal BOM form—single BOM, multi-BOM, single BOM/multi-views and are relational databases still the optimum for BOM handling[91]?

Thus, at its core, it concerns the simplification of the three structures, E-BOM, M-BOM, and BOP[92]. This is almost a political task, as here we are in trench warfare between MRP and SysLM, whereby, in recent years, a trend has also resulted to transfer functions of MRP both according to SysLM and according to MES. The ideal would be to also place

---

[90]DMU Digital MockUp.

[91]http://beyondplm.com/2021/02/27/bom-management-single-bom-multi-bom-multi-views-how-to-decide/, (retrieved 2.22.2021).

[92]Here, we refer to an interesting blog from Jos Voskuil, "PLM is changing and affects us all," in particular "Learning from the past to understand the future—some E-BOM / M-BOM clarifications (1–7)." https://virtualdutchman.com/2020/08/23/learning-from-the-past-to-understand-the-future-some-E-BOM-M-BOM-clarifications-7/, (retrieved on 8.31.2020.

the two work groups together organizationally as well, according to the motto "design to assembly." An important simplification, particularly for complex product structures, is introduced by Schichtel by a division of the product structure into [87]:

- Generic structure (also referred to in variant systems as modular product structure or breakdown structure) and
- Instantiated individual parts and assemblies (concrete BOM and parts).

The structure of a product (P) is initially classified in logical/generic groups (G1–G6). This so-called product classification represents the skeleton of a product. The actual individual parts (E1–E6) or assemblies (Z1, Z2) are added below the generic leaves (logical nodes) and extend the generic structure to the complete product structure (Fig. 4.42).

This separation represents a drastic simplification of the product structures in certain use cases. It is used for quantitative complexity (a high number of components), for example in machinery, plant, airplane construction, and shipbuilding, as well as for qualitative complexity (high number of variants) in automotive.

The following discussions are held in the community and should lead to an overall optimization of the BOM process:

Reduction of the number of parts lists, or which parts lists are actually needed
The following measures are discussed in order to reduce the number of BOMs by at least one structure:

- E-BOM and M-BOM identical
  This will be applicable in simple application situations; however, for many companies currently with parallel manufacturing, it hides the disadvantage that E-BOM and M-BOM cannot be uncoupled and changes to the M-BOM are imposed on the E-BOM.
- It is possible that we could assume a flat E-BOM generated automatically from the CAx systems, similar to Fig. 4.41, and make the configuration management and merge to the mechatronic BOM in the M-BOM.
- In contrast to the traditional conversion of the E-BOM in the M-BOM, the BOP is created from the E-BOM on the level of parts and assembly-ready modules with systems support (→ drag and drop). This requires tight coupling with the MRP system in regard to the resources and a transfer of not only the M-BOM but also the BOP into the SysLM system. The M-BOM, which continues to be required for logical purposes, is automatically derived from the BOP. E-BOM and M-BOM are automatically synchronized and know from one another which parts are contained or changed (→ reconciliation) (see Fig. 4.74).

In the last solution, change effects are identified, synchronized, and managed through the mutual referencing of E-BOM and M-BOM.

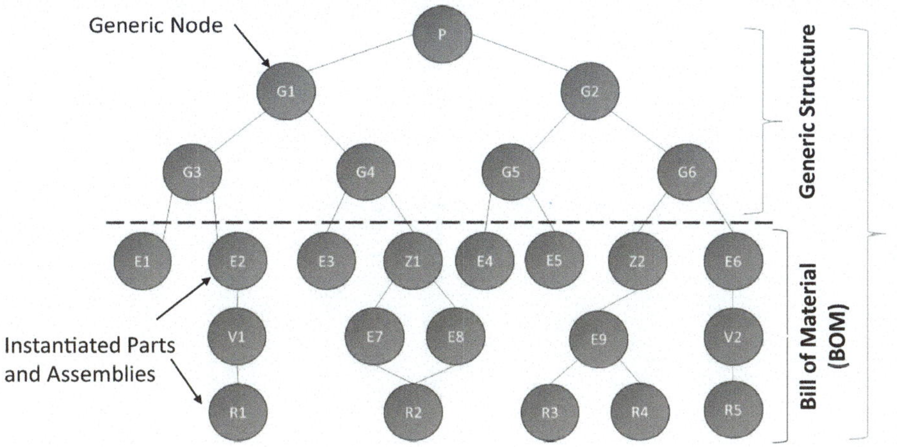

Fig. 4.42 Separation of the product structure into a logical/generic product structures and instantiated parts and assemblies [87]

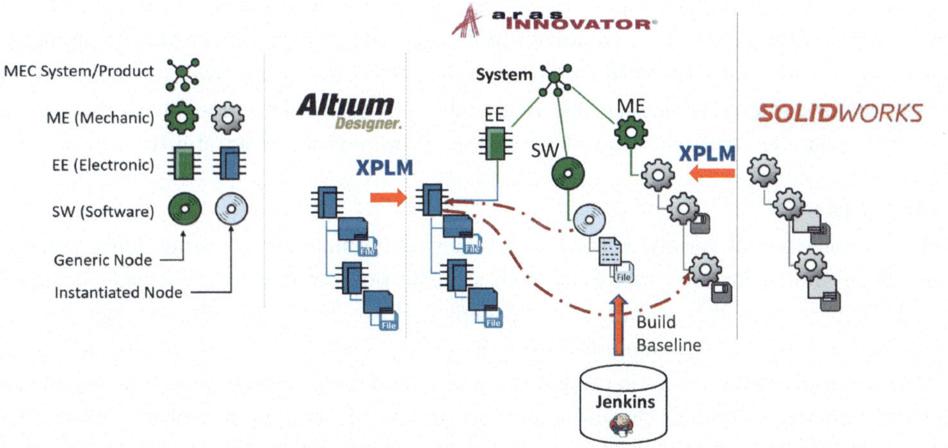

Fig. 4.43 Basic principle of uploading CAx structures into mechatronic BOM and build up an interdisciplinary link [99]

Merging the discipline-oriented parts lists into a Mechatronic BOM

As the discipline-oriented CAx structures for E-BOM formation according to Fig. 4.40 play a significant role, in the first stage, the two mechanical and electronic disciplin oriented BOMs are linked to generic nodes for mechanics and electronics under a mechatronic/cybertronic overall system node. The same occurs with software according to Fig. 4.33. The process is relatively easy to implement which is simple to automate in the CAx coupling (Fig. 4.43).

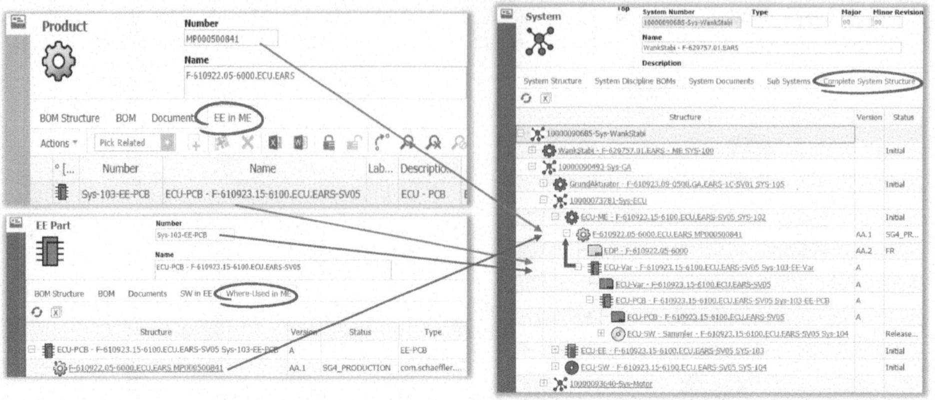

Fig. 4.44 SysLM customizing for uploading and linking discipline-oriented structures into a mechatronic BOM [99]

The cross-references and links between electronics and mechanics, and software and electronics must be created in the second stage. Figure 4.44 shows the transformation of the discipline-oriented structure into mechatronic BOM, as outlined in Fig. 4.43. In Part Item, a menu *EE in ME*[93] is customized. With drag and drop, the electronic component is pushed in there. In the BOM menu on the right side of the figure, you can see the result. This creates a traceable mechatronic BOM which in turn is a prerequisite of this product lifecycle phase to create the Digital Thread and thus guarantee traceability.

Variant BOM

The complexity and variety of modern products is permanently growing. Only when we are successful in meeting this trend with suitable product development and production methodologies can the overall costs be kept under control. Suitable product structuring and design as well as manufacturing, assembly, and procurement processes built on these are prerequisites. On the product side, the systematic construction of a variable and modular product structure through construction kits, as well as measures to reuse components, can make a significant contribution. Variant building blocks are a long-known working technique in the industry. Its importance for a company's cost structure has further increased under the framework conditions mentioned. The propagated platform and reuse concepts in the automobile industry are an expression of this working technique. The creation and management of this variants is of considerable importance for development and scheduling, as the number of possible product features can be extremely high, for example in the area of automotive approximately 10^{30}. The rapid increase of the solutions—mathematically triggered by the faculty rule—can be seen in Fig. 4.45, with the example of a driver's seat.

[93]Connect electronic (EE) with mechanic (ME).

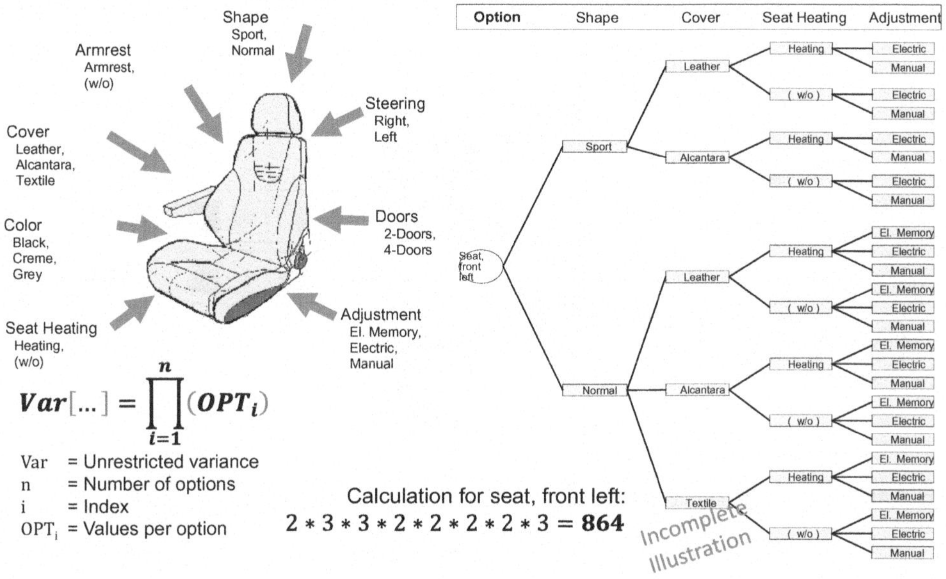

Fig. 4.45 Complexity of a simple variant (driver seat) [108]

Variant configuration systems represent the current state of the art technology [108]. Instead of an explicit definition of all product variants, only a variant generation logic is managed in this type of BOMs. This contains all modules, for example individual parts, from which all conceivable product variants can be derived from rules. The selection of required modules/individual parts and the determination of the amount information for a certain product variant are made by a configuration logic. The concrete approaches marginally vary between the providers of the variant BOM systems, but all reproduce the configuration logics in formal rules. The derivation of a concrete product variant on the basis of a rule-based variant BOM occurs through the evaluation of the configuration logic based on order-specific information (→ options or features). These are usually characteristics of the desired product, for example color or materials. The order-specific information thereby defines the configuration of a product variant. Figure 4.46 shows the fundamental construction of a variant definition which forms the prerequisite for an automatic evaluation and instruction of special order-specific variants by a configurator.

The *Variability Item* encompasses:

- *Feature Families* (FF), for example color or material,
- *Features (Options) (F),* for example black, beige, white or leather, textile, alcantara,
- *Rules/Constraints*, for example material = leather AND color = beige, and
- Validation is a test mechanism which tests the completeness and clarity of the rules.

In the feature families, we decided between two types:

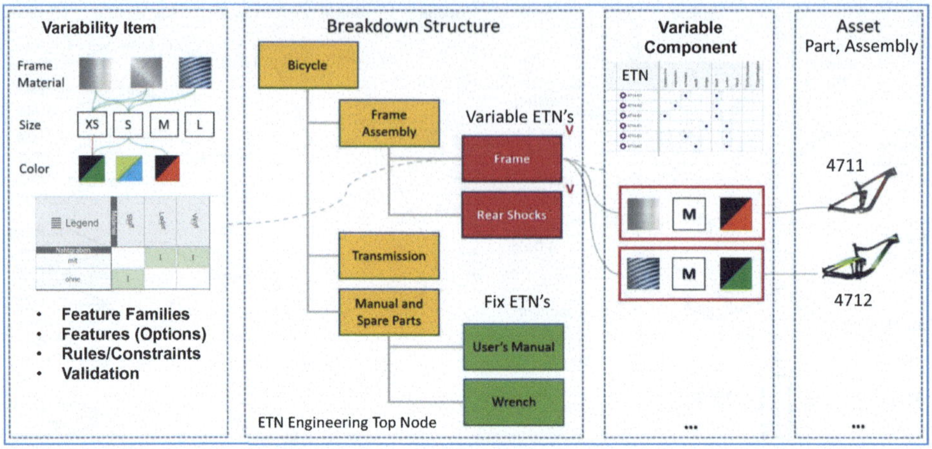

Fig. 4.46 Fundamental concept of a variant definition (according to Aras)

- Feature families that classify the part or the assembly. Here, the ratio between FF and F is always 1:1.
- Feature families that identify an usage conditions; for example, the black leather headrest is used for both the black leather and the alcantara seats. Here is the ratio between FF and F 1:n.

The *Breakdown Structure*, often also referred to as the generic or modular structure, is the basic framework of the rule-based variant BOM. This structure is identical for all instantiated variants and corresponds essentially to the generic/logical product structure according to Schichtel (Fig. 4.42). There are either variable, inc. optional, components or fixed components, which occur in every variant on the lowest level of the breakdown structure (→ leaves). The Variable Component assigns every variant alternative on the leave level to a concrete part or assembly with an item number. Figure 4.47 is to be taken from a concrete example from the automobile industry that is implemented in Fig. 4.48.

Feature families, features, and rules are defined for the armrest in steps one to three. The variant building blocks from 1 to 3 are validated in the fourth step. In the fourth step, we can see that leather has been selected, and therefore, only the color alternatives black, beige, and white are possible. The yellow matrix is generated from the existing rules in the fifth step. The item numbers can be inputted manually or generated by a number generated on the left side. The integration in the technical organization and strategic product planning must also be guaranteed in the variant systems. This means the

- variant building blocks (→ variability item feature families, features, rules, and breakdown structure) must be able to be revisioned and subject to an ECM process, for example, in order to depict model years of products or changes in the break down structure or a new feature is added.

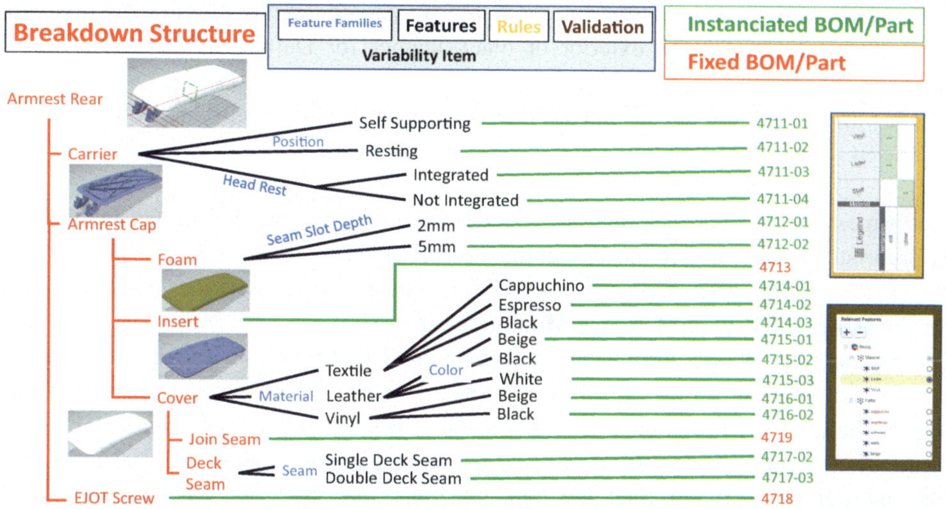

Fig. 4.47 Concrete example of a variant construction from the automobile supplier area

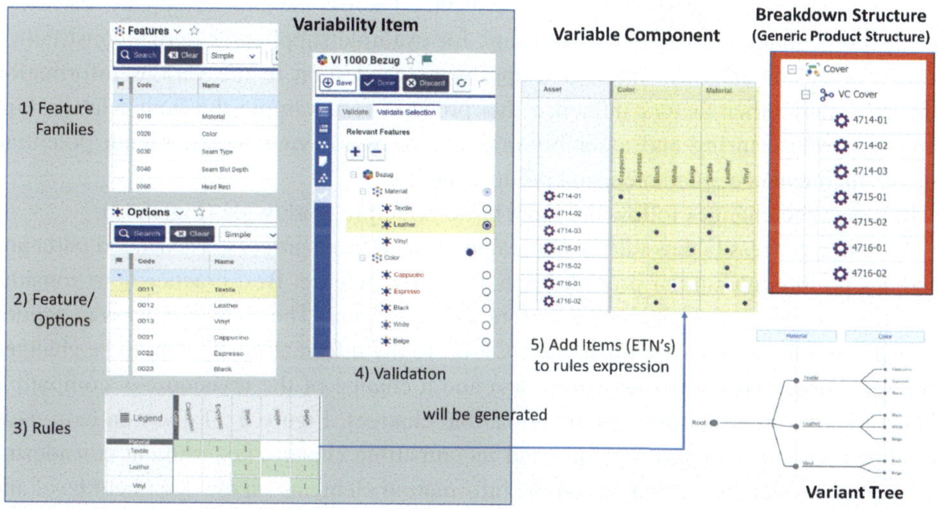

Fig. 4.48 Implementation of an example in a variant system (basis Aras)

- The variant building blocks require a link to super-ordinate information objects, for example product line families, i.e., BMW 3 series, and models, i.e., sedan, convertible, touring. This information supports the inheritance and reuse of feature families, features, and rules.

- Feature families and features must enable a semantic differentiation, for example black interior and black exterior or black leather for Daimler C class or BMW 5 series.
- The variant building blocks and all their components must be reusable.
- The BOP should be generated in parallel with the BOM.

Companies often forget that such a variant logic has to be established and maintained. Appropriate capacity and management must be provided.

Engineering Release and Change Management

The engineering release and change management (ERM/ECM) is a significant process for product creation in order to fulfill the objectives traceability, transparency, and support of the quality management. Moreover, these processes form the foundation of the configuration management (CM). Critics maintain that the change process itself indicates inefficiency in the product development. Some may agree with this, but in general, the development, planning, and production of complex products is not an exact science, rather an iterative process which is, of course, afflicted with errors and possibilities toward permanent optimization. A release and change process is not an isolated event, rather a sequence of corresponding steps distributed across the entire company. Part steps can also be domiciled outside the company, for example suppliers, external construction offices, or customers, who must give approval. In addition to the various information affected, many other factors influence this process, for example the item relationships and use, manufacturing and assembly-oriented points of view, costs, saving potential, quality, maintenance, guarantee, and product liability.

Releases and changes follow the same basic principle worldwide (Fig. 4.49).

A change process is generally comparable to the release process, as changed parts and modules go through similar procedures as for a new release. Best practice is to consider the release as the first change and therefore save a process. The change process is defined through the allocation of a status/maturity level to the information element to be changed through change processes (→ workflows) and a change of the revision (→ compatible change) and/or item number (→ incompatible change). Figure 4.50 gives an extremely simplified summary of a compatible and incompatible change, i.e., without considering the changes to also be carried out on the information elements on the document level and on all other affected information elements (→ affected items). There is also a simplification of the fact that, generally, every element to be changed is used in several superordinate modules that must of course be included for every change.

The digits (1–8) each symbolize an item number, and the letters (a–f) represent the revision/version. We differentiate between compatible and incompatible changes. In compatible changes, parts and modules can be exchanged forwards and backwards if they are identical in regard to form (→ installation space), fit (→ links, connections), and function. For compatible changes, the revision of the changed information element is incremented and the item number retained. At the next higher level, the information

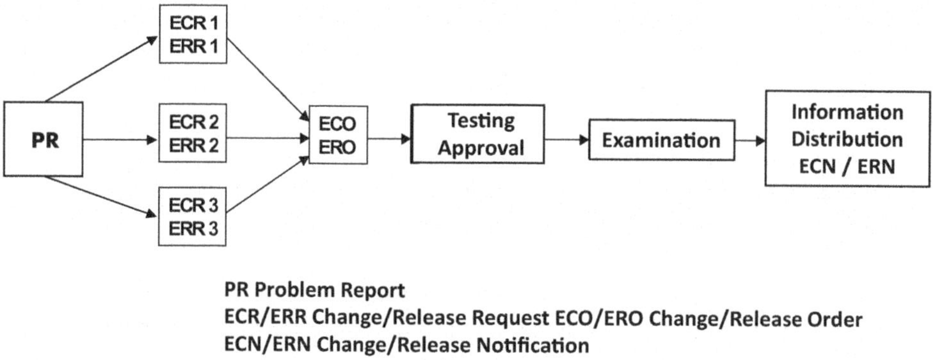

Fig. 4.49 Globally recognized release and change basic principle

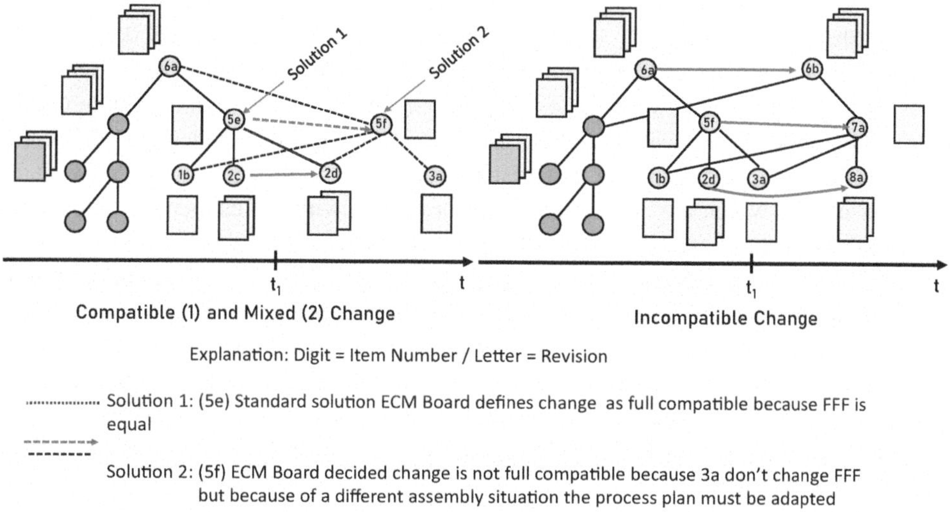

Fig. 4.50 Simplified implementation of a compatible and incompatible change

element is simply linked without change. For incompatible changes, the information element to be changed is allocated a new number and the revision begins again on the lowest value. The change is continued upwards in the product structure until compatibility is achieved.

Solution 1: With a compatible change (according to Fig. 4.50), the following has changed within module 5e:

- 2c was replaced by the compatible part 2d and
- 2d was directly connected to 5e. No change on this level!

The new product configuration at time > t1 is 6a → 5e → 1b, 2d

Solution 2: 3a was additionally inserted. 3a is a thin paper seal and does not change form, fit, and function but it changes the process plan because of an additional assembly operation. The Change-Management Board decided as a compromise to introduce 5f at the next higher leveland link it directly to 6a. This example shows very wellt hat the ECM process does not always consist of clear black and white decisions.

The new product configuration at time > t1 is 6a → 5f → 1b, 2d, 3a

▶ In change management based solely on revisions, there is always a clear assignment of an item number and of a revision on the time axis. An identical item number with different revisions may not exist at this point in time.

SysLM systems support the change processes extremely efficiently. Depending on whether the change occurs before product release or afterwards, the processes can be adjusted to various "sharpness." All forms (→ PR[94], ECR, ECO, and ECN) are actively depicted. The change processes are designed interactively through so-called workflows (Fig. 4.51).

We can recognize the vertical and horizontal queries of the linked Configuration Items (CIs) under 1. Vertical (→ where used) shows the use of the element MP2954 to be changed in the super-ordinate parts list positions. The structure browser horizontally connects all CIs which are connected as object classes with the element to be changed across all lifecycle phases. The real affected elements (→ AIs) must be selected from this. The selection process of CI to AI should be supported by AI methods in the future. However, at the moment, it is already a huge step forward that all CIs can be displayed. This is entered in the impact matrix after the selection process (2). The ECO is subject to a workflow which is displayed under 3. We can recognize the current status in the workflow (4). In the appendix is a detailed example of a change that includes a part (MP2966), its parent assembly (MP2954), and its associated 3D CAD model (Fig. A14 and A15 in Appendix).

The following discussions are being held in the community and should lead to an overall optimization of the change process.

Stronger Administration of Change and Configuration Management through Additional Introduction of a Serial Number

For products where damage tracking is compulsory for legal regulations, the series or serial number is used in addition to the revision (typical in automotive, aerospace and defense, as well as in transportation). Every instantiated, delivered product (cf. Digital Twin Sect. 3.4) contains this sequential serial number and can therefore be clearly identified. Through the clear assignment of an item number to a serial number, a close temporal connection can be raised; i.e., an item number can exist in the same time frame with various revisions, if it is clearly identified by a serial number Fig. 4.52. Just imagine that

[94]PR Problem Report.

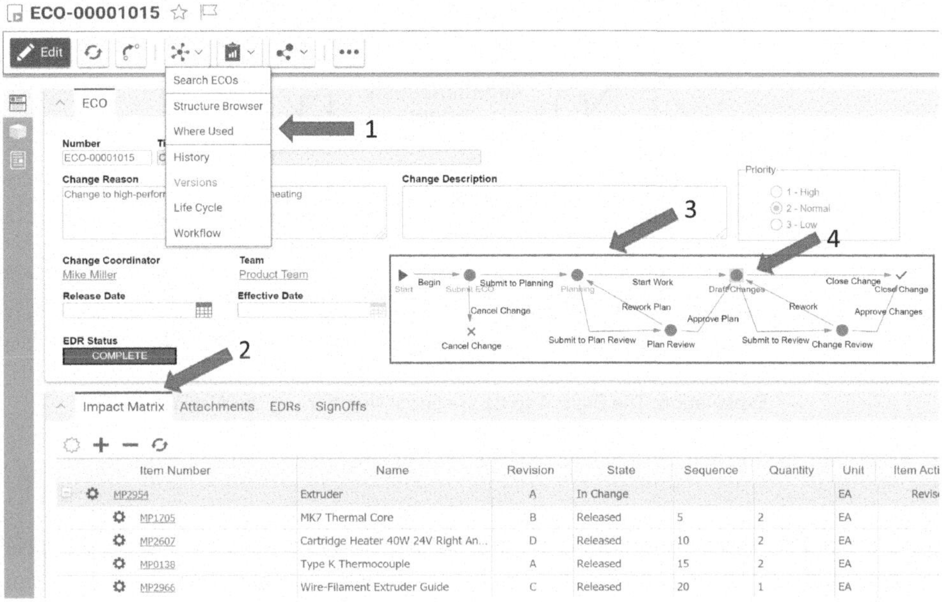

Fig. 4.51 Example of an ECO (basis Aras)

1b and 1c were revision statuses of an airplane turbine and two orders were made for a batch of A320 by Air France and Lufthansa. The orders can be processed simultaneously. Air France still receives the 1b turbines (in a batch different revision statuses should be avoided as far as possible for maintenance reasons), and in the LH order, 1c is already assembled. The assignment of the respectively built-in revision is guaranteed via the serial number on A320 level [34].

The principle of the serial number will become more and more important due to the growing legal significance of traceability and the increasing use of the DT. It is under discussion in some branches whether the serial number can not only be introduced on the uppermost product level, but also on the level of relevant supply components. This would make tracking software statuses easier, for example. On the other hand, the uses resulting from the service-oriented business models of a DT automatically lead to a serial number from which a DT can be clearly identified and linked to the physical twin—the real product.

Implementation of Interdisciplinary Change Processes

Again, a deep chasm, but this time not between engineering and production or between SysLM and MRP, but between the discipline hardware (mechanic and electric/electronic) and software development (case). Figure 4.53 clearly shows the difference between the various applications and disciplines in the change process.

Here, too, company-wide regulations and processes are to be established before the introduction of a SysLM system.

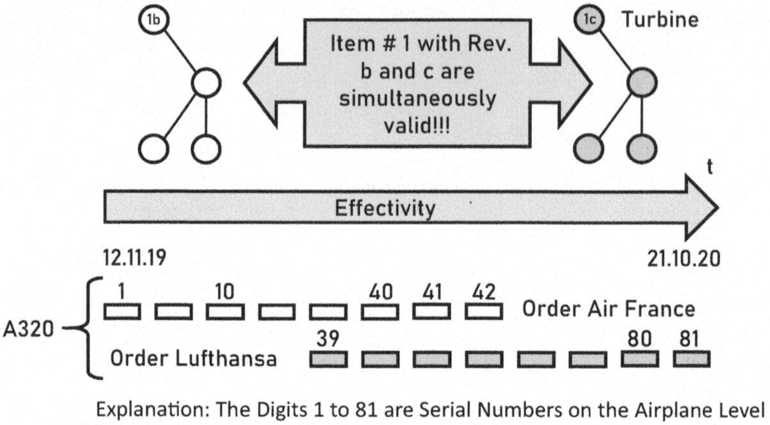

Fig. 4.52 Introduction of a serial number enables item numbers with different revisions

Two Change Processes for Engineering and Production or a Super-Ordinate Process
The orchestration of a company-wide solution also plays a weighty role in change management—and here again, it becomes political. Typical in the current situation:

- An engineering-oriented ECM process which relates to the CAx documents and the E-BOM and
- A production-oriented MCM[95] process which relates to M-BOM, production planning, production resources, production, and logistics.

Here, too, as with M-BOM and BOP, precise consideration should be given as to whether or not an implemented dominant ECM process on an engineering backbone which includes all information elements of the lifecycle, including M-BOM and BOB and synchronized with a simpler MCM in MRP, is the better and more integrated solution. This breaking point between engineering and production is so serious that here, too, the opportunity for a change should be taken. Figure 4.54 shows the concept of an automotive supplier for a so-called Engineering Cockpit, that the higher-level Engineering processes (→ ECM, CM, data exchange) in a SysLM backbone orchestrated and in addition to the engineering applications also SAP is superimposed. The typical landscape of existing old legacy systems on the authoring and TDM level which are connected by innovative, federated, and linked data integration is highly recognizable.

Configuration Management

Configuration Management (CM) based on Engineering Change Management
CM techniques are the formal consequence of well implemented change management. They were originally used in aerospace, the military, area and generally for the

[95]MCM Manufacturing Change Management.

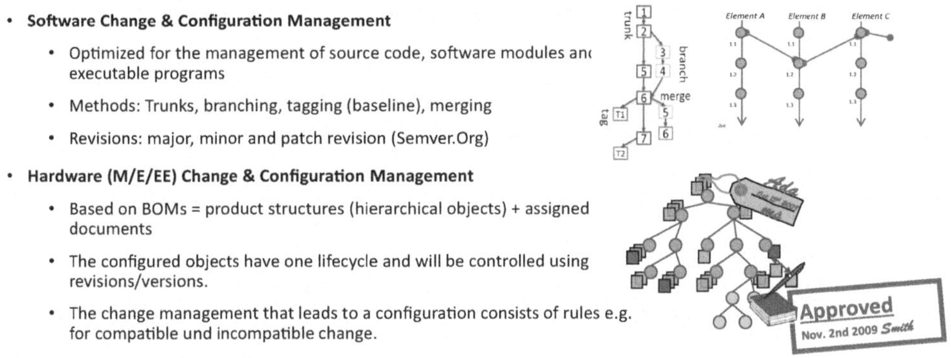

Fig. 4.53 Diversity of the change process depending on applications and disciplines

development of complex products with a high product liability risk and the necessity to support and maintain these products throughout the entire operating period. In the meantime, however, more and more modern companies beyond aerospace are using CM in order to monitor the complete lifecycle of their products [34].

Configuration Management (CM) according to EN ISO 10007:
CM is a management discipline which is applied across the entire lifetime of a product in order to ensure transparency and monitoring of its functional and physical features. Ingredients for CM are well-established change management with clear rules for revisions and possibly a unique identifier by serial numbers. The current state of a product (→ delivery or operating status) must always be reconfigurable. CM is a control and regulation process that ensures the correct assignment of Configuration Items (CI) of an enterprise.

Compliance with the CM guidelines further enables the fulfillment of requirements from standards, such as ISO 9001 and EN ISO 10007. CM therefore delivers and important contribution in the scope of certification measures and traceability scenarios in the event of damage. A configuration is a description of a product at a certain point in time or in a defined delivery or maintenance status. The description contains all the relevant documents for manufacture, assembly, quality control, and maintenance, for example the product structure including any software components as well as all necessary documents, for example text descriptions, CAD models, calculation results, NC programs, and office documents. CM systems consist of a standardized set of tools and rules in order to clearly identify and control amounts of information belonging to a product configuration during the entire product lifecycle [34, 63]. ISO 9001 describes this connection as

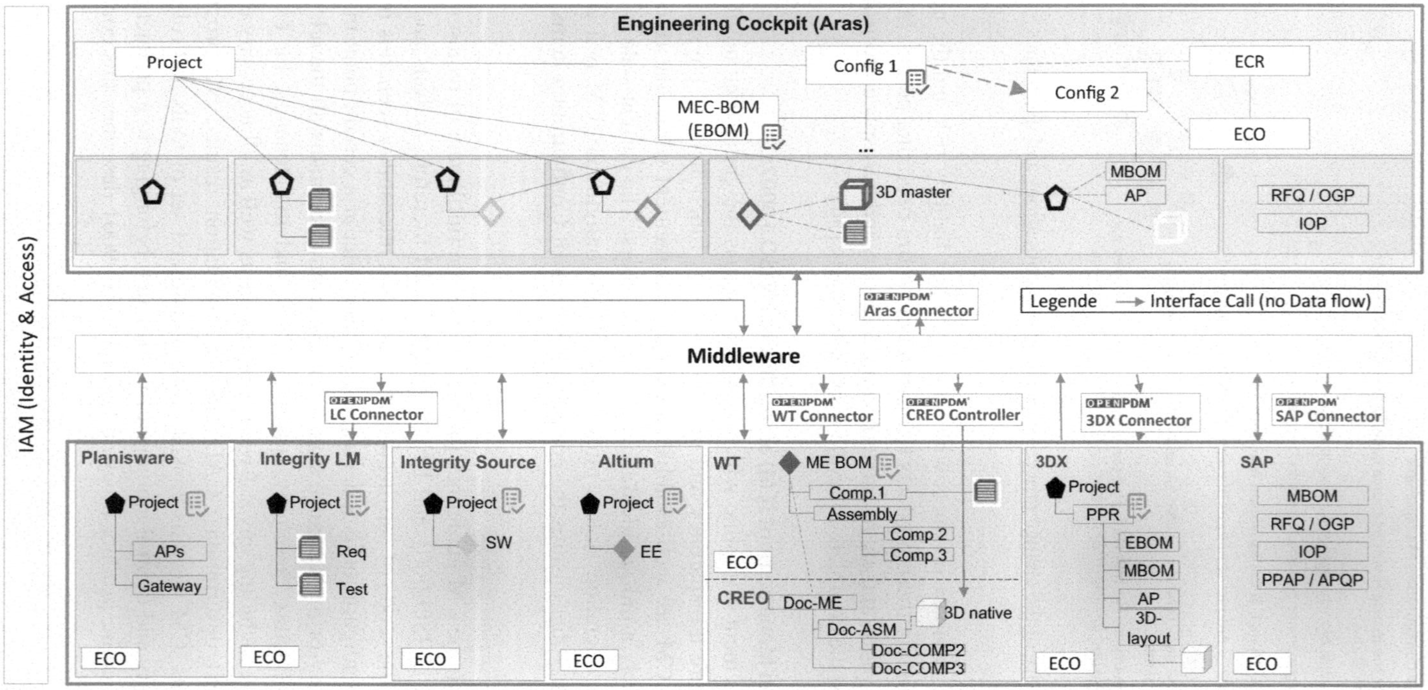

Fig. 4.54 Super-ordinate engineering backbone for the implementation of company-wide engineering processes [99]

follows: "Where it is appropriate, the supplier must introduce and maintain processes for the clear assignment of the product to the associated drawings, specifications, or other documents throughout all phases of production, delivery and assembly."

A significant feature of the CM of the product is the so-called *Configuration Item* (CI).

▶ **Configuration Item (CI):**
Under the control of CM, a CI is something which should be identified and traced throughout the entire product lifecycle. It can be any information element (→requirements, functions/behaviors, individual parts, assemblies, E-BOM/M-BOM, simulation results, software program, BOP, production resources, etc.). CIs are all information elements which must potentially be observed in the engineering change management and configuration management process. If they are actually affected by the change, then we refer to an affected item (AI).

A further term for supporting the identification is the so-called configuration baseline. In general, this determines a "frozen" product description status in an internal or external transfer, for example the functional baseline (FBL) of feasibility to product definition, the design baseline (DBL) of project definition to development, and the formally most important product baseline (PBL) at the start of production. Changes to a baseline can only be carried out after examination and approval by internal and/or external configuration control boards (CCB).

Parts and modules to be serialized must already be determined in an earlier project phase and, if necessary, confirmed by the customer or operator. Serialization means higher costs, so that careful consideration must be made between traceability and costs. The serial number at the top product level or on the assembly and part level is a second identification number in addition to the item number. In contrast to the item number, the serial number describes exactly one instance of a product. An item number can be the same for multiple instances. The relationship between serial and item number is 1:n. Marking of the parts and modules with item and serial number and further contractually determined markings are often required by the standard, for example MIL-STD 130 or contractually from the customer (→ part marking).

Configuration Management based on Variants

In general, there are different possibilities to depict a variant logic on IT systems. The method building blocks for variant definition were already presented in the sub-chapter BOM (Fig. 4.46). In variant-rich products, it is important to understand configuration management as a result from ECM and variant management (Fig. 4.55). There is only traceability, for example, in automotive, if the customer's selected options can first be determined by the vehicle serial number (→ VIN vehicle identification number), for example ceramic brakes, and then the built-in real ceramic brake 4711 with the revision status D in the physical twin triggered by the option. In the case of a DT concept, both

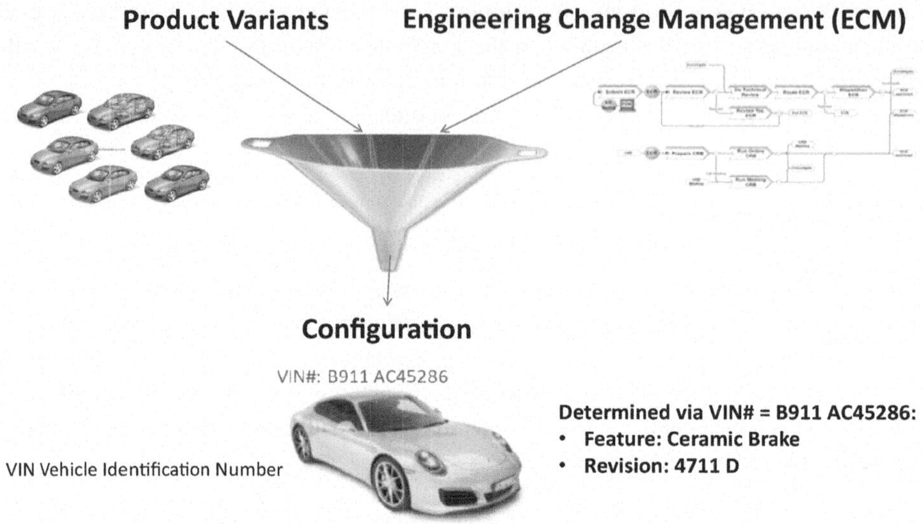

Fig. 4.55 Configuration as a result of variant and change management

the features and the revision status of built-in components subject to tracking can be determined.

4.2.2.2 Integration of the Product Lifecycle Phases

Today's innovations are shifting the limits of what is possible. System design is continuously developing and requires the seamless integration of software, mechanical, and electronic components in the scope of interdisciplinary design, but parallel to this also the integration of the various work areas along the product lifecycle, in particular between product development and production planning, production and operation. The good old "design for X" (→ DfX) from the 90s is being picked up again. The term stands for the efforts in product development to simultaneously take into consideration the often contradictory requirements on a product to be developed and to find the best possible compromise between them [17, 97]. DfX means firstly design for excellence, a general term which is used in engineering in order to serve as a placeholder for various construction objectives. In reality, the term DfX is better through of as design for "X," where the variable X can be exchanged with a one of many values, depending on the respective objectives of the company. The most frequent interpretations of X are assembly (DfA), costs (DfC), logistics (DfL), manufacturing (DfM), reliability (DfR), maintainability/service (DfS)[96]. A continuous, iterative spiral of changing requirements, designs, and results from the digital analysis and physical test is decisive for the development of complex products and systems. However, existing organizational silos of data, processes, tools,

[96]https://www.creativemechanisms.com/blog/design-for-x-dfx, (retrieved (12.6.2020).

and experts are a huge obstacle for the introduction of system thinking (Sect. 3.5.3) and MBSE (Sect. 3.5.4). In order to design the product development seamlessly, effectively, efficiently, and interdisciplinary, the borders between mechanics, electrics, and software as well as between requirements, system modeling, design, and analysis, simulation and production planning as well as production must be overcome. Simultaneously, a complete, precise, and traceable and configuration managed digital recording of the product and production definition must be captured as a complete Digital Model with a Digital Thread of all product and production data across the entire product lifecycle.

No company in the world as of today has implemented this kind of complete horizontal integration; rather, only fragmented approaches have been implemented. However, it is precisely from the experiences of these approaches that we can learn. The concepts are there, and the first partial prototypes have been realized. An ideal horizontal integration is presented in this chapter, and the part areas of an integration are explained. The process of an overall integration in front of us is precisely like the progress in the last 50 years of use of optimization solutions of engineering, in terms of evolution. The engineering-relevant processes, methods, and IT solutions are present (see Chap. 3). They must now be applied to all phases of the product lifecycle. Figure 4.56 shows the theoretical ideal of a complete horizontal integration. The green arrows represent forward transferred information, the red arrows provide feedback for the upstream areas. The vertical integration in the individual phases of the product lifecycle must, of course, be implemented and the partial models must exist.

Methodological prerequisites for such an approach are system thinking – whereby system thinking is not only on the interdisciplinary between the disciplines, rather also on the various work areas along the product lifecycle – and model-based systems engineering. The first will internalized the ideas of the interdisciplinary and overarching thinking and MBSE connects the phase-specific partial models to a digital overall model. The higher the level of completeness and integration of the partial model, the more efficient the traceability through the Digital Thread will be.

Subareas of an overall integration with partial integration concepts and their features in SysLM are listed below. The green and red connecting arrows from Fig. 4.56 provide the guidelines of an information forwarding and return from the respective phases and applications of the product lifecycle. Before the integration of the partial models, an overarching and relevant foundation is still available: the program and project management.

Program and Project Management

Managing new product development projects and programs is complex and comes with many challenges. Program and project management simplifies product development by enabling organizations to link new product development with engineer-to-order processes. By connecting deliverables to project tasks, project management improves completion status and control exponentially (Fig. 4.57). As a result, organizations achieve greater success through the combination of proven project management, approaches, stage-gate methods, and comprehensive risk management. Projects can be bundled up

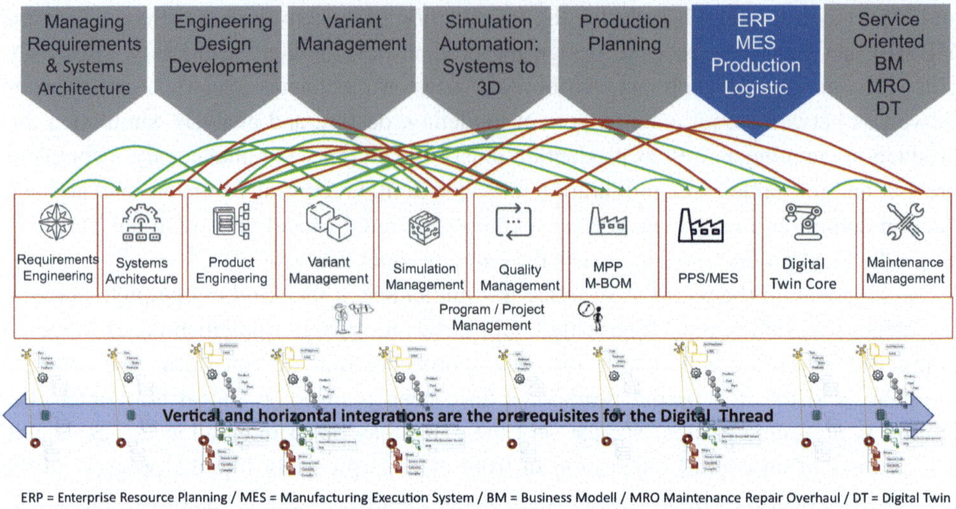

Fig. 4.56 Concept of a complete implementation of a horizontal integration

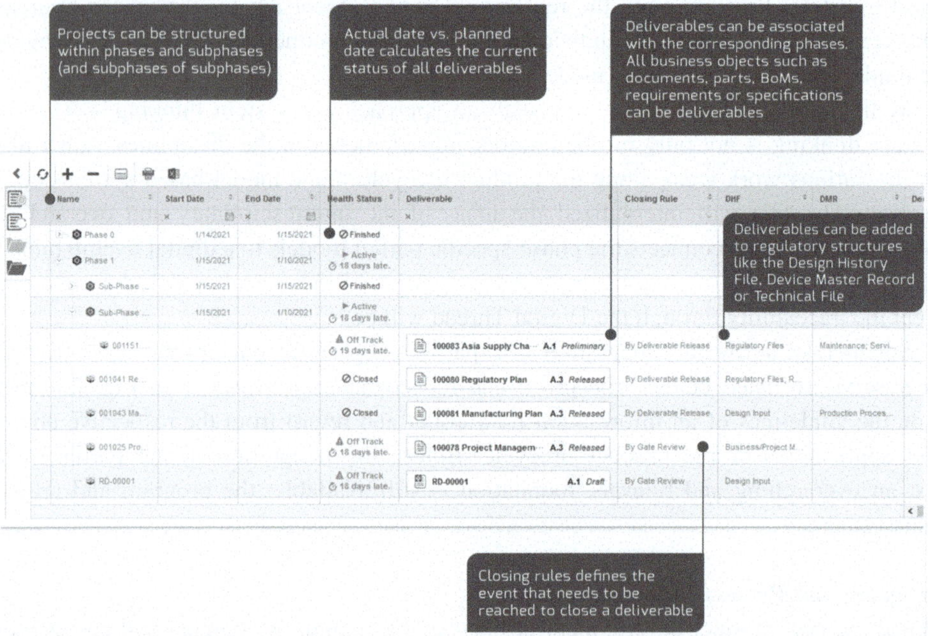

Fig. 4.57 Deliverable matrix supports an integrated project management. *Source* Minerva

into programs. On top of this, you can leverage engineering collaboration to communicate in real time with both the project and the program teams.

With features like Gantt charts, project collaboration workspaces, and project templates based on PMI[97] principles, as well as forward, backward and milestone scheduling, interactive dashboards and intuitive user-friendly score cards System Lifecycle Management becomes more powerful (Fig. 4.58).

Intelligent Requirement Management (RM)

A concrete product development process begins with requirements. In order-oriented production, these come directly from the customer and in consumer goods from the market translated by marketing. It is all the more important to pay precise attention to completeness, clarity, and feasibility here. It has already been indicated in Sect. 4.2.1.1 that a purely textual description of the results in complex interdisciplinary products is only necessary but not sufficiently. Furthermore, in an extension, requirements engineering embeds the requirements in system modeling, in order to connect them with use cases, stakeholders, functions, behavior, and logical blocks. Figure 4.59 clarifies what is meant by intelligent requirement management (cf. Sect. 4.2.1.1).

The numbers in Fig. 4.59 represent the significant features in requirement management and mean in detail:

1. Requirements are hierarchically structured and contain texts and passive graphics (general standard).
2. This is a significant feature for overarching integration, in particular to virtual and physical simulation (→ V&V). Parameters can be extracted from the text of a requirement and
3. managed as independent information objects and used in other applications, for example system architecture and simulation.
4. Shows the intelligence of the system. Texts, lists, and graphics are intelligent and linked information objects that can originate from other applications. If these objects or other applications are changed, the linked objects are shown in the requirements and can be to "affected items" (→ reconciliation).
5. The tabs created links:
   - Related parts: if no system architecture is used, direct link to physical parts and assemblies,
   - External links: with these, OSLC[98] or REST[99] can construct links to legacy systems, for example Doors and
   - Satisfied by: is the direct link to the system architecture, for example to functions behaviour and logical elements.

---

[97]https://www.pmi.org/

[98]OSLC Open Services for Lifecycle Collaboration.

[99]REST Representational State Transfer.

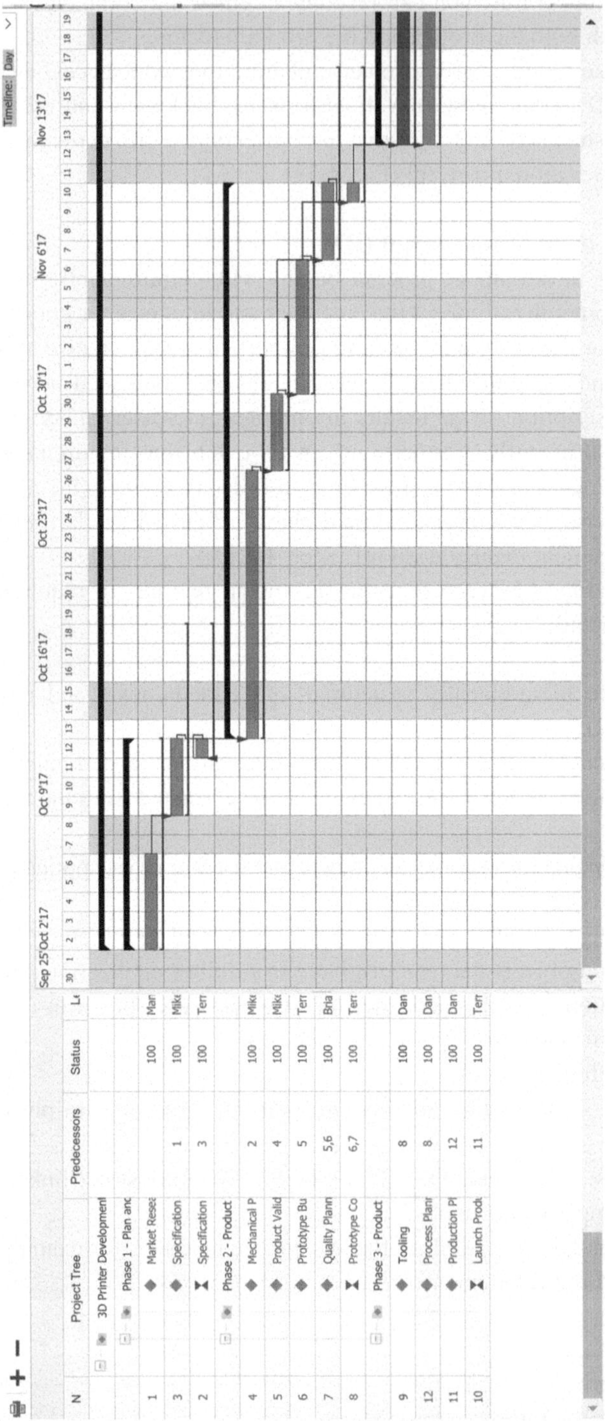

Fig. 4.58 Graphical support for project management—the GANNT diagram. *Source* Aras

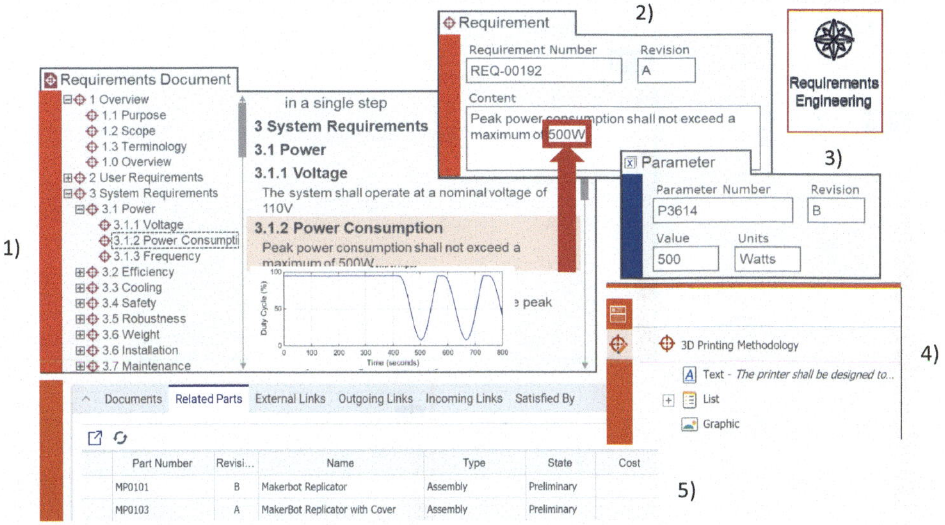

Fig. 4.59 Intelligent requirement management

Figure 4.60 shows the requirement REQ-000000072 in detail and the link to the function FN-00000044 of the system architecture.

Requirements can be entered into the SysLM system in very different ways:

- The requirements are entered and managed directly in the RM module of SysLM,
- The requirements are actively entered and managed in a legacy system and linked by SysLM via OSLC/REST (→ non-redundancy); the legacy system is the master.
- The requirements are transferred into SysLM with the standardized, XML-based exchange format ReqIF[100] (→ redundancy). Which system is the master?
- The requirements are entered directly in SysLM in the system architecture module and managed with the overall system model.
- There exists a hierarchical functional distribution of requirements. General and higher level requirements in SysLM and on a more concrete and detailled level discipline-oriented requirements f.e. in ALM.

The Dominant Role of System Architecture (SA) for Interdisciplinary Product Development

The system model (cf. Sect. 3.5.6 and 4.2.1.2) is the link between requirements (R), the system architecture (F, L), the physical components of the E-BOM (P), and the simulation model and test cases in the early phase, which is extremely relevant for interdisciplinary product development in particular. Figure 4.61 and 4.62 show links to the system

---

[100]ReqIF Requirements Interchange Format.

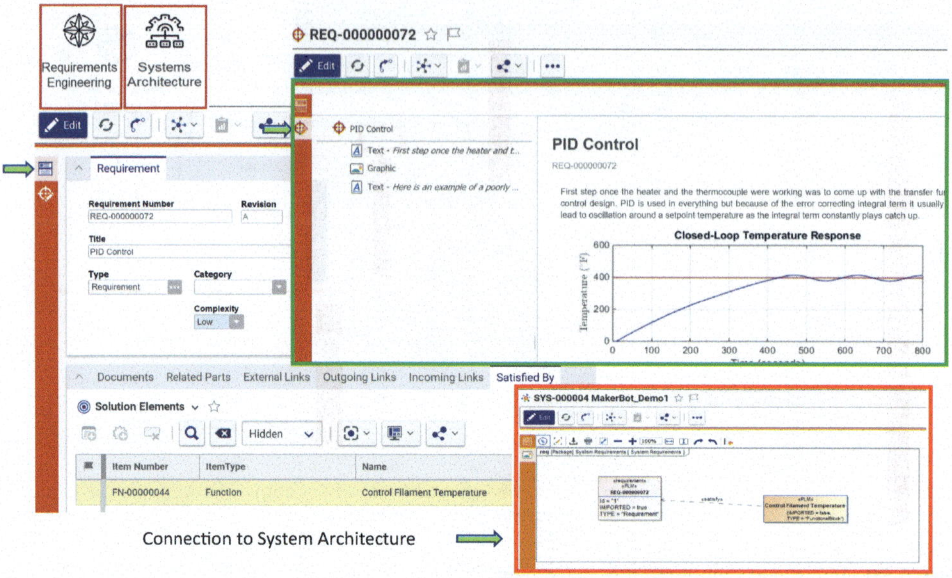

Fig. 4.60 Graphic design and linking of a requirement with the system architecture

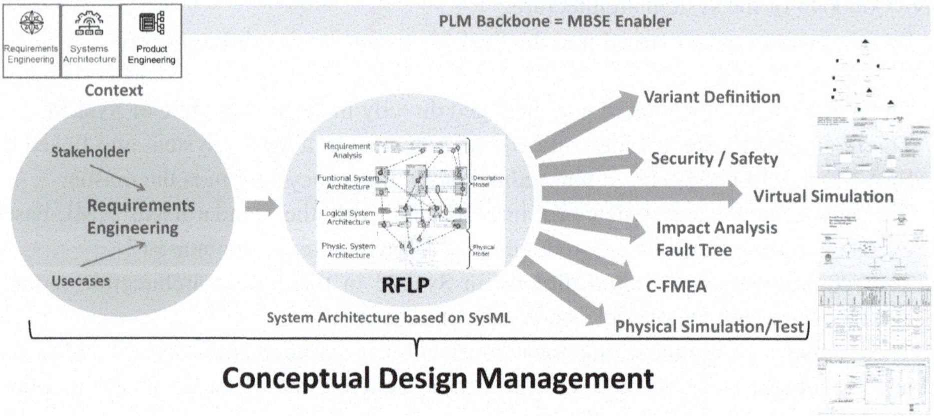

Fig. 4.61 Conceptual design management

architecture to further relevant applications in order to consistently design the engineering process and thereby optimize it.

The risk when applying the system architecture is that – similar to the calculation departments 20–30 years ago – there is the threat of a total isolation within the development process. The function of the system architecture should not be a service function according to the motto "input in, output out," rather an organizational and process-technical embedded organizational unit. The system architecture itself offers many connections to other applications:

- Variant and product line engineering (PLE),
- Analysis of functional security (→ security and safety),
- Connection with virtual and physical validation and verification (→ V&V),
- Impact analysis with fault tree, and
- Concept FMEA[101] (→ C-FMEA).

In both images, it is clear how important the connection of requirements via the system architecture to subsequent applications is, in particular for simulation and quality management. Companies such as pure systems[102], EnCo[103], and PLATO[104] provide part solutions for the integrations presented. We have already presented how requirements reach the SysLM system. Figure 4.63 clarifies the administrative function of SysLM. Depending on the level of detail of the integration (Fig. 4.10), SysLM only recognizes the diagrams, all the individual RFLP[105] hierarchies, or even the F/L networks via port management.

A further significant—but unfortunately barely automatable—link is the connection of the logical elements (→ system elements or logical breakdown) with the physical elements (→ parts, E-BOM). The simplest solution is to integrate the physical elements in the system architecture and to use the graphic or tabular possibilities of the SysLM system. As the logical elements generally have a generic character and the physical element instantiated parts and modules, the product structure method of Schichtel [87] could be applied (cf. Fig. 4.42). In variants this occurs automatically, as the instantiated engineering top nodes are hung on the lowest elements of the generic product structure (→ leaves) by the configurator (cf. Fig. 4.46, 4.47 and 4.78). An alternative support would be a semi-automated link via naming conventions. For example, the logical element is called "anti-roll control element" and all physical components contain this as a portion of the name with a specific supplement in the designation.

As the integration of the system architecture is not yet in wide industrial use, and the associated process is still new and not very stable, the highest degree of flexibility should be guaranteed in the integration in SysLM. Open points are (cf. Sect. 4.2.1.2):

---

[101]FMEA Failure mode and effects analysis.

[102]https://www.pure-systems.com/

[103]https://www.enco-software.com/

[104]https://www.plato.de/

[105]Actually, the term RFLP should be extended to include simulation and testing (->RFLPST) according to Fig. 4.56

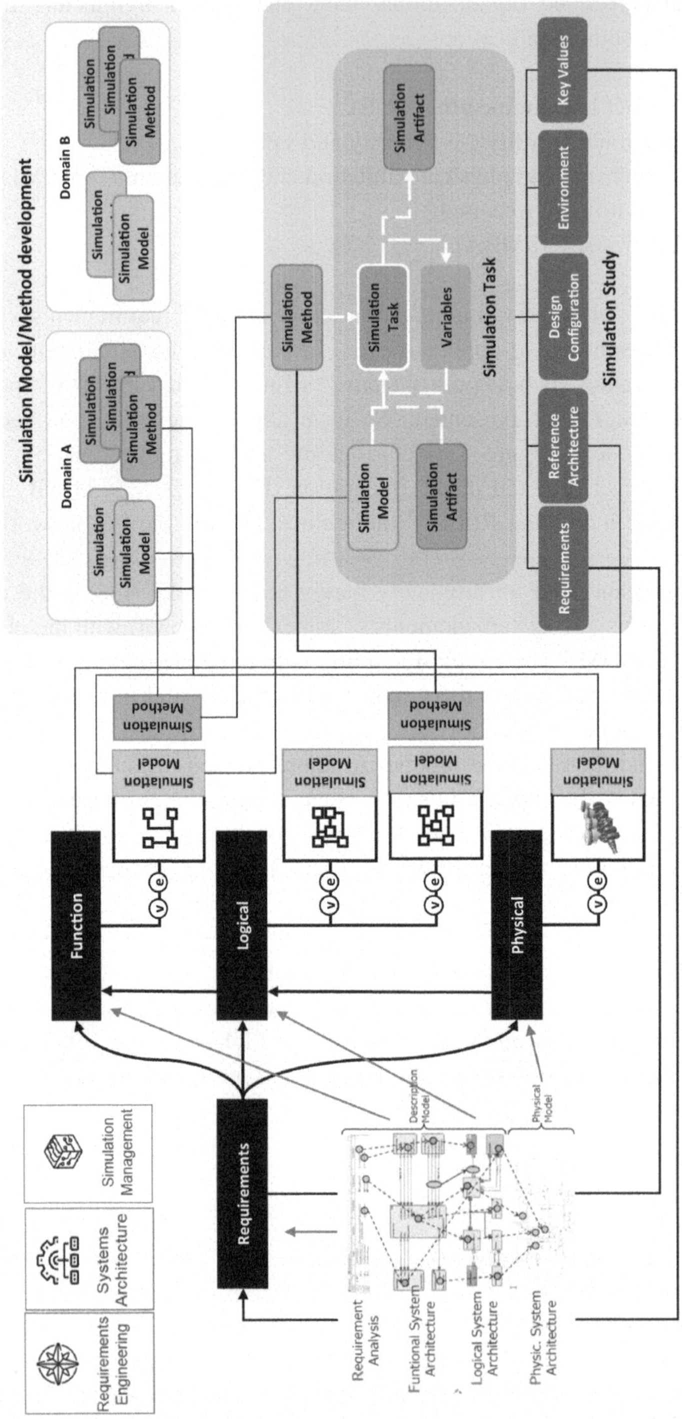

Fig. 4.62 Connection of the system model (RFLP) with the simulation model. *Source* Aras, Matteo Nicolich, und Malcolm Panthaki

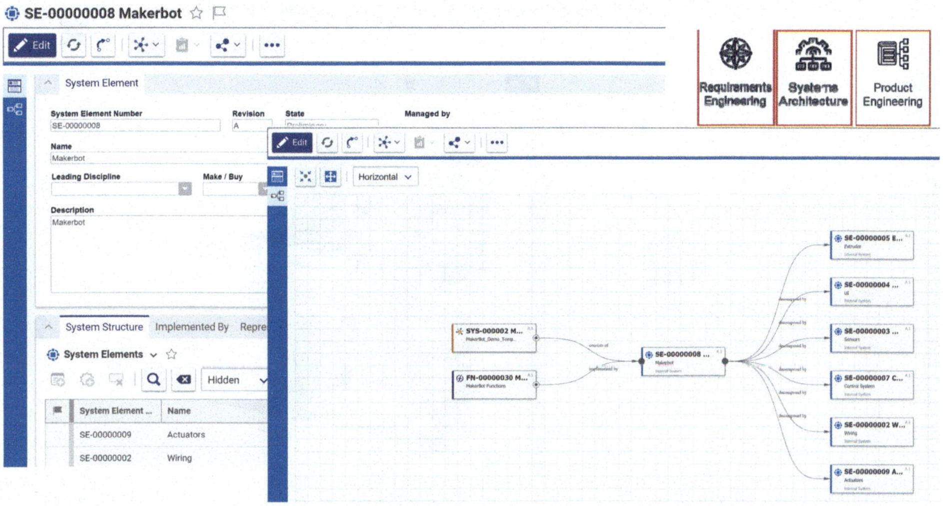

Fig. 4.63 Administration and visualization of the system architecture in Aras

- Integration directly or via TDM. In this case, what is the functional distribution?
- Level of detail of the integration (cf. Fig. 4.10),
- From when and on which level—SysLM and/or TDM—does engineering change management take place and how sharp is it set (cf. Fig. 4.11)?

The orchestration of the system architecture integration in SysLM is one of the heaviest in vertical and horizontal integration. The applications up to now have shown that users must be guaranteed a high level of autonomy and as little administration as possible in this early phase. From a certain maturity in the system architecture, this is then continued by the SysLM change and configuration management.

Product Engineering (PE) the Spine of Product Development

PE is at the heart of development and construction. It includes the original PDM functions with which companies can define and manage information during the design process. This application supports parts, assemblies and products, various types of product structure (→ BOM), standard configuration management and change management, permitted supplier, and manufacturer lists with corresponding libraries as well as types of technical documents. The management and communication of constructive changes remains a central challenge for every company in product development due to the ever-increasing complexity. SysLM contributes toward designing a clearer, more collaborative, and transparent process. Product change histories are automatically collected and recorded in the scope of the product development. Engineering change management (→ ECM) (cf. Sect. 4.2.2.1) is applied in SysLM systems to all information elements at various levels of detail, whereby the Digital Thread of the product is maintained and guaranteed.

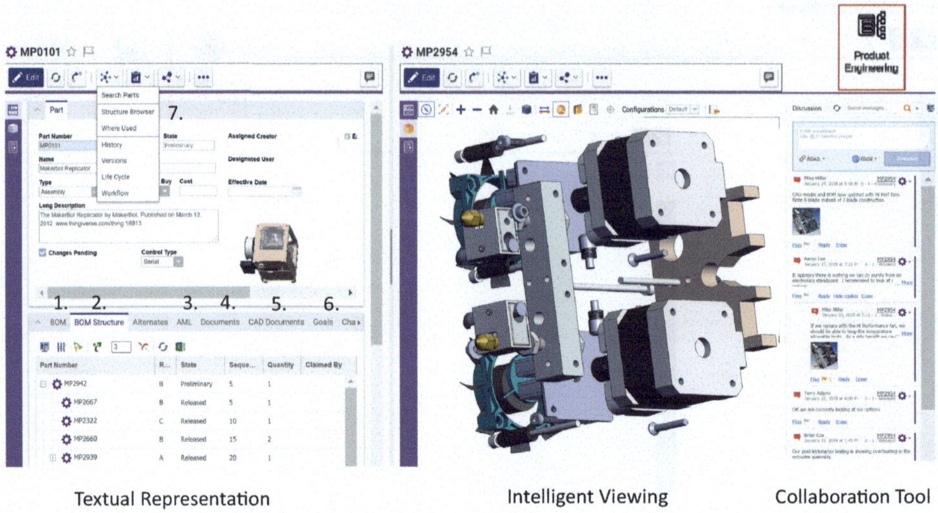

Fig. 4.64 Central functions of product engineering

The engineering master – represented by the part number MP0101 – with all links is represented in Fig. 4.64. The screen is split into three. In the middle is an exploded representation of a dynamic viewer (cf. Sect. 4.2.1.3), and on the right side, there is a collaboration tool (→ "engineering Twitter"). The left side represents information and references that belong to the engineering master.

1. The two level BOM,
2. Structure BOM across all levels,
3. AML[106] parts,
4. General technical documents, for example office documents,
5. CAD documents (recommended: native and lightweight format for viewing),
6. Goals, contained weight, costs, volumes in estimated, calculated, and measured values, and
7. Other links (Fig. 4.65):
   - Where used, horizontal link across the product lifecycle to further linked information elements in the PLC and processes (→ Digital Thread) (red),
   - Structure browser vertical resolution including all linked documents → Digital Thread) (orange),
   - Revisions/versions (blue),
   - Workflow in the event of changes (green),
   - Status/lifecycle, and
   - History.

[106] AML Approved Manufacturer List.

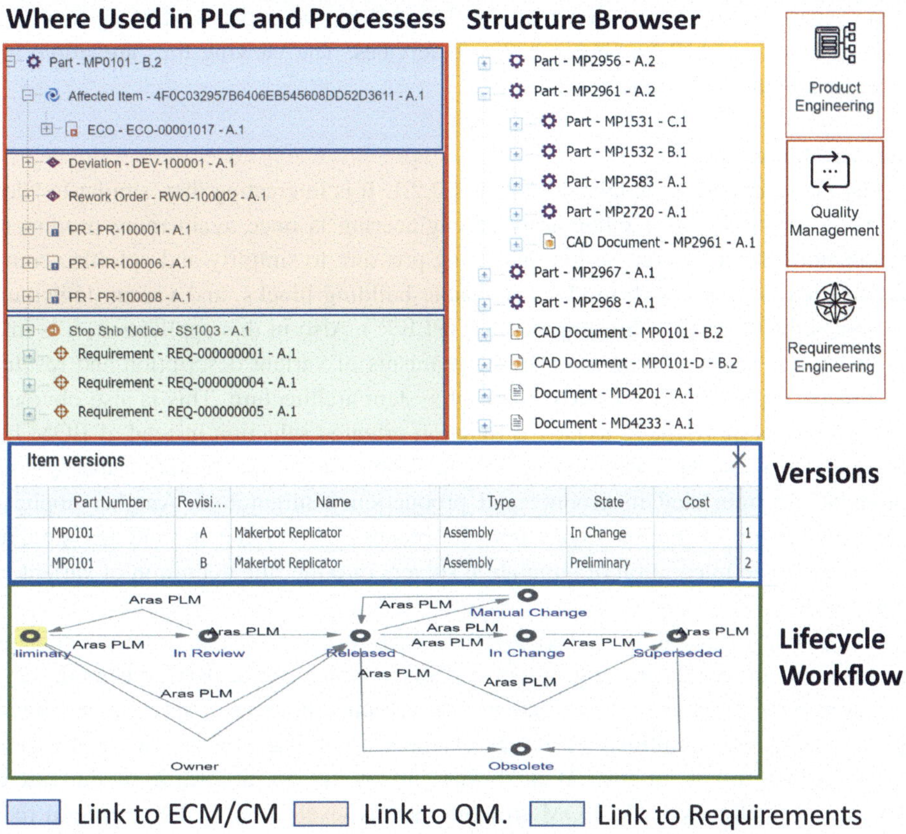

Fig. 4.65 Parts master links horizontal and vertical

Of course, all CAx integrations and visualization techniques from Sect. 4.2.1.3 are necessary in order to link the parts master with its documents and to visualize these in lightweight formats.

Significant demands on PE are:

- Support of all product structure types and views (BOM and where used list),
- Display all vertical and horizontal CIs connected with an information element. With this, the vision from Fig. 4.38 is implemented (→ horizontal and vertical browser),
- Display all uses of an information item in processes (→ ECR, ECO, CAPA, etc.) in a process browser,
- Existence of an ECM/CM concept applicable to all information elements and easy to configure,
- Integration in an engineering communication and collaboration tool, and
- Integration of intelligent lightweight 3D viewing technologies including functions such as cut, explosion, dimensions, tolerances, and depict of various configurations

via a date or serial number which can be used by the entire SysLM system as well as by MES and MRP as integrated web services. The viewing files are stored in the SysLM vault (→3D master, Fig. 4.15).

Variant Management

Variant management is explained in Sect. 4.2.2.1. It is interesting that, similar to "design for X," a relatively old methodology of engineering is once again more prominent in digitalization strategies due to the increasing pressure to simplify and optimize complex products and systems with modular, reusable building blocks, and variant logics under the new name of product line engineering (PLE[107]). Also in this application, it is important not to use them in isolation. The mechanisms of variant description and resolution are already applicable in the early phase of system architecture. This is also obvious, as the customer is more likely to have a function-oriented selection instead of BOM based decisions. It would be consistent on this to develop sales configurators that are semi-automatically translated in product and production configurators. Another application is the integration of variant logics into the process of deriving instantiated simulation models up to the integration of simulation solvers into the rule definition of variant logic (Fig. 4.66).

It is insightful that in a product system based on a variant logic, the manufacturing and assembly processes are subject to a systematic which can be derived from these.

Figure 4.67 highlights the output of the variant configurator through the input of options using the simplified customer example from Fig. 4.47[108]. Since the generic product structure (→ Logical Breakdown) already set up, according to the assembly planning, it is already an M-BOM structure at this level. This is also a general recommendation for far more complex variant logics; the leaves of the generic product structure should be in sync with the assembly process. If differences still result between the development and assembly view below the engineering top node (ETN) level, i.e., on the level of the variant-free part and assemblies (parts numbers 4711-02, 4712-02, 4713 etc.), it is recommended to work here retaining the parts numbers with a view concept, i.e., 4711-02 E (engineering) and 4711-02 M (manufacturing).

The next step is to derive an instantiated BOM with a specific part number (Fig. 4.68). This number (→ 123456789) was generated from the number generator and should correspond directly to the MRP number (1). If the breakdown structure node (→ armrest

[107]PLE actually comes from the field of software development; however, it is now applied to all disciplines.

[108]The instantiated variants are not derived via a fully automatic configurator, but manually for better visualization of the process. The interpretation of the rules in complex cases is actually much too slow. However, Aras allows the manual validation of the variant logic in advance. In later productive operation, the rules are compiled and the variant configurator generally works automatically after inputting the options.

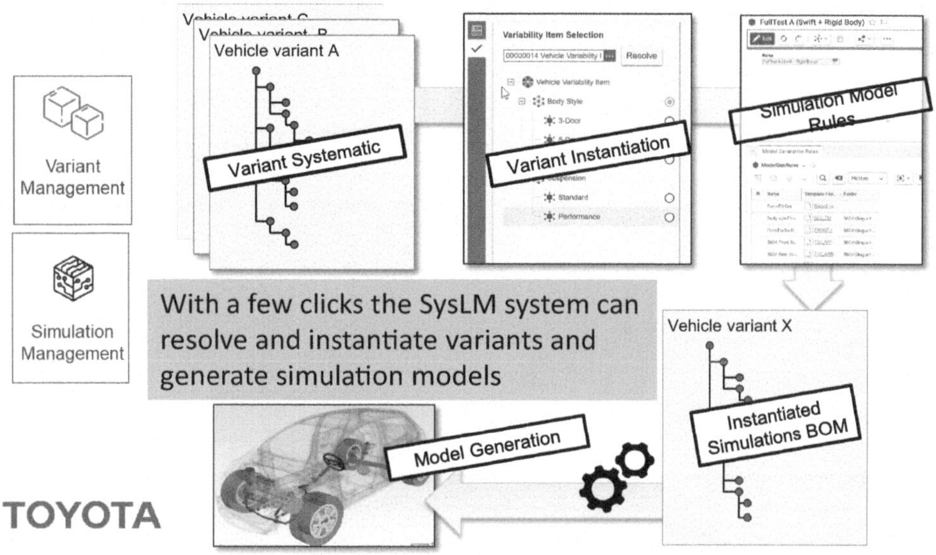

Fig. 4.66 Variant methods applied to product instantiation and the selection of the simulation models [76]

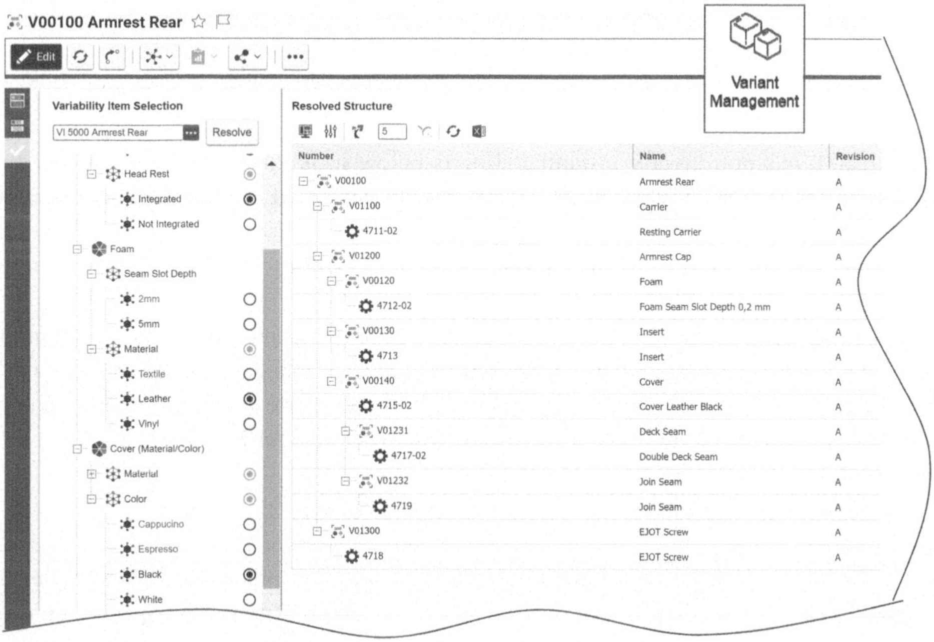

Fig. 4.67 Output of the SysLM variant configurator

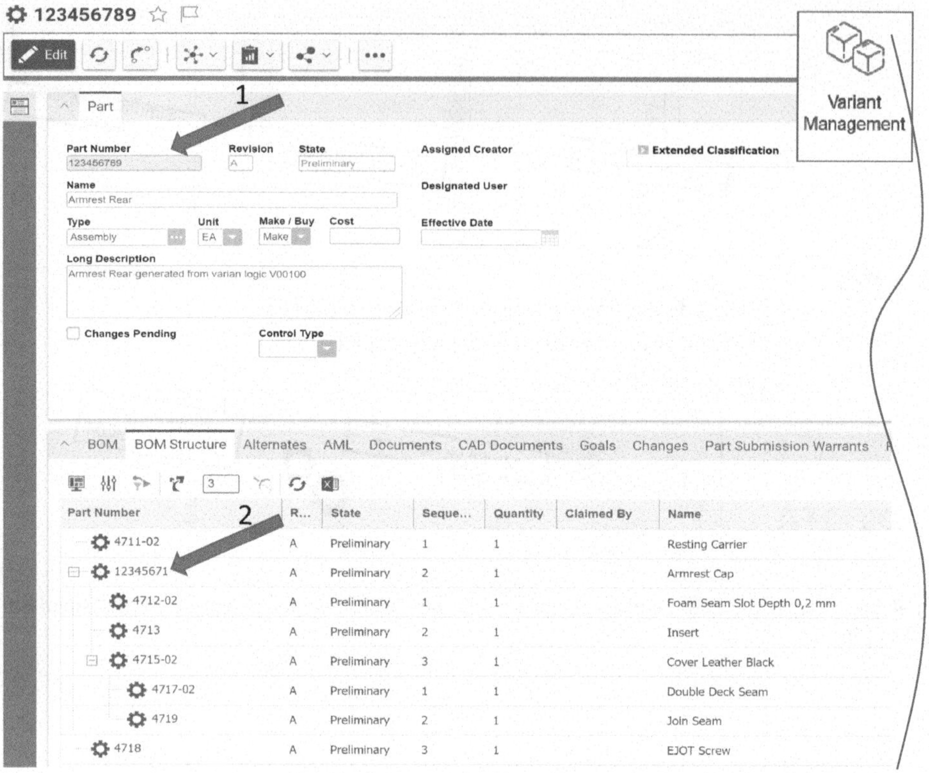

Fig. 4.68 Derivation of an instantiated BOM with a generated part number

cap V01200) has not directly instantiated parts below it, it can be decided whether the node is eliminated or an instantiated assembly with generated part number is generated (→ 12345671) (2).

A potential positioning of the variant system regarding MRP/MES is shown in Fig. 4.69.

Of course, other solutions are also conceivable; for example, the variant configurator is covered by an independent configuration lifecycle management (CLM) system[109] or completely in MRP. The presented solution, however, corresponds to the philosophy represented in this book, to transfer functions from MRP to SysLM and MES (cf. section MRP/MES).

Significant requirements on a variant solution in SysLM are:

- The variant logic must be subject to change management; otherwise, for example, no model years can be depicted in the automobile industry. This variant-specific change management affects all elements of the variant logic (→ feature families, features,

[109]Typical independent CLM systems are Configit (https://configit.com/de) and Tacton (https://www.tacton.com).

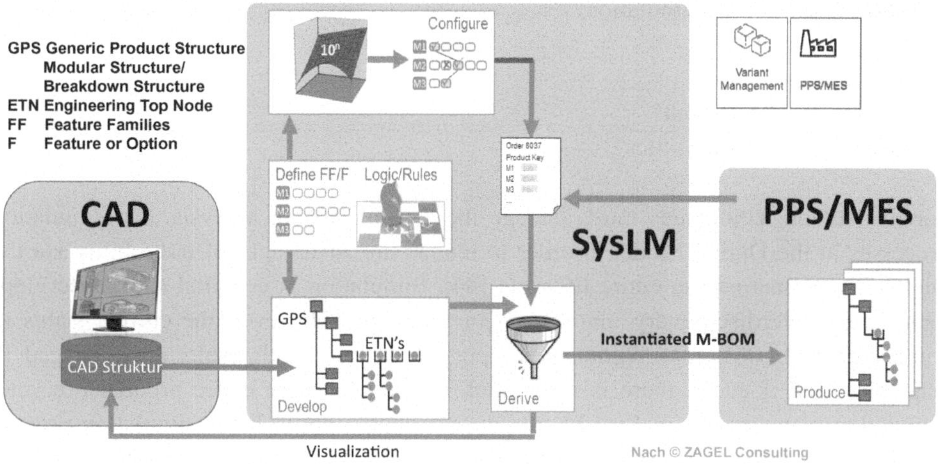

Fig. 4.69 Positioning of the variant system to MRP (according to Zagel Consulting)

rules, and breakdown structure). It must be synchronized with the standard ECM (→ changing a part or assembly at the ETN level via ECM also affects the variant logic).

- It must be possible to define portfolios, program lines (→ 3 series BMW) and model lines (→ 3 series sedan).
- Elements of the variant building blocks must be reusable in portfolios, programs, and models, for example the variant logic of the engine and powertrains.
- Features/options must be assignable with semantic meanings, for example black in regard to leather seats, black for textil seats and black as an external color, and most important features families and features must be able to assign programs and models to give them a unique semantic mapping.
- The variant logic can be applied to any information elements of the product lifecycle beyond the parts/assemblies, for example to functions, logical elements, MPPs, and technical documentation.
- It must also be possible to represent the instantiated product solutions determined by the variant configurator in a lightweight viewer. It is also necessary that the SysLM system can file the topology in the variant logic (cf. Fig. 4.17)

Bringing Simulation Management out of the Silo and integrating it into the Digital Thread

A company must be confident that all inputs and outputs incorporated into a particular simulation are traceable. The 5 Ws should be answered with every simulation[110] [75]:

[110]https://www.nafems.org/community/the-analysis-agenda/simulation-data-management/

- Who carried out the simulation?
- Why did they do it?
- What was the result?
- When was it carried out?
- Where are the relevant files?

For this reason, companies must manage the system design, analysis, and simulation processes in the Digital Model in order to realize the advantages of traceability via the Digital Thread across the entire lifecycle [84]. Simulation is essential for the development of the interdisciplinary innovations of tomorrow. However, the current status in companies looks different. Simulation inputs, processes, and results are managed in fragments, and if at all, there is no Digital Thread along the entire product lifecycle. Simulation management in SysLM introduces a new approach for simulation process and data management (SimPDM) that connects simulation inputs, processes, and results and guarantees precision, repeatability, and access to simulation processes for teams throughout the entire lifecycle [23]. Furthermore, multi-physical and multi-fidelity simulations are enabled. An open SysLM platform creates a Digital Thread and data model independent of the CAE tool that guarantees consistency, continuity, and traceability of product data in the entire company. Details of the vertical integration of SimPDM and the underlying data model with the elements

- Simulation study,
- Simulation model,
- Simulation task, and
- Simulation artifact.

can be taken from Sect. 4.2.1.4 The example of a simple calculation of a front axle component here shows the continuous process of the simulation. Figure 4.70 shows the call up of a calculation task (→ simulation task) with the textual and graphical result representation [76].

A calculation is carried out multiple times under variation of the parameters which build on one another. The results of the various simulations (→ SIM000007.1, SIM000007.2, SIM000007.2.1, etc.) are shown in a table. The simulation versions can be represented graphically (Fig. 4.71).

It can be graphically shown how which requirements and functions are validated and verified by which simulations and their results at the end of the simulation process. The reference to the simulated parts and assemblies as well as the simulation model derived from this are also represented (Fig. 4.72).

In this way, the Digital Thread of requirements across system architectures to constructive parts and assemblies and their simulation is created and kept up to date. According to Fig. 3.13 in Sect. 3.4.1, the virtually tested parts and assemblies as well as the simulation models resulting from these concerns an instantiation from the Digital

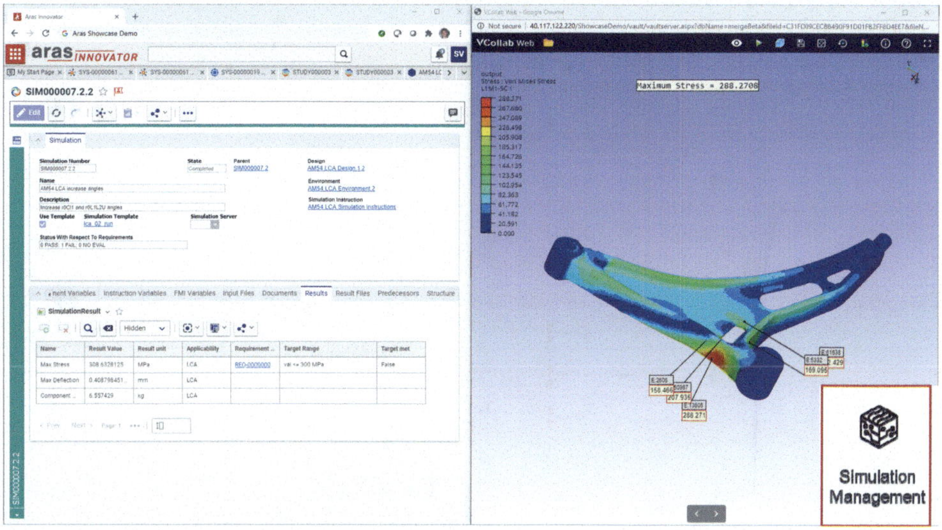

Fig. 4.70 Calling up a simulation task and display of the results [76]

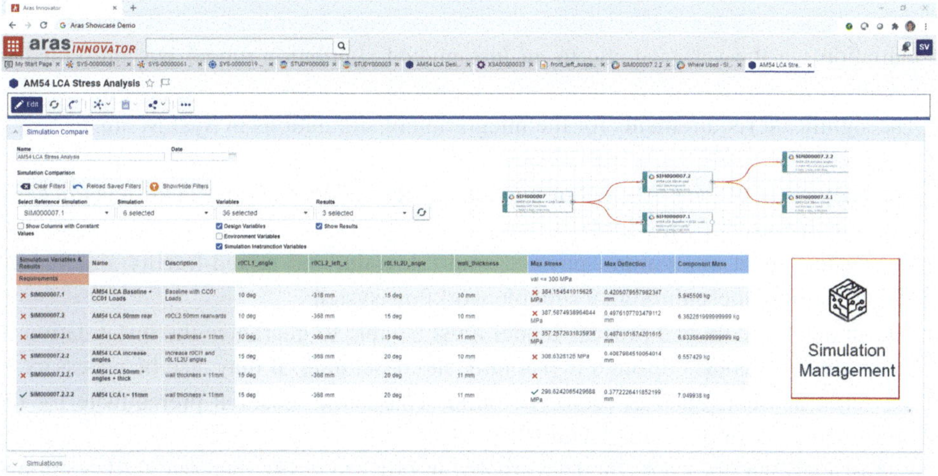

Fig. 4.71 Comparison of the results from various simulations (basis Aras)

Model (→ simulations twin[111]), for example a crash simulation for a vehicle with sunroof, 6-cylinder engine, and 5 doors and parallel to this for a vehicle without sunroof, 4-cylinder engine, and three doors. Reminder: The generation of such vehicle

[111] As there is not yet a prototype, we cannot yet refer to a DT Sect. 3.4.1, Fig. 3.13.

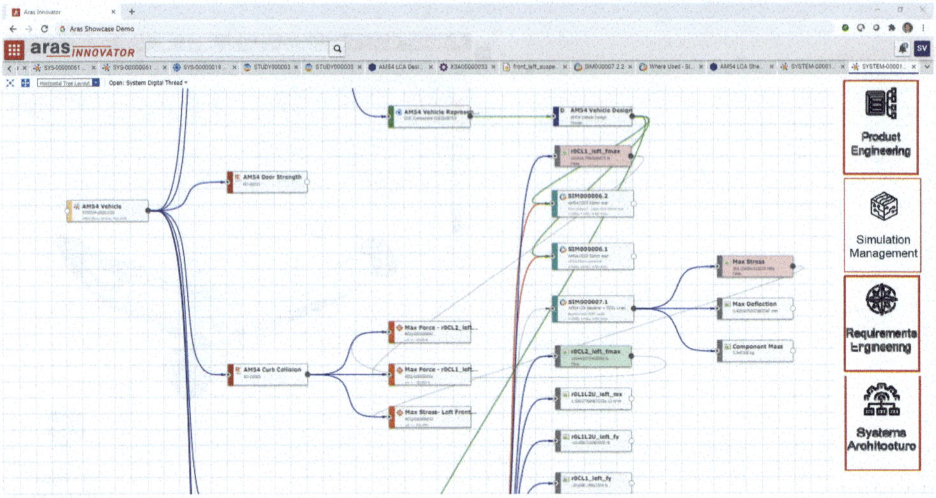

Fig. 4.72 Representation of the connection of requirements, simulation, and tasks and results (basis Aras)

instantiation via the variant logic to an instantiated simulation model is represented in Fig. 4.66.

The significant requirements for the integration of simulations in SysLM are:

- The integration of the CAE authoring systems always occurs via a TDM level in complex simulations (→ SimPDM).
- The system must allow multi-physics, multi-domain, and mixed-fidelity simulations and orchestrate these through a suitable workflow system.
- The simulation data and process model must be able to guarantee the instantiation of the simulation models through revisioning and baselining. It may also be necessary to use serial numbers such as prototype build. This is the essential prerequisite for the implementation of the Digital Thread.
- It must make an open data model available in order to easily integrate CAE solutions from partners (Fig. 4.37).
- In the future, test management must be integrated in the solution. This would make a virtual and physical validation and verification possible (→ V&V).

Quality Management (QM)

Companies use quality management (QM) to plan, organize, and control the quality of work processes, products, and services in the long-term and consistently. QM ensures consistent, high-quality output service of a company and thereby the success[112] and defines all activities that ensure product and process quality in the scope of a QM system. This happens through continuous quality planning, quality control, quality assurance, and quality improvement. A current survey of upper management of large producing companies resulted in 83% that QM must be a significant application embedded in SysLM [64]. The first approaches to a structured quality management were based on the Japanese philosophy of life, Kaizen. Translated the term means "change for the better." The American physicist William Edwards Deming derived the principle of the "continuous improvement process" (CIP) from this in the 50s. This fundamentally concerns continuous teamwork on small production and work process improvements within a company that overall lead to a significant increase in quality. The most common quality management models are ISO 9001 and the EFQM model:

- The internationally valid standard ISO 9001 has a detailed criteria catalog that companies can use to realize quality management as a continuous process. Companies can be certified accordingly[113].
- Total quality management (TQM)—in contrast to ISO 9001—does not only concern technical processes. Instead, the relationship to the customer is the focus: "What is the benefit for the customer?" is therefore the central question posed for every change in a process. Companies that work with TQM are organized in the European Foundation for Quality Management (EFQM). The so-called EFQM model was created as a counterweight to ISO 9001.

There are also industry-specific standards which are generally stricter (→ automotive, aviation, defense, medical technology, telecommunications); the state sometimes makes certification obligatory. A QM system (QMS) within SysLM should connect quality with system architecture, construction and development, production planning, and company decision-making processes as well as change management. Furthermore, QM offers interdisciplinary teams and the extended supply chain high-performance closed-loop functions for the identification and management of risks to increase the quality, to fulfill customer requirements, and to comply with ecological, safety, aerospace and defense, or medical regulations. QMS includes quality planning and the actual quality system in order to equip organizations with comprehensive QM capacities. QMS enables quality specialists to manage the product quality through the entire product lifecycle. Quality planning consists of the functions for proactive quality management that is traditionally carried out during the product development. The functions for failure mode

[112]https://www.rechnungswesen-verstehen.de/lexikon/qualitaetsmanagement.php, (retrieved on 9.23.2020).

[113]https://www.rechnungsoftwareesen-verstehen.de/lexikon/qualitaetsmanagement.php, (retrieved on 9.23.2020).

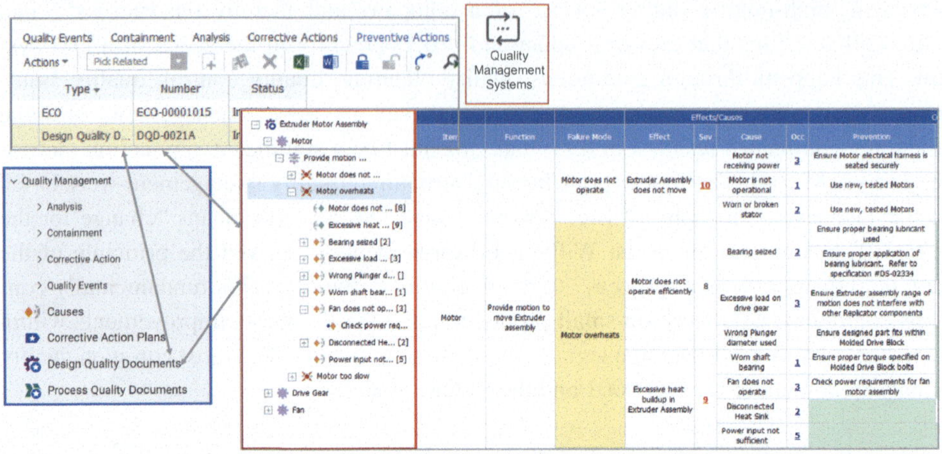

Fig. 4.73 Quality planning with FMEA incorporated in the change management (basis Aras)

and effects analysis (FMEA) to manage risks improve quality and achieve environment-related, official, safety technical, and other forms of conformity belong to the planning features. These functions enable the use of design for Six Sigma (DFSS) for risk control and reduction of critical product risks, the use of FMEA processes to achieve conformity with legal requirements and quality standards, including ISO-9001, TS/16949, AS9100, FDA-QSR, the management of libraries with critical features, error mode, effects, and control mechanisms. Moreover, ISO compliant processes to promote the cross-function collaboration as well as the creation of dashboards, scorecards, KPI metrics, and reports for tracing the project can be used (Fig. 4.73) [5].

The FMEA application corresponds to Excel, which is so popular in quality planning, but has a much higher intelligence. For example, every SysLM information element is able to form the breakdown structure (red). For example, a concept FMEA can be formed from the system architecture on the basis of a logical or functional breakdown. Design FMEAs use extracts of the BOM structure and process FMEAs manufacturing and assembly plans (→ MPP). The functions of the QM embedded in SysLM can be taken from the blue area.

The quality system should react quickly to problems and communicate change requirements company-wide. It includes CAPA functions (closed-loop corrective action/preventative action) and RCA functions (root cause analysis), in order to accelerate the containment and analysis of problems as well as the tracking of affected elements. Identified problems are analyzed with a variety of tools for cause analysis, including error tree analysis, fishbone analysis, and five reasons (→ 5 whys). In order to ensure that problems can also be traced and your teams are up to speed, there are 8D reports[114] and CAPA metrics (% closed, % due). Metrics and status are conveyed to company management with integrated reports and aggregated data.

The significant requirements for the integration of quality management in SysLM are:

- Application of the QM methodology to the entire product curriculum vitae and its essential information elements, in order to thereby enable, for example a concept FMEA, design FMEA, and process FMEA.
- Integration of typical SysLM cross-section functions such as ECM, visualization, collaboration, and workflow.
- Aggregation of metrics and conclusions to management, and
- Direct transfer of problem reports (PR) in ECR.

Manufacturing Process Planning (MPP)/Bill of Processes (BOP)

The manufacturing and assembly planning in SysLM provides an integrated approach and closes the gap between product development and process planning. The creation of the MPP from the E-BOM and the automatic derivation of the M-BOM as well as the automatic real-time comparison of E-BOM/M-BOM is already almost a BOM evolution in the right direction (Fig. 4.74) [24].

In contrast to the traditional conversion of the E-BOM in the M-BOM (a), the BOP is created from the E-BOM on the level of parts and assembly with systems support (→ drag and drop) (b). Only assemblies with an already M-BOM character should be used for this mapping. This requires tight coupling with the MRP system in regard to the resources and a shift of not only the M-BOM but also the BOP into the SysLM system. The M-BOM, which continues to be required for logistical purposes, is automatically derived from the BOP (c). E-BOM and M-BOM know from one another which parts are contained or changed (→ reconciliation).

Change and configuration management are fully applicable for M-BOM, process plans, work instructions, and other elements. By mutual referencing of E-BOM to M-BOM and vice versa the effects of changes on both types of BOMs can be identified and synchronized. One BOM fewer through automatic generation means that time-consuming manual processes are prevented. Moreover, graphic and parallel process plans are supported. This means that various factory locations can create and manage their respective resource and supplier contingent local changes and automatically compare between one another according to the same principle as the E-BOM/M-BOM.

---

[114] "An 8D report is a document that is exchanged between the supplier and customer (but also internally) in the scope of quality management for complaints. 8D here stands for eight obligatory disciplines (process steps) which are required to process a complaint." https://de.wikipedia.org/wiki/8D-Report, (retrieved on 9.24.2020).

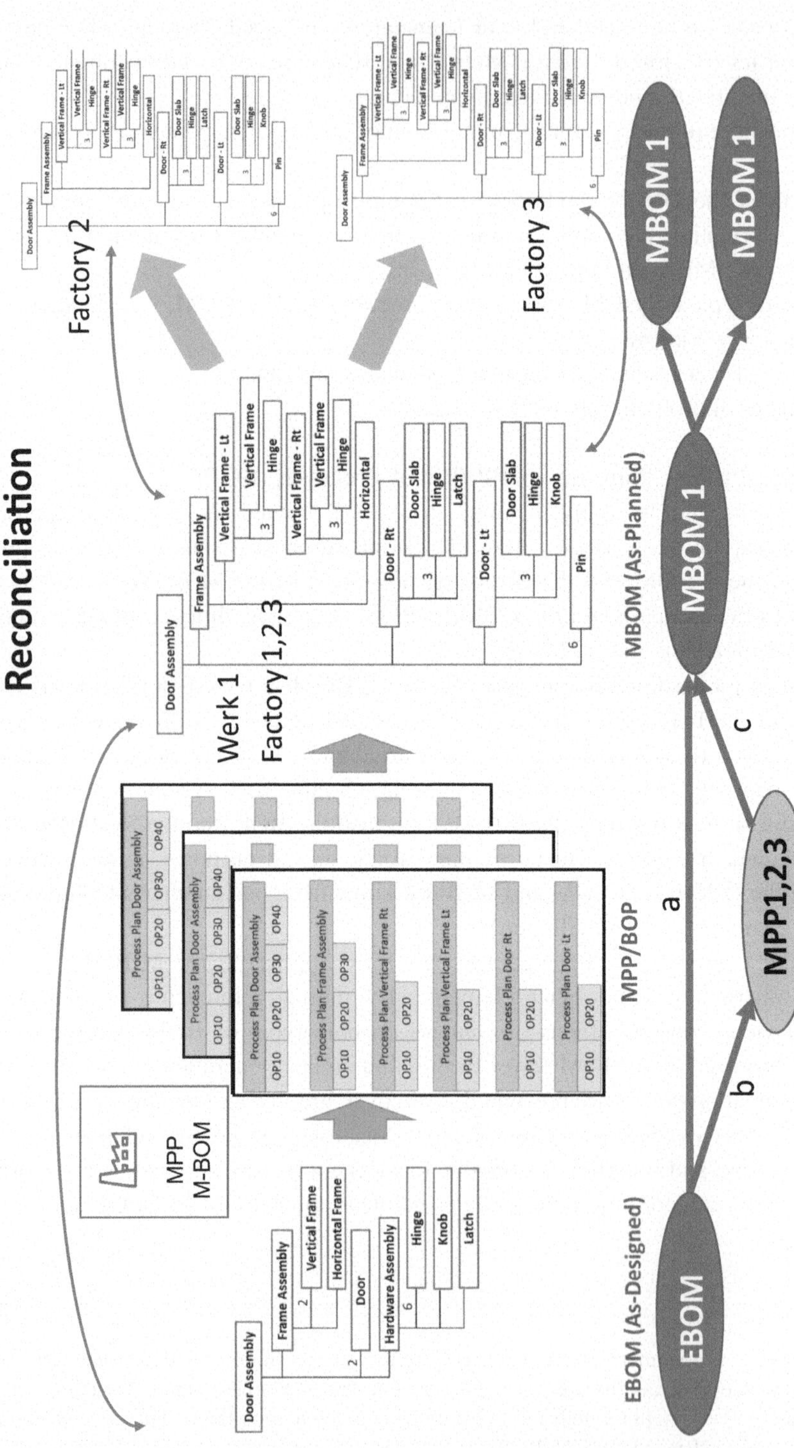

Fig. 4.74 Creation of BOP/MPP from the E-BOM and derivation of the lead M-BOM as well as site-specific M-BOMs (supplemented according to Aras)

The global management of complex engineering processes which stretch across the requirements, system architecture, product development down to manufacturing is now effortlessly possible. In addition, the information elements of production planning are automatically taken on in the Digital Thread. These capabilities offer a completely new level of organization, integration, synchronization, and traceability along the product lifecycle.

MPP items can be accessed from the folder named Process inside the Content menu (Fig. 4.75). This folder contains six Item Types: Graphics, Locations, Machines, Process Plans, Skills, and Tools:

- Graphics—The Graphics item is inserted in a Work Instruction under an Operation or a Step.
- Locations—The Location item represents a Manufacturing factory where the product or assembly is manufactured. The Location Item is related to the Process Plan Item.
- Machines—The Machine item represents an asset/resource used on the shop floor, necessary to perform an Operation. The Machine Item is related to an Operation Item.
- Process Plans—The Process Plan item is the top-level Item that details how a product or assembly is manufactured.
- Skills—The Skill item represents the shop floor technician capability that is necessary to perform an operation. The Skill Item is related to an Operation Item.
- Tools—The Tool item represents an asset/resource used on the shop floor, necessary to perform an Operation. The tool Item is related to an Operation Item.

The individual areas of the screen are:

- Workbench from development parts and assemblies with an already M-BOM character that are used 1:1 in MPP and M-BOM (red).
- The work area to create the MPP. The development parts are pulled up with drag and drop from the workbench (blue).
- The factory locations regarding deviations, for example different resources and supplier situation, are entered here (orange).
- Functions and input objects for MPP, for example parts/assemblies, documents, phantom, process steps, resources, and skills. A phantom is a module that does not exist in the E-BOM; rather, it is need due to a special assembly and storage situation (green).
- The area for assigning graphic to the process plans is located on the right (brown).
- We can switch between the views MPP, M-BOM, and E-BOM using the vertical bar on the left.

Figure 4.76 shows the comparison and the displays between E-BOM (red) and M-BOM (blue). Quantity differences, splits in assembly groups, and existence only in M-BOM or E-BOM are displayed in the column "reconciliation." The synchronization of both BOM types and the change management occurs on the basis of this information.

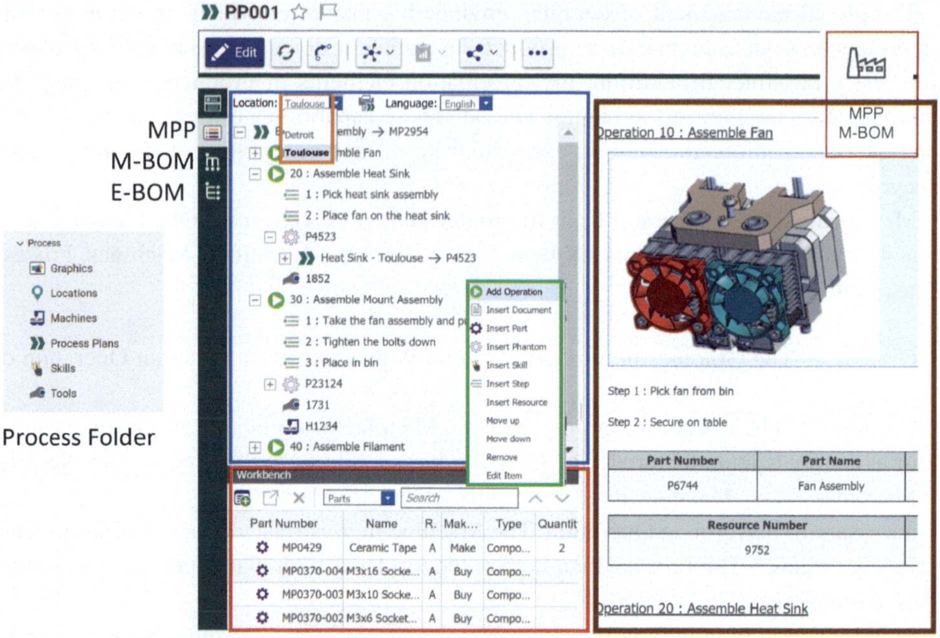

Fig. 4.75 Principle of the connection of E-BOM, process plan, and M-BOM (basis Aras)

The reconciliation status provides useful information of the part and cannot be edited. Reconciliation Status is calculated based on how a part in the E-BOM is consumed in the M-BOM. The information is calculated and displayed as icons. The following icons can appear:

- Partially Consumed Quantity—This icon tells the user that the quantity of a part in the M-BOM is less than the quantity of the same part in the E-BOM. The system checks all places in the M-BOM (that is opened in the MPP tear-off window) where a specific part is consumed, adds the quantity of the part in all the positions, and compares against the total quantity of the part in all the positions in the E-BOM.
- Overconsumed—This icon tells the user that the quantity of a part in the M-BOM is greater than the quantity of the same part in the E-BOM. The system checks all places in the M-BOM (that is opened in the MPP tear-off window) where a specific part is consumed, adds the quantity of the part in all the positions, and compares against the total quantity of the part in all the positions in the E-BOM.
- Quantity Split—This icon tells the user a part exists in more positions in the M-BOM than it exists in the E-BOM, and the total quantity of the part in the M-BOM = total quantity of the part in the E-BOM (for all the positions).

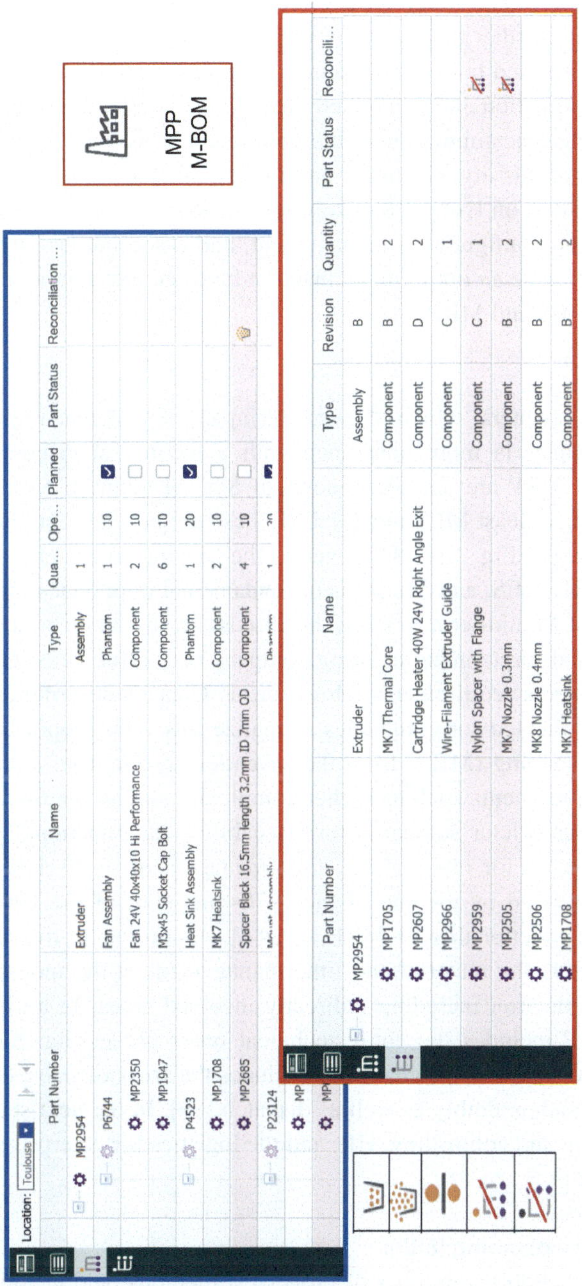

Fig. 4.76 Automatic comparison of E-BOM and M-BOM with display of the differences (basis Aras)

The MPP application concerns the shift of MRP functions to SysLM and MES, as already repeatedly mentioned. Similar trends were already referenced in the variant configuration (cf. Fig. 4.69). The functions are still relatively new in SysLM systems and must be extended with further functions, for example line balancing, material flow simulation as well as surface and buffer space analysis, resource planning, run time and manufacturing cost determination and completed through the integration of partner solutions, for example for the graphic simulation of the production processes and ergonomic investigations. At the end of this organizational and functional change process, the numerous, fragmented—often based on Excel—part solutions in SysLM under a common database, a common change and configuration management, and a continuous Digital Thread are summarized and guarantee an optimum connection between development and production in connection with MES and MRP.

MRP and MES

Unlike the previous chapters, which showed examples of concrete implementations in SysLM, this sub-chapter is more concerned with a theoretical discussion level. This is because MES and MRP are not components of SysLM today. If SysLM vendors are thinking of integrating at least MES into SysLM, these projects are only in the concept or implementation phase (cf. Fig. 2.10). However, as the way the orchestration and functional division between MRP, MES, and SysLM is of fundamental importance for the orientation of an operational SysLM mid-term strategy, this discussion should be brought up here.

The core task of a discrete manufacturing company is to offer a cost-effective product tailored to the customers requirements. However, as Chap. 3 describes, it is becoming more and more complicated for companies to realize this. The supporting concepts for Material Resource Planning (MRP) from the seventies are far from able to correspond to the increasing requirements such as higher complexity and an interdisciplinary nature on one hand and ever-stricter standards for traceability and functional security on the other. MRP as a concept for material and time management of discrete manufacturing industry was characterized by the end of the 1970s and start of the 1980s. Schomburg defined an extended functionality of MRP [72, 88]. According to this, MRP is a concept which summarizes the material and time management in the industry and encompasses the entire production including indirectly involved areas. In an initial extension of the concept, MRP includes the entire technical order processing. Starting with the offer processing down to the dispatch, it thus includes the areas of sales, design, purchasing, manufacturing, and assembly as well as dispatch [89]. In the next stage, the MRP is extended by financing and controlling, HR, and the logistics for enterprise resource planning (ERP) [65, 70].

▶ **Enterprise resource planning (ERP):**

ERP should make available a number of processes, methods, and techniques for effective "planning and control of all resources for the procurement, manufacture, sales, and invoicing of customer orders for production, trade, and service companies" [90]. ERP systems are data and function integrated, company-wide applications for the management, control, and evaluation of the business management process. ERP systems extend MRP systems with the holistic financing and accounting as well as human resource and asset management [56, 89, 90].

This makes it necessary to not only serve internal company production and order processing processes, but also requires the inclusion of the entire value chain along the entire supply chain across several companies. This results in a further function complex: Supply Chain Management (SCM) [65, 88, 89]. These functions are partly extended in MRP; however, due to the sluggishness of the often 40-year-old (or older) basic structure of these systems, an independent market has developed from this. The development of the market layout in the areas PLM/SysLM, CRM[115], and MES[116] was similar to this. Below is a limitation to these three systems. There were several attempts (→ a common BOM) to integrate the functionality of PLM/SysLM in MRP. These failed due to the lack of openness and flexibility of the data model, the user interfaces which were difficult for engineers to adapt, and the anachronistic methods of change and configuration management (→ restricted revisioning of the parts master[117]) as well as almost complete ignorance of the ever more important disciplines of electronics and software in particular and the new application area, system architecture, and simulation.

Manufacturing execution systems (MES) are understood to be the connecting link between machine controls of the operative level and the ERP planning tools of the management level. On one hand, it is necessary to prepare and compact the detailed (often to the second) data of the operative level in such a way that it can be used as a decision-making foundation in ERP. On the other hand, the specifications of the ERP (e.g., cycle times) should also be forwarded to the machine controls. A particular challenge lies in the different time horizon, in which the systems act. While the processes of automatic manufacturing are in the seconds range, ERP systems generally focus on mid- to long-term planning. The information flow between MES systems and the operative level is oriented on the technologies of machines and systems and is real-time capable. An MES can therefore depict the materials, machines, tools, and other capacities used online at any time. This data enables planners, masters, dispatchers, and also plant managers to react to problems and make real-time alternative decisions. Additionally, the information transferred from the ERP is prepared and provided in a manufacturing appropriate form for this [70, 72, 88]. In summary, an MES based on the widespread and recognized VDI[118] guideline 5600 can be defined as follows [106].

---

[115]CRM Customer Relationship Management.

[116]MES Manufacturing Execution System.

[117]In reality, operational solutions are found again and again, for example, depicting the well-organized ECM in PLM/SysLM on MRP through the extension of the item number through the revision because many MRP systems has restricted solution for revisioning parts and BOMs.

[118]VDI German Engineers Assoziation (Verein Deutscher Ingenieure).

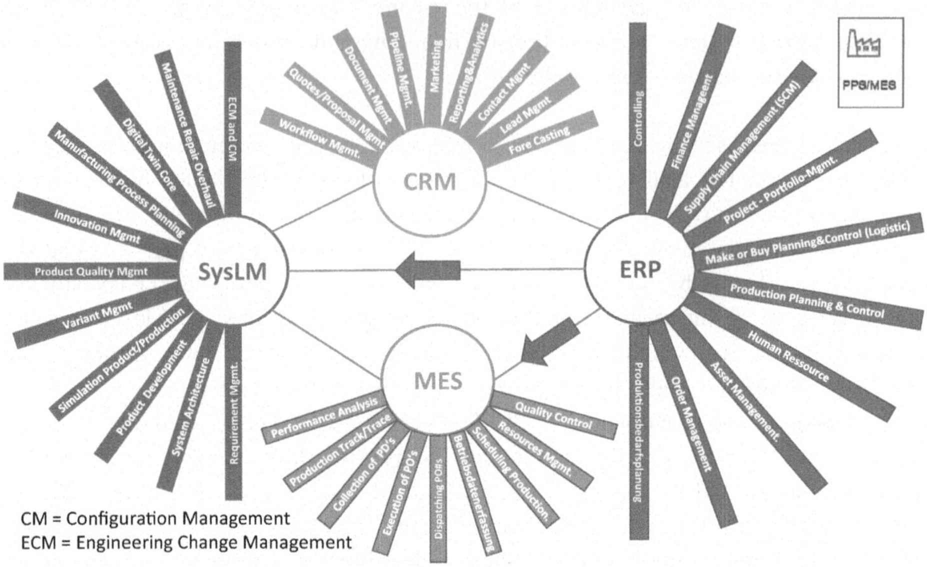

Fig. 4.77 Shifting functions from MRP to MES and SysLM [69]

Manufacturing Execution Systems (MES):
MES describes process-oriented operating manufacturing management systems that extensively support the tasks detailed planning and management, operating equipment and material, information and HR management, data recording, performance analysis, and quality management.

It has already been indicated in various sections of this book that there is currently a rethink taking place in the functional distribution between SysLM, MRP, and MES. Fig. 4.77 indicates these changes. A similar shift is described in [85].

Overall, in the development of a company strategy, all involved functional and organizational units should impartially and openly discuss, analyze, and decide on the functional distribution between the three systems. It is important that, following this, no fraction "wins," rather that there is a deep integration between engineering and production processing and the Digital Thread is extended by MES and ERP. The following plays a role in this orchestration:

- The qualitative complexity and the interdisciplinary nature of the product,
- The functional and technical performance capacity of the underlying systems, and
- The company strategy to implement new approaches such as MBSE, System Thinking, Digital Thread and Digital Twin, company-wide, and consistently.

It may well be that in simple, predominantly mechanical product environments, the ERP system plays a stronger role, as the production and logistics processes are more dominant than the engineering process (→ case a). The same of course also applies in reverse, generally for mechatronic and cybertronic products, which received updates over the air in the scope of a Product Service System functional extension (→ case b). The two extreme forms of different coupling possibilities are represented in Fig. 4.78.

Digital Twin Core (DTC)

Within SysLM, the *Digital Twin Core* creates and manages the precise digital representation of an individual physical asset. At the time of the generation, the DTC initially represents the serialized *as built* or *as delivered* BOM. The mechanisms of reduction and extension presented in Sect. 3.4.1 (Fig. 3.12, 3.14 and 3.17) are allied to this. This creates the foundation to continually synchronize changes in the PT over time in the DT and to manage a Service BOM which is always up to date. This DT configuration reflects a certain component of a product or system in the company, for example an individual vehicle according to the VIN[119] or an airplane according to tail number [33]. The DTC provides a low-code editor for modeling, adapting, and extending highly developed twin representations of the configuration. Users or operators of a service-oriented business model receive an extended visualization in the after-sales area in combination with the dynamic and graphic product navigation, in order to see the DT and PT relationships in the overall information structure of the Digital Model along the PLC (Fig. 4.79).

Connectors to existing legacy systems—such as ERP or CRM, SCM, MES, and others provide possibilities for derivation and persistency of configuration twin data from other sources and guarantee the digital traceability through the Digital Thread across the entire product lifecycle. Open APIs enable connections with IOT cloud services and data lakes in order to merge configuration twins with time series sensor data for analysis. Then, the connection with the SysLM simulation management enables the combination of DTs in simulations with real loading data for insights into wear, load effects, and other effects which result from the running operation of the product of production system.

At this point, a customer should have a say on the topic of Digital Twin

"As the manufacturers continue to digitally transform and the complexity of the products, supply chain, and operation increase, manufacturers and product operators use DTs in order to optimize the product and plant quality and improve the customer experience. The capacity to manage a plant configuration while it is changing over time is a fundamental element of an effective strategy for DTs," says Jeff Hojlo, program director for product innovation at IDC.

[119]VIN Vehicle Identification Number.

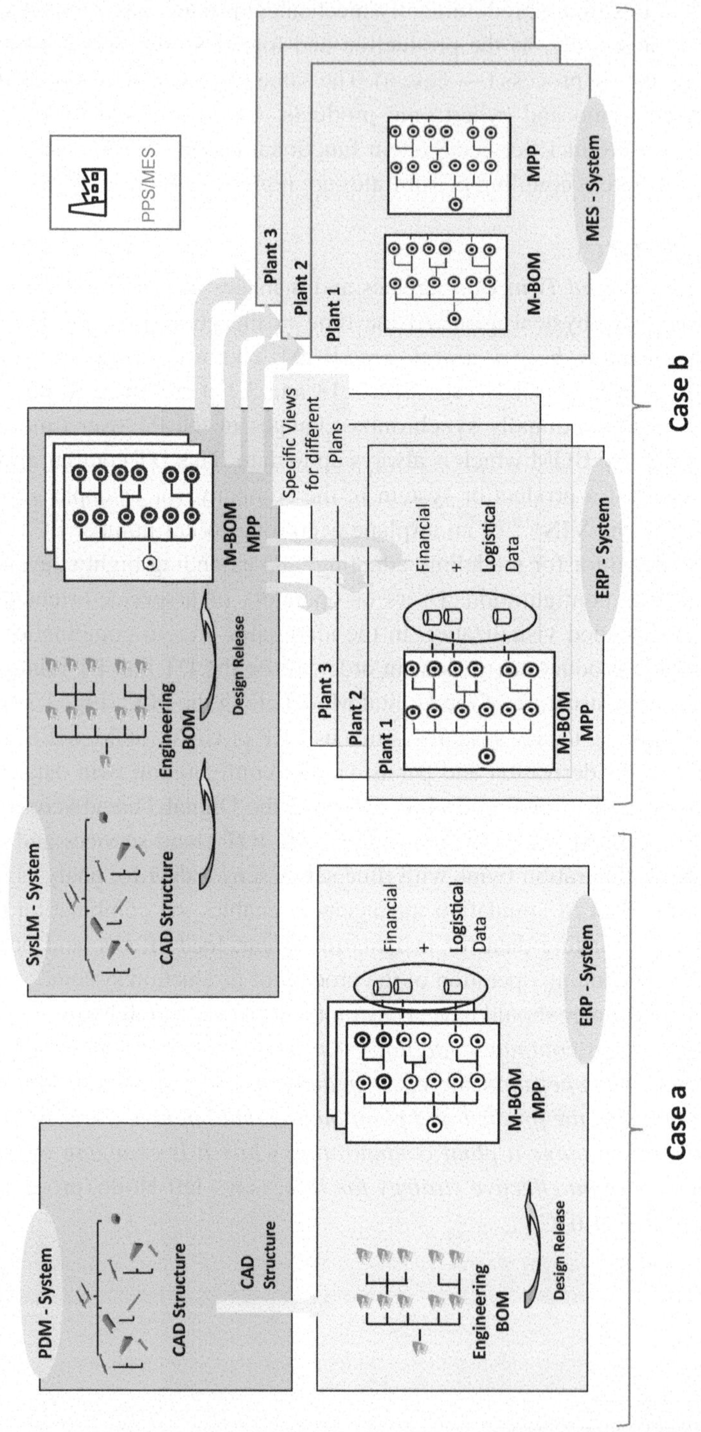

Fig. 4.78 Two extreme possibilities of the PDM/PLM—or SysLM-ERP integration

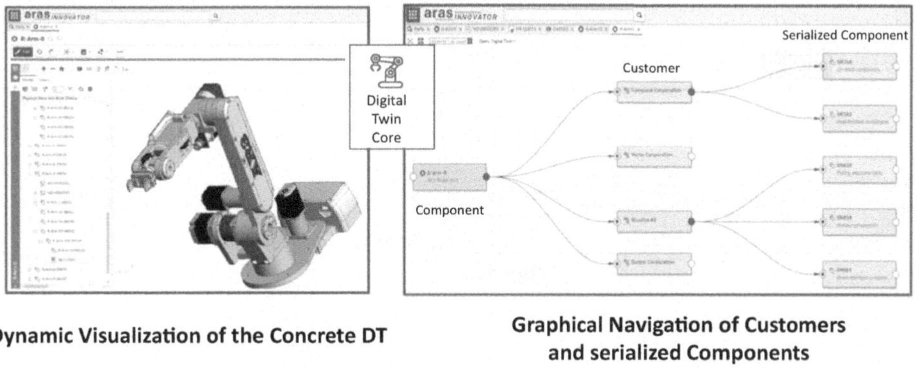

Fig. 4.79 Typical SysLM functions for visualization applied to the DT (basis Aras)

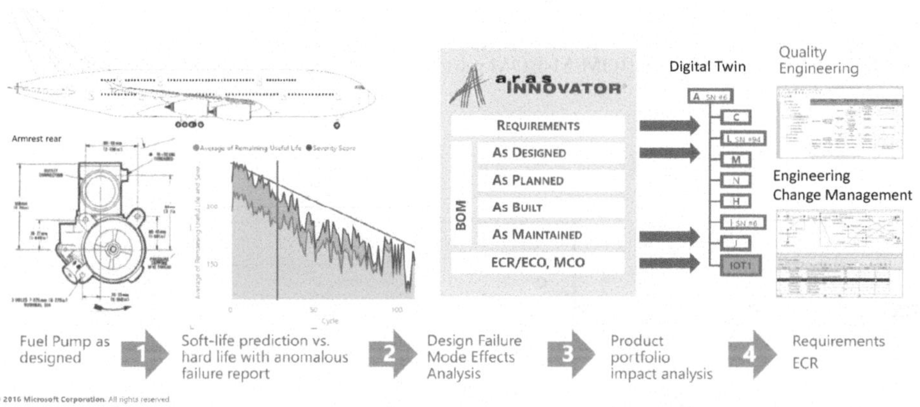

Fig. 4.80 Service-oriented business model on the basis of a DT integrated into SysLM [39]

An example of such a service-oriented business model (→ here predictive maintenance) can be taken from Fig. 4.80 [39].

Back again to the simplified example of the variant armrest (cf. Fig. 4.47). This demonstrates the transfer from an instantiated BOM as output in serialized DTs. In an initial step (Fig. 4.81), the armrest with the part number 123456789, revision A, is prepared for a series production. In the "control type" (1), serial is selected, and another alternative would be batch. The series to be planned has 222 armrests (2), for which serial numbers 2000–2221 are reserved (3). A manufacturing order is created in a second step (Fig. 4.82). As the breakdown structure of the variant logic was already designed assembly-oriented, there is a direct transfer into the M-BOM (1). In the serialized as built BOM, we can see that in addition to the armrest, the carrier was also serialized. The E-BOM can also be represented (3).

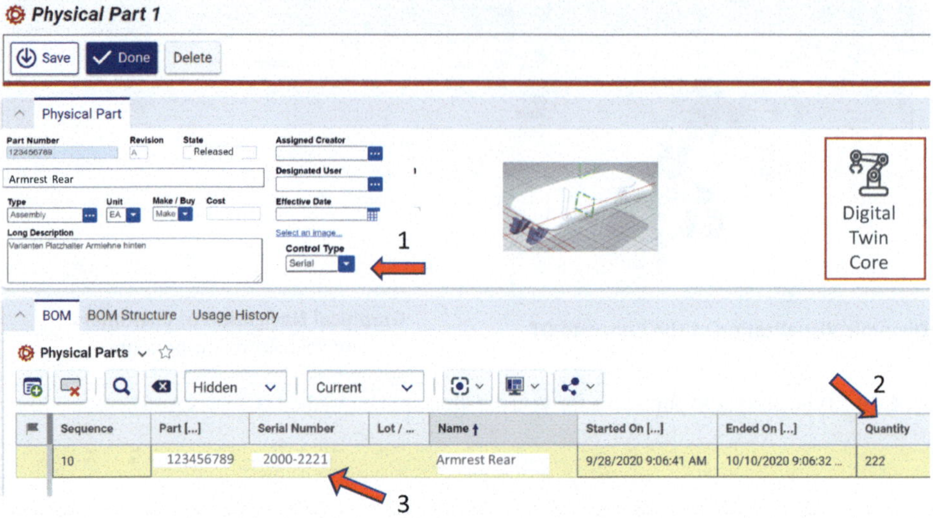

Fig. 4.81 Preparation of an E-BOM/M-BOM part or assembly in a serialized DT (basis Aras)

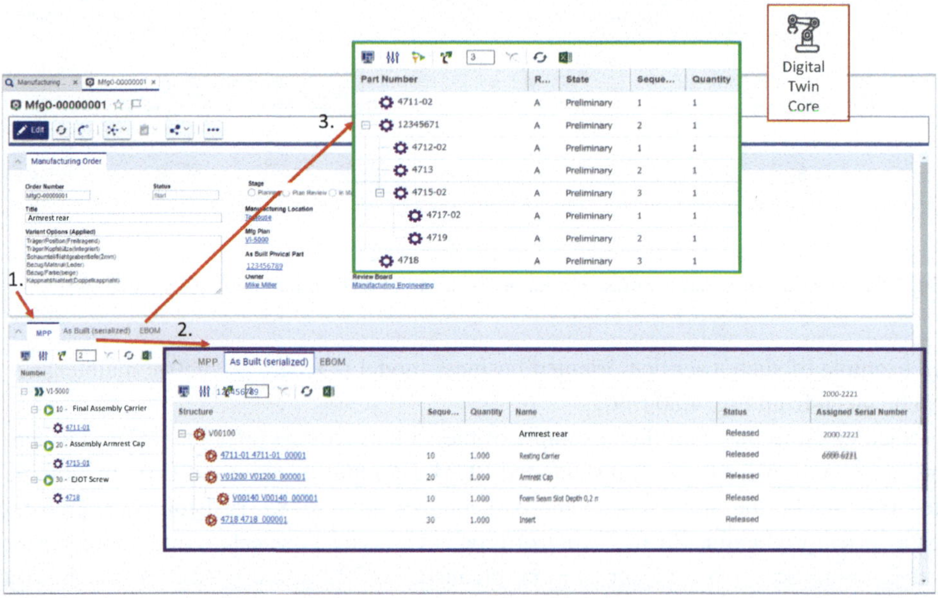

Fig. 4.82 Creation of the production order for the serialized armrest

Michael Grieves first mentioned the concept of the DT in 2002 [45]. Today, machine intelligence, connectivity to the cloud, and the performance of SysLM enable us to provide all information belonging to an instantiated DT via the Digital Thread, an unprecedented potential for large-scale implementation of the DT technology for companies in many fields. Even if it now looked as if SysLM controls the entire process, in the coming years, hybrid concepts between SysLM, ERP, and MES will represent reality. It is important to network these systems as redundancy-free as possible via linked data and, above all, to orchestrate them intelligently (cf. Fig. 4.77).

Maintenance, Repair and Overhaul (MRO)

The term MRO was first used by the Department of Defense (→ DOD) of the USA; however, it is now used across areas. Today, MRO includes the exchange, tests, measurements, and repairs which are required in order to keep or return components to an operative state. It incorporates all measures as well as the procurement of materials and workforces. MRO is often referred to as indirect procurement, as the purchase of auxiliary and operating materials, direct materials and tools is required for the implementation of many maintenance tasks [104]. Due to these dispatch and logistical activities, MRO systems have also implemented a subset of MRP functions. MRO includes three standard types of maintenance (cf. Sect. 4.1):

- Preventive maintenance (PM),
- Condition-based maintenance (CBM), and
- Predictive maintenance (PdM).

and has three main tasks:

- Planning and documenting,
- Executing work orders, and
- Financing and logistics (→ capacity and material provision, storage, invoicing)

Typical maintenance methodologies are FMEA, root cause analysis, lean Six Sigma, and SCADA[120]. The latter has gained a high importance through IOT, smart products, and service-orientated business models. MRO is very closely related to the topic DTC/DT and Digital Thread. The functions QM, PE, and the cross-section functions ECM, dynamic and graphic visualization, and engineering collaboration are also heavily integrated. SCADA is a computer control system which is used to monitor products and systems. It uses many sensors, data communication, and extended management in order to monitor systems for UpKeeping and is often integrated in a DT concept. Below, once again, a simplified example serves as an explanation (Fig. 4.83).

[120]Supervisory Control and Data Acquisition.

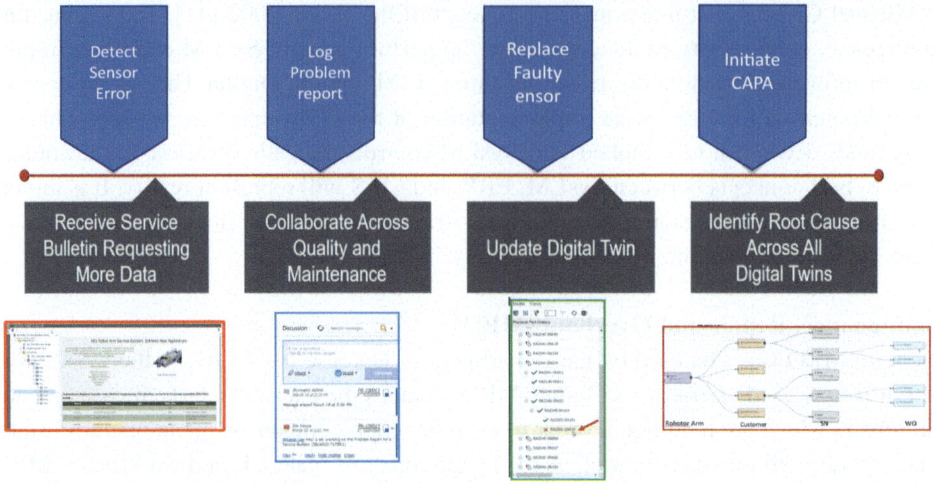

Fig. 4.83 DT in maintenance and service (basis Aras)

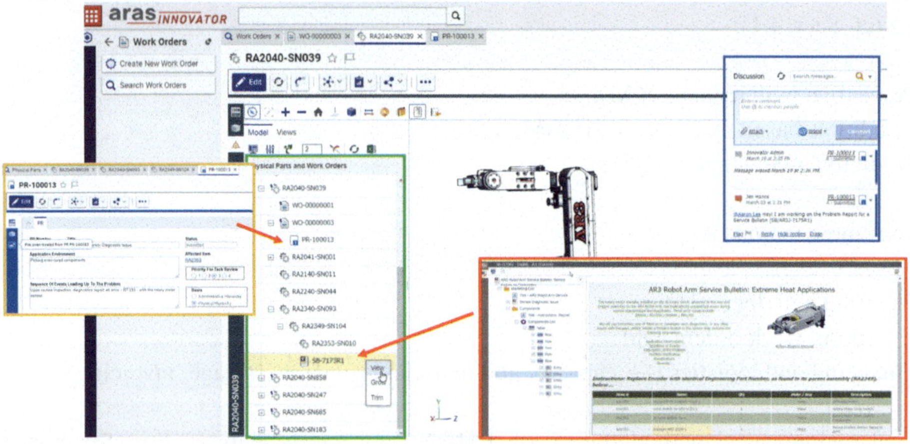

Fig. 4.84 Start of a maintenance even with problem report and service bulletin (basis Aras)

A customer problem was recorded by service and a problem report PR 100013 was created (orange), as well as a service bulletin (red) (Fig. 4.84). An encoder with the item number RA2353 with serial number SN010 has failed in the robot arm with item number RA2040 with serial number SN039 due to overheating. The current overall structure of the instantiated and serialized robot arm with all information elements is visible in the browser (green), for example also the problem report and the service bulletin. The processing engineer is in close contact with product development, simulation, and quality management via the discussion panel (blue).

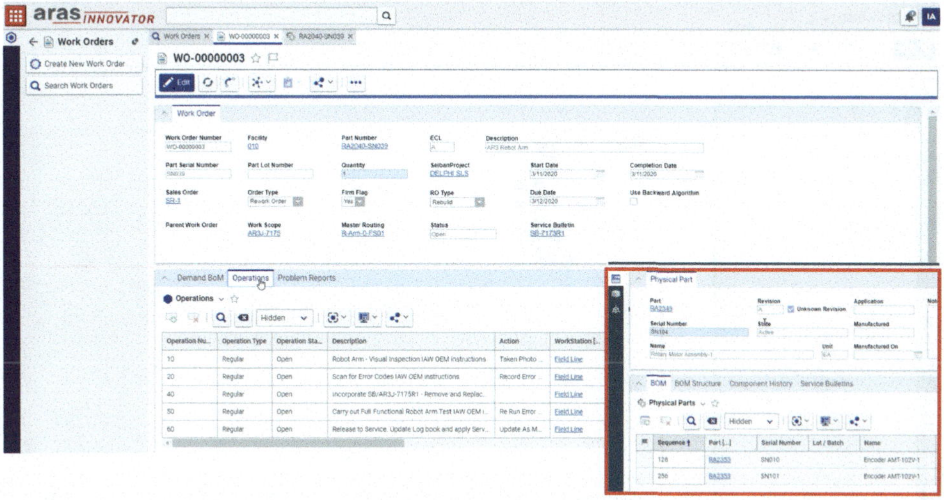

Fig. 4.85 Issue of a work order with the individual work steps for the repair (basis: Aras)

After a discussion with his colleagues, a work order is issued which specifies the individual work steps and estimates time and costs. After executing the work, the exchange of the new encoder with the serial number SN1010 in the robot instance is documented (red) (Fig. 4.85). The maintenance order is now closed.

All work order steps incl. the change in the instantiated product structure are documented and represented both in the browser (green) and in the graphic navigator (red) (Fig. 4.86).

The example was extensively related to the planning and documenting portion of MRO. Of course, a complete MRO system must offer many more functions[121] (Fig. 4.87).

Other SysLM Cross-Section Functions

Through the application of a temporally consecutive run along the product lifecycle (Fig. 4.56), the more cross-section-oriented applications of an SysLM system came up a little too short. Several functions such as ECM, engineering collaboration, graphical navigator, and dynamic viewing are demonstrated in the scope of various applications. Figure 4.88 gives a final overview of the management cockpit (orange), project management (blue), technical documentation (red) as well as viewing and engineering collaboration again (green).

---

[121] In 2018, Aras bought one of the leading MRO systems in the USA and transformed the software directly on the Aras platform. A MES integrated in SysLM is planned from the components of the applications MPP and MRO (cf. Figure 2.10).

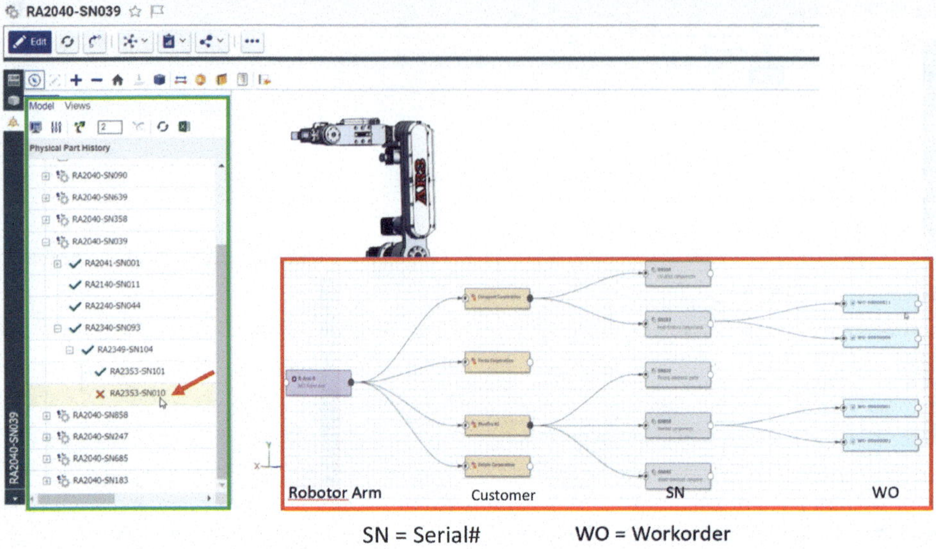

Fig. 4.86 Documentation of all work steps after completion of the maintenance order

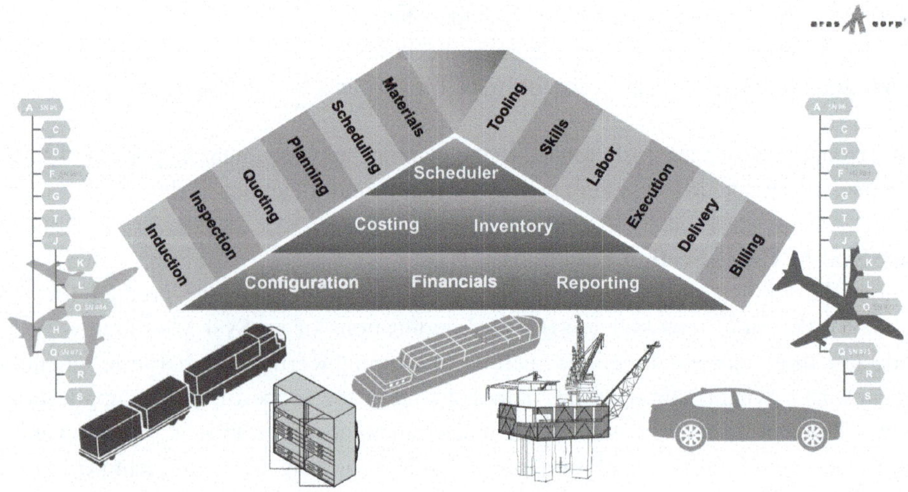

Fig. 4.87 Overview of MRO functions. *Source* Aras Impresa

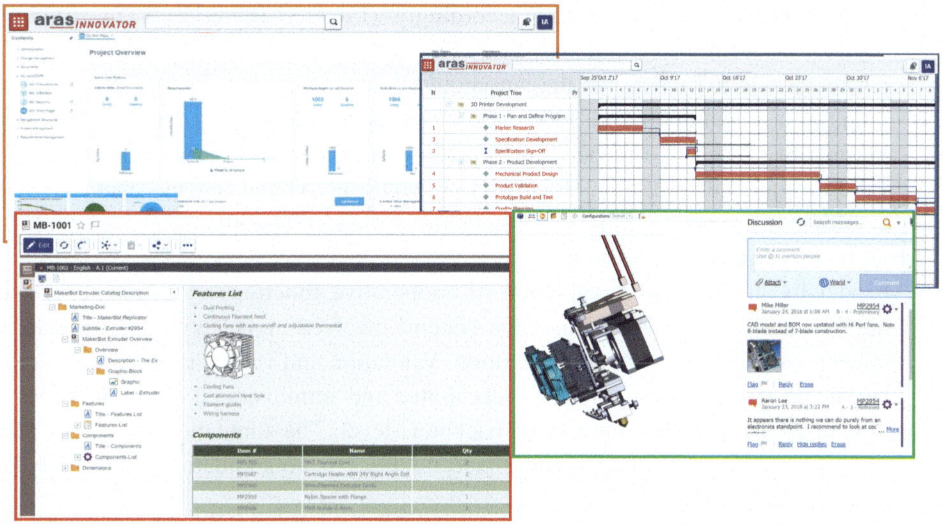

Fig. 4.88 SysLM applications management cockpit, project management, and technical documentation

4.3 Technical Requirements on SysLM

The necessity for the functional extension to PLM was already mentioned in Sect. 2.4, and in particular also for a modern WEB-based and cloud-capable technology and architecture. Significant prerequisites of the digitalization of the products and services as well as of the vertical and horizontal integration of the engineering processes are:

- **Interdisciplinarity leads to stronger requirements for quality and traceability**
  - With IOT[122], IOS[123], Industry 4.0 and the Industrial Internet, more and more interdisciplinary networked cybertronic/cyberphysical products and systems are coming to the market [9, 16, 19, 61, 94].
  - In the scope of service-oriented business models, the operations of the product are increasingly in focus (→ DT, Digital Thread, AI, and big data analysis) [71, 83, 98].
  - Embedded software and electronics partly have more than 50% of the value of a product with an increasing tendency.
  - The high interdisciplinary nature leads to an obligation to guarantee quality, traceability, and the functional security within the product lifecycle. Standards and

[122] IoT Internet of Things.

[123] IoS Internet of Services, as the latest value creation the abbreviations are summarized to IOTS.

recommendations were defined accordingly (ISO/TS 16949[124], VDA 5005[125], AIAG[126] Traceability Guideline) [46]. The Safety Integrity Level (ASIL) according to ISO 26262 exists for the automotive branch. Security is considered in the scope of a guidebook (SAE J3061[127]), which is currently being further developed toward ISO 21434 (→ cybersecurity). More standards and recommendations are valid for medical (→ISO 13485, MDSAP, MDR, FDA) and aerospace and defense (→ ISO 9001). Many of the above standards refer to ISO 9001.

- **Modern Engineering Processes**
  - Especially in the virtual validation of cooperative functions of autonomous and highly automated systems, numerous systems involved in, behavior and simulation models must be taken into consideration. Validation and verification through simulation and virtual testing must be orchestrated and administrated on the TDM level (→ SimPDM) and the engineering backbone level. The simulation should support multi-physics, multi–domain, and mixed-fidelity (→ co-simulation) as well as the new standard FMI[128].
  - The increasing interdisciplinary product and system development requires new processes, methods, and adapted IT tools for these for the engineering. MBSE, System Thinking, and agile processes[129] gain a strategic importance for the digitalization and optimization of the engineering [8]. Super-ordinate interdisciplinary system architectures and system models, among others, on the basis of the SysML[130] language is gaining increasing importance [28, 29, 35, 77].
  - Development is increasingly distributed and must be supported accordingly using the modern federation and collaboration techniques of SysLM. Cloud solutions support this process. Data security must be guaranteed here. Leading providers such as Microsoft Azure offer additional levels of data security FedRAMP (Risk and Authorization Management Program), ITAR Compliance, and DFAR 252.2047012 for DoD[131] contract companies.
- **New Application Functions**
  - In the 90s, there was a positive discussion about parametric construction in the CAD environment. At that time, it was "only" about parametric change of the

[124]In conjunction with ISO 9001, ISO/TS 16,949 defines the quality management system requirements for the design and development, production and, when relevant, installation and service of automotive-related products.

[125]Recommendation describes processes and procedures for the traceability of vehicle components.

[126]AIAG Automotive Industry Action Group.

[127]Cybersecurity Guidebook for Cyber-Physical Vehicle Systems.

[128]FMI Functional Mock-Up Interface https://fmi-standard.org/, (retrieved on 9.27.2020).

[129]SAFe 5.0 Scaled Agile Framework https://www.scaledagile.com/safe-50/

[130]SysML System Modeling language (not to be confused with SysLM!).

[131]DoD Department of Defense.

geometry. In the meantime, this discussion is being continued, not in relation to just geometry any more, but parameters as a super-ordinate, continuous guardrail for the entire engineering process. The parameters result from requirements management and are transferred on to the following phases of the product lifecycle.

- The integration of authoring systems must be reconsidered. Today, they run predominantly as desktop tools, and in a future cloud installation of authoring systems, the couplings will look completely different.
- Authoring systems are becoming more and more intelligent, and customers require new capabilities in integration, for example the integration of new, lightweight and intelligent visualizations (→ dynamic/shattered view), VR[132] and AR[133], more generative design technologies and electronic design automation (EDA) under the direct integration of simulations and AI [15]. This should also enable the intelligent support of 3D printing technology.
- Development and production planning will grow closer together. The redundancy between an E-BOM, an M-BOM, and the BOP[134] will be intelligently triggered. The integration to MRP will be less significant compared to integration in MES[135] in the future.
- Engineering processes (release, change, and configuration) will cover all disciplines, all phases of the product lifecycle, and the supply chain.
- Artificial intelligence (AI) and reinforcement learning[136] only played a minor role in the strict PLM context up to now, and if at all, then more on the interface digital/physical twin in the evaluation of field data. Variant, configuration, change management, and traceability only remain controllable with the increase of quantitative and qualitative complexity with greater automation and application of AI. AI design, simulation, and in particular decision processes support and take over detailed work as a sort of super-assistant [14, 67]. AI can be positioned within the SysLM system or also as an external service, for example from the cloud provider. This also appears logical if we can assume a separation of data retention and application in the future via intelligent cloud concepts. How about an Amazon-style: "Other engineers have used the following solutions in this case, etc." ?
- As opposed to MRP, PLM has hardly offered any decision-making support for management up to now. SysLM must become the central management dashboard

[132] VR Virtual Reality.

[133] AR Augmented Reality.

[134] BOP Bill of Processes = manufacture/assembly plan.

[135] MES Manufacturing Execution System.

[136] Reinforcement learning is based on the idea of formulating problems as a Markov decision-making process, where an AI agent (a specialized algorithm) learns a control point in order to always select the best-possible action for a given system state.

in order to manage and control the engineering processes (→ statistics and analytics, governance).
  - The provision of various access systems such as secure external access (SEA), discretionary or domain access control (DAC), and mandatory access control (MAC) [1]. It would be optimal if there were cloud services that offer these services.
- **Software Technology**
  - The software technology has drastically changed since the end of the 90 s, particularly under the influence of the Internet. The objective is: a farewell to the monolithic aircraft carrier to the federated, flexible frigate (→ cloud, REST[137], RDF[138], WEB services, multi-tenant databases). This trend goes on continuously [96].
  - An open question is the architecture: central core with distributed and linked applications, total decentralized with multiple cores or a hybrid solution[139]? After researching with REST/RDF at the Institute of VPE at the University of Kaiserslautern with a SysLM system and several legacy systems (→ ALM, TDM, MRP), the result was a good solution to display all networked information like a non-redundant data warehouse. But in a fully decentralized environment, no solution was found to realize a holistic engineering backbone to support global ECM/CM processes. The approach has been changed to a minimally redundant solution. All core information elements contain a reduced required record: URL, item number, revision, version, and name. The name is only needed for the better comfort of the users. With mouse over you can view all connected data in the legacy systems. With a double click you jump directly into the legacy system. On this basis, a fully synchronized ECM/CM process was implemented software technically [99]. On this basis, a centralized and hybrid SysLM core system with minimal redundancy and decentralized internal and scalable internal (SysLM) and external functionalities based on linked data (REST/RDF) was designed (Fig. 4.89). The linking of the CI items could be implemented through this linked data concept along the entire PLC and across all system boundaries (cf. Fig. 4.38).
  - Scalability problems will be a strong management problem in the future. It is difficult to predict the growth rate of applications, memory capacity use, and bandwidth. The possibility to scale the cloud quickly and handle unexpected fast grown or seasonal demand shifts has become a major advantage of public cloud services, but can also become a burden if it is not correctly managed[140]. There is

---

[137] REST, Representational State Transfer, describes a programming paradigm for distributed. systems, in particular web services.

[138] RDF - resource description framework - is a fundamental building block.

[139] http://beyondplm.com/2020/10/03/learn-why-decentralized-version-of-truth-is-the-future-plm-paradigm/, (retrieved on 3.2.2021).

[140] https://blog.turbonomic.com/blog/on-technology/cloud-scalability-scale-vs-scale (retrieved on 9.26.2020).

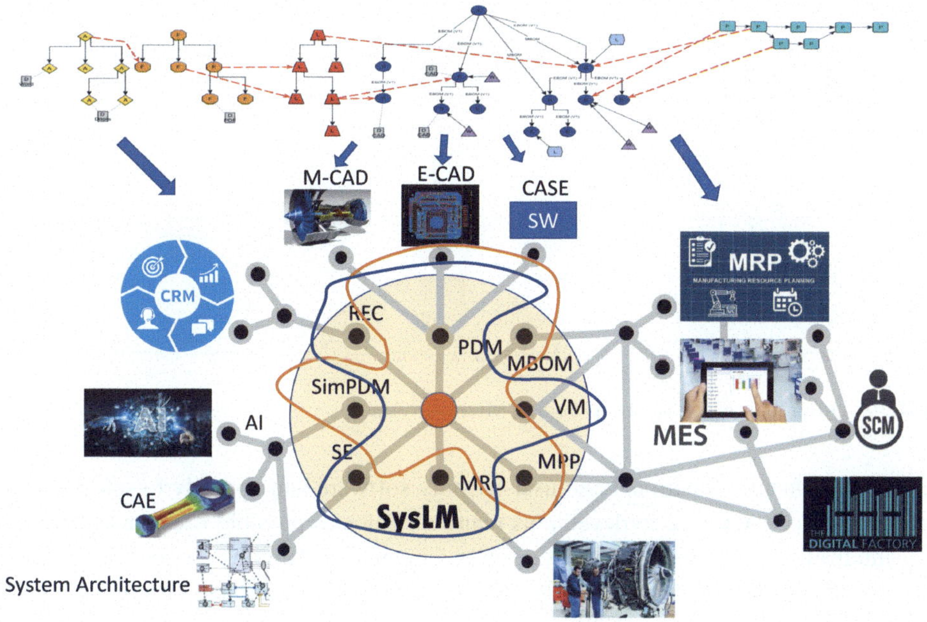

Fig. 4.89 Scalable hybrid SysLM solution with display of the CIs in an ECM process

a differentiation between vertical scaling (→ adding resources) and horizontal scaling (→ scale-out is usually associated with distributed architectures). The latter generally sets extreme changes in the programming technology. For many old PLM systems, this would mean at least partial reprogramming. For SysLM, horizontal scaling is a must [78]

- In order to obtain the mentioned elastic scaling (→ vertical and horizontal) and therefore performance and optimum deployment, further software technologies are required (→ micro-services, containering, convergent, and hyper-convergent systems). Micro-services architectures are the foundation for the development of digital workloads and future system landscapes. Through the convergence of container technologies such as Docker and scalable cloud infrastructures, the system architectures can first be designed, developed, and operated which promises the highest level of agility and scalability with simultaneous robustness.Cloud follows a completely open philosophy with "declarative APIs." For example, Amazon: Amazon does not force anybody to use its own payment service, so there is PayPal, and it is cleanly and openly integrated. Cloud computing goes hand in hand with SaaS[141]. If the software manufacturer is not paid for the license of the software but for its faultless operation, the interests shift enormously. The provider has the incentive

[141]SAAS Software as a Service.

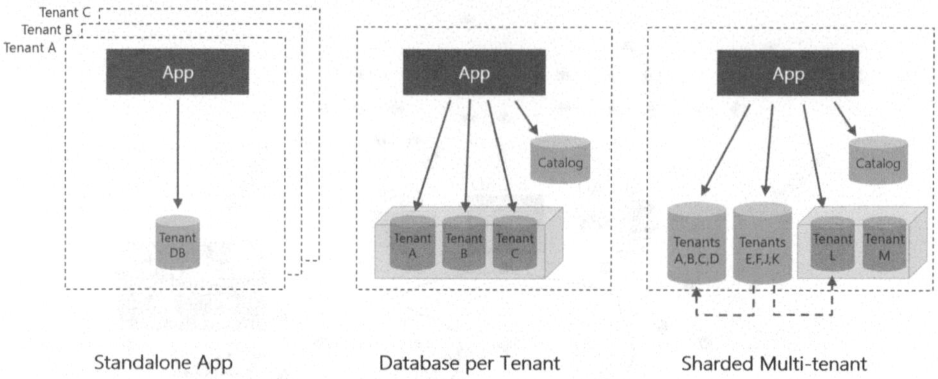

Fig. 4.90 Different tenancy solutions *Source* Microsoft Azure

to manufacture functional and user-friendly software and not only to sell licenses. It also does not have to develop as many modules as possible itself and retain the intellectual property on these. It generates income by building the most cost-effective, best solution and often includes open source libraries. SaaS brings objectivity to software development [78].

- PLM providers have been working with relational databases for years, which has also proven itself for the standard functionalities. There are SysLM functions, for example visualizations, evaluations, in which parallel NonSQL databases, for example graph DBs, are suitable (→ polyglot DB's). Time series databases that are specifically developed for the monitoring of mass data from measure values from feedback from sensors can be optimally used for IOT/IOS applications. They can record, analyze, and save millions of points per second (→ InfluxDB).
- When cloud starts to become important PLM provider soon started to migrate from its client–server routes with the launch of single-tenant SaaS environments. But it was recognized that this temporary solution was not sufficient for elasticity, scaling, and load balancing of the PLM solution[142]. "Multi-tenant architecture is a foundation of all global web applications—Facebook, Twitter, Google. All these solutions are multi-tenant. For the last 15–20 years, there was a growing interest from enterprise application developers in delivery of multi-tenant enterprise applications. Such software like Salesforce, NetSuite, and some others are development using multi-tenant architecture approach[143]." Figure 4.90 shows three samples differ in the underlying database tenancy model used. The first uses a single-tenant

[142]https://www.whichplm.com/multi-tenant-plm-platform%E2%80%8Bs-a-breakdown/, (retrieved on 3.2.2021).

[143]http://beyondplm.com/2018/09/14/multi-tenant-architecture-next-plm-backbone/, (retrieved on 3.2.2021).

application with an isolated single-tenant database. The second uses a multi-tenant app, with a database per tenant. The third sample uses a multi-tenant app with shared multi-tenant databases[144].
- Why do most PLM systems have such old-fashioned user interfaces? Referring to a graphical user interface (GUI) seems presumptuous in most cases. Conversely, the operation is often reminiscent of traditional MRP systems. Here, an example should be taken from smart phones and apps, because the typical SysLM user, in contrast to the typical MRP, is an occasional user and has very particular requirements for the operation of the SysLM system. It cannot be assumed that SysLM only uses a joint GUI concept. It would be better to design the GUIs appropriately for applications and only keep a few fundamental elements consistent across all applications.
- A problem that has existed for years, which has an extremely negative influence on the running costs, is the permanent tracing and separately upgrade of the first time customizing in every new PLM version. Today's software technologies must automatically provide solutions via repositories and difference analysis.
- However, there remains a break in the wonderful new world of software technology: The connectors to the authoring systems which are in the majority of desktop tools. If the authoring systems also migrate to the cloud (→ such as OnShape), the integrations via REST/RDF and linked data can be provided much more easily but differently.

- **The New Business Models of System Providers**
  - The partly three times higher PLM/SysLM implementation costs relate to the one-off license costs, and the permanent costs in order to adapt the company-specific customization of every new PLM version are becoming less and less acceptable. The objective is to drastically reduce the total cost of ownership (TCO).
  - The market is demanding new business models from PLM providers, for example transaction-oriented payments, which guarantee and implement system operation, maintenance, upgrade of the basic version and the automatic upgrade of the company-specific customization.
  - Even if acceptance in the companies is still reserved, it is assumed that the future SysLM systems will be provided as SAAS cloud solutions.

[144] https://azure.microsoft.com/de-de/blog/new-multi-tenant-patterns-for-building-saas-applications-on-sql-database/, (retrieved on 3.2.2021).

4.4 Validation of the Maturity of Digitalization in Engineering[145]

As already explained, companies are under ever-increasing pressure to develop innovative products and save costs, to increase the quality of their products, and reduce product introduction times. The implementation of suitable processes, methods, and IT tools in engineering is the foundation for mastering the challenges of digitalization. The development of cybertronic products and their service systems requires interdisciplinary collaboration, integrated solutions, and processes as well as accessibility and traceability of data and information across the entire product lifecycle. For this reason, companies should continuously make an effort to bring their digital competencies up to date according to their requirements [8, 40].

Moreover, their own strengths and weaknesses should be determined for strategic product planning and internal and external potentials should be identified [41]. In this way, future required actions can be derived as well as possibilities to achieve effectivity, adaptivity, and profitability. These action requirements can represent the foundation for product innovations on one hand, and on the other hand, the determined deficits can demonstrate the required changes in the engineering processes within a company. This makes it clear that in such potential finding, not only products but also processes, methodologies, technologies, IT tools, organizational framework conditions, skills, and strategies must be considered [91]. In order to recognize a possible action requirement, the digital maturity level can be estimated using a *digital maturity assessment.* Improvement potential can be discovered on this basis and a subsequent digitalization strategy and an individual digitalization roadmap developed, in order to drive forward, implement, and ultimately validate digital transformation within a company [68]. The implementation of a digital maturity assessment necessary for this therefore becomes a significant prerequisite for the digitalization of engineering. As preparation for potential finding, the initial situation of a company must be recorded at the start. In order to carry out as detailed a maturity recording as possible, firstly, fundamental company features such as company size, sectors, organizational structures, and already existing strategies must be recorded. Then, a so-called maturity model is used to determine the maturity in order to determine the current status of the company.

[145]The validation of the digitalization and the underlying maturity model was developed in the scope of a research project and through PHDs at the Institute VPE at University of Kaiserslautern by Mrs. Mona Tafvici and Mr. Philipp Pfenning.

4.4.1 Maturity Models

Maturity models are often used as an instrument to record the maturity level of an organization or a process in relation to a certain target state [91, 100]. The actual status of organizations can be evaluated based on a maturity model using various criteria. Action recommendations can then be derived based on this actual status. According to Becker et al. [11], a maturity model consists of a series of maturity stages for a class of objects. Here, it represents an accepted, aimed for or evolution-typical path that the observed objects go through in discrete steps [74]. These objects typically concern organizations or processes. The lowest level here stands for the initial status, which is characterized by the fact that the organization hardly demonstrates any capabilities in the observed area. The highest maturity level on the other hand stands for a mature status. Maturity models refer to the stage theory, which states that elements of a system go through a series of different development phases over time. Here, it must be possible to differentiate between, and empirically prove, the properties of the various phases and it must be possible to trace what causes an element of a system to move up to the next development phase. All properties of the current phase must first be fulfilled for a system to move up a development phase. The origin of these models lies in the quality management maturity grid developed by Philip B. Crosby [26]. This concerns a five-stage matrix with which the organizations can evaluate the maturity of their service and quality management processes. Then, at the end of the 1980s, the capability maturity model (CMM) appeared, which is designed to reveal improvement possibilities in the software development process [49]. This model was further developed by Ahern et al. [6] to the capability maturity model integrated. In addition to this model, there is the model for software process improvement and capability determination (ISO/IEC 15504 [53]), which was developed by the international standard organization ISO into the most widespread maturity model. Automotive SPICE is a domain-specific variant of the international standards from the German Association of the Automotive Industry (VDA). The purpose of Automotive SPICE is the evaluation of the capability of the development process of control device suppliers in the automobile industry [48]. Maturity models are also used in other areas in order to analyze the development of more complex systems, for example in the analysis of human consciousness [59] or the classification of national economics [60].

Conception of Maturity Recording for Digitalization

The general features, for example company size, sector, organizational structure as well as existing strategies, are well suited as a benchmark at the end of the maturity analysis, as they make it possible to compare to similar companies with similar challenges [73]. Maturity models usually have three to six maturity levels which respectively reflect the evolution stages of digitalization in the company. Here, the lowest level corresponds to an analog approach and the highest level to complete digitalization. At this stage, it is important to precisely define the individual levels of maturity of a maturity model in advance in order to be able to make a clear assignment. Next, the digital capabilities to be investigated are determined. Here, several dimensions should be taken into consideration, for example processes, IT systems, and employees, as, on one hand, several dimensions are integrated into the engineering processes and activities, and on the other hand, a digital transformation is most effective if there is a company-wide agreement on changes. The conception of the maturity model of Engineering 4.0 was founded on the ^{VPE}SystemDevelopment$_{Methodology}$ (cf. Sect. 3.6.1) [31, 32]. Several capability indices are defined in the area of process optimization, IT systems, and the integration of the systems. In the area of employees, parallel methodological competence was validated. In order to highlight the capabilities that are most significant for the company under investigation in particular, the criteria are set up differentiated on several levels, for example in the implementation of the SysLM. Various depth of detail can be given as required, for example the consideration of a SysLM system for data management and its connection to authoring and TDM tools or a more detailed examination of the individual functions along the PLC and processes applied by SysLM (→ ECM, CM). Of course, the priorities of digitalization in different areas vary from company to company. Thus, in an electronics company, a well-developed and embedded E-CAD system is more decisive than in machine construction companies, for which a reliable M-CAD system is more important. For this reason, it is sensible to introduce a scoring model that enables companies to evaluate the individual capabilities according to their importance and thus allow them to flow into the final result. Here, either a precise definition of the individual stages of the priorities can be formulated or a mathematical model drawn upon. Thereafter, an evaluation of the maturity level takes place that can be well visualized in a network diagram. On this basis, the analyzed company can estimate its current status with the help of benchmarks and through the comparison with other companies in the same sector of the same company size or with the same existing strategies.

Implementing the Maturity Model for Digitalization of Engineering

The *Maturity Model* consists of four maturity levels: Explorer, Beginner, Advanced, and Expert. These levels and their value range are shown in Fig. 4.91. ML_{total} describes the

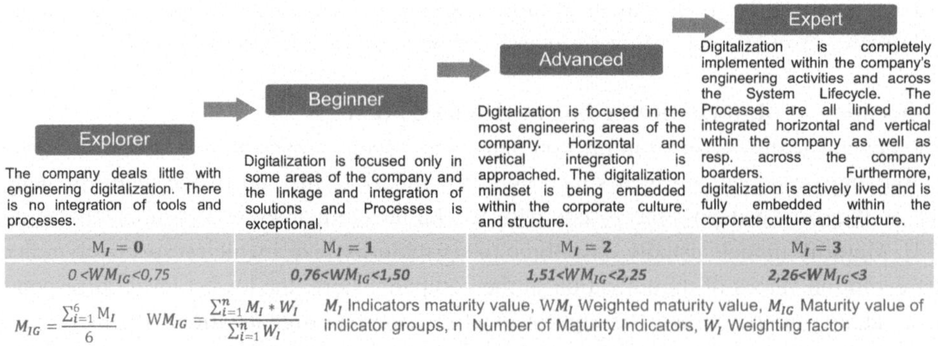

Fig. 4.91 Maturity levels and values [101]

total maturity level regarding digitalization of the companies engineering and is calculated by the arithmetic mean of the achieved maturity level of the four general dimensions (→ people, organization culture, organization structure, and data) and the superior dimensions (→ horizontal and vertical integration and digitalization of products and services) (cf. Fig. 4.1).

The dimensions digitalization of the products and services as well as digitalization of the processes and tools in engineering represent the objectives of the digitalization of engineering. The other dimensions form the foundation to achieve these objectives. The exploitation of potential and realization of requirements which are associated with digitalization in the areas of the human, organizational culture and structure, and data play a significant role here and form the foundation of digitalization. Within every dimension, maturity level criteria and patters are introduced and defined that can be used to determine the maturity level of the respective dimension. For this, the characteristics of the possible levels are given.

The status assessment of the vertical and horizontal integration takes place via integration matrices. All the relevant part functions in engineering and their integration stages (→ from 0–3) along the product lifecycle are depicted in the first matrix. The existence of the respective solution and its level of integration to the TDM level and SysLM backbone are queried. In addition, the grade of harmonization and tool landscape are queried (Fig. 4.92).

In the second matrix (Fig. 4.93), the horizontal networking and coupling of processes and data between the product lifecycle phases are queried (→ horizontal integration). It is easiest to imagine this matrix if you remember the green forwards and red backwards arrows from Fig. 4.56 that represent the horizontal integration.

Here, again, the evaluation should be in stages from 0 to 3 (cf. Fig. 4.91). Moreover, information is queried about the following topics (Fig. 4.94):

Vertical Integration	**Engineering relevant applications**	Digital Solution not existence	Digital Solution implemented	Solution is integrated in Team Data Management or a local application	Solutions are integrated in the SysLM Backbone	Weighting 0: Not relevant, 1: Rather less relevant 2: Neutral, 3: Rather relevant, 4: Of highest relevance	WM_{IG}
Requirement Management	Requirement Definition				3	4	
	Use Cases Modeling			2		3	
	Stakeholder Modelling	0				1	
	Interfaces Modelling	0				1	
	etc.	0				0	2.0
System architecture	Function Modelling		1			4	
	Logical Modelling		1			3	
	Behavior Modelling	0				0	
	etc.	0				0	1.0
Design and Development	Mechanical				3	4	
	Electrical		1			1	
	Electronical		1			1	
	Software			2		2	
	Service design	0				0	
	Generative Design	0				1	
	etc.	0				0	2.0

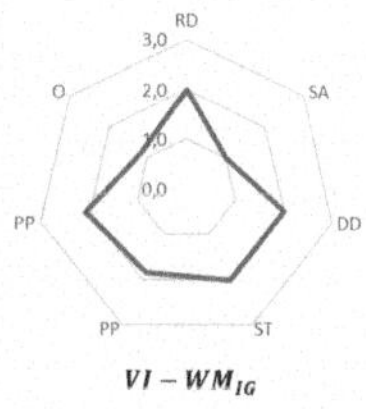

Fig. 4.92 Querying the maturity level of the vertical integration (simplified) [101]

Horizontal Integration	RM	SA	DD	ST	PP	P	O
Requirements Management (RM)		1	1	2	0	0	0
System Architecture (SA)			0	0	0	0	0
Design and Development (DD)				3	1	1	0
Simulation/ virtual Tests (ST)					1	0	0
Production Planning (PP)						3	0
Production (P)							1
Operation (O)							

RM not integrated to SA = 0, RM and SA are semi-coupled requirements are combined to functions = 1, Requirements are also integrated to logical and behaviour Modelling=2, RM is fully integrated into the system model. Complete RFLP models are built = 3.

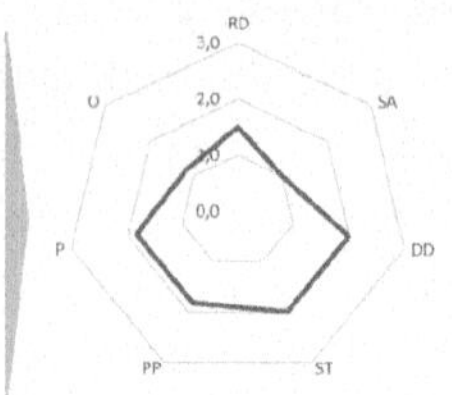

Fig. 4.93 Querying the maturity level of the horizontal integration (simplified) [101]

- HR development (→ are the employees trained and taken along on the journey?),
- Organizational culture (→ digital management, agility, innovation, and learning, etc.),
- Organizational structure (→ flat structures, CDO, DevOps, customer success management),
- Data and technologies (→ business analytics, cloud, Digital Thread, AI, etc.).

Recognitions from the Practical Application of Maturity Models

The aimed for target status of an Engineering 4.0 solution of a company depends on many factors, for example the company strategy, the company size, industry-specific requirements, or legal framework conditions. An evaluation of maturity level analyses carried out shows that industry-specific focuses are set. Thus, in aerospace and defense, a higher priority can be recognized in systems engineering, while in machine engineering, it tends to be more capabilities, such as virtual commissioning, which are relevant [79–81]. Due to the various framework conditions, an Engineering 4.0 solution should always be observed individually and tailored for the respective company. Every company does not have to implement every capability at the highest development level. Rather, it is necessary to identify the relevant capabilities, prioritize these and put them in a chronological sequence. To this end, approach models are increasingly being established that allow capabilities and use cases to be oriented to the company strategy and business model with the help of a maturity model. As a higher level of maturity is often associated with additional costs, in practice, generally one business case is required. Here, the calculated costs are compared to the potential benefit potential. This allows management to compare the digitalization project with other projects and make decisions on this basis.

In summary, it can be determined that a maturity model represents a powerful instrument for defining, validating, and implementing company-specific concepts for Engineering 4.0 approaches.

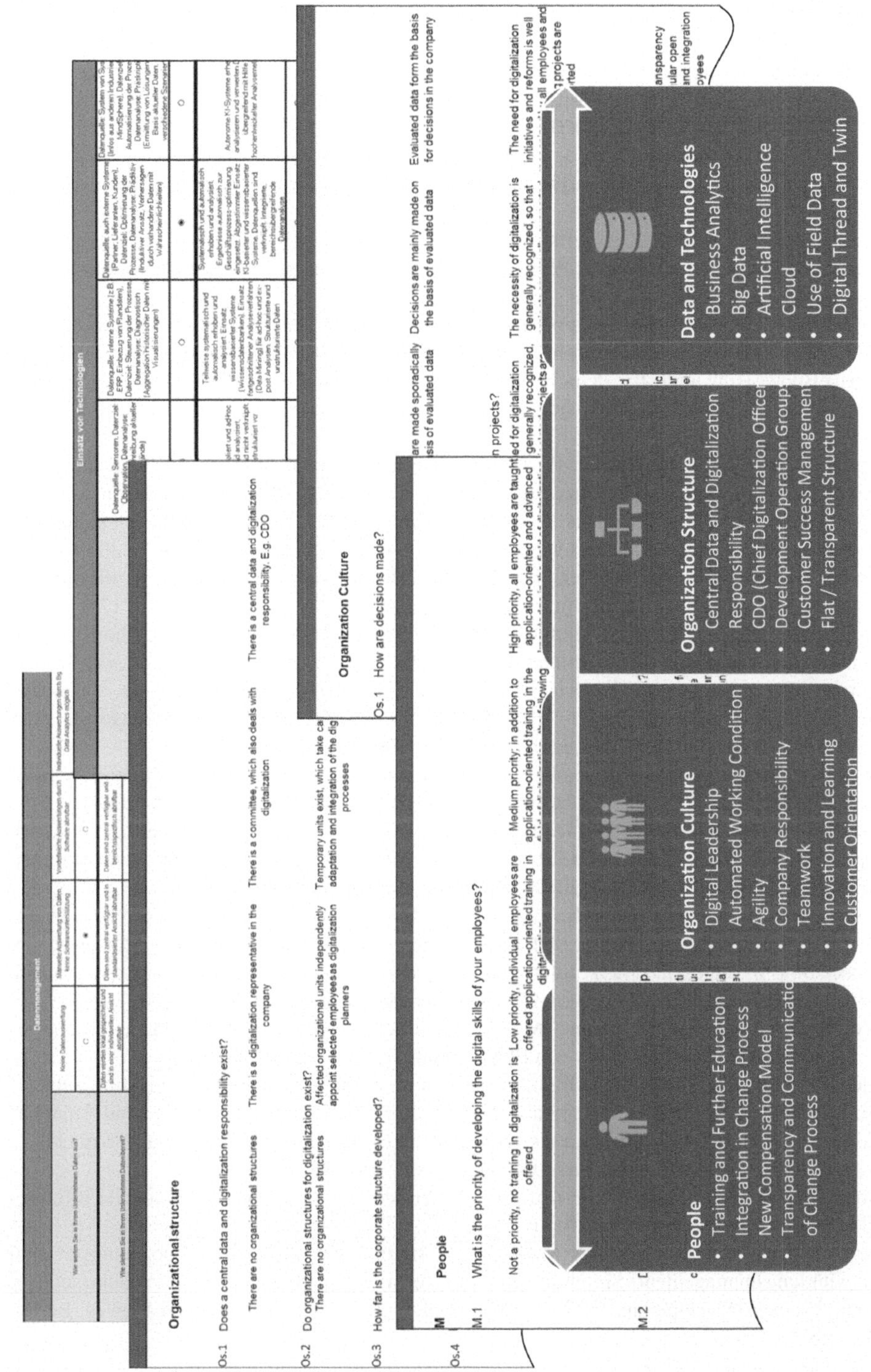

Fig. 4.94 Further queries for the maturity level of the digitalization [101]

References

1. Aras eBook: Making the connection: The how-to's of connecting suppliers, partners, and manufacturers, White Paper. https://www.aras.com/en/resources/all/sea-making-the-connection (2018). Retrieved on 9.27.2020
2. Aras eBook: The engineer's guide to the resilient platform, White Paper. https://www.aras.com/de-de/resources/all/ebook-engineers-guide-resilient-platform (2020). Retrieved on 9.26.2020
3. Aras eBook: S Blueprint for digital transformation: meeting the challenge of product complexity for global manufacturers, White Paper. https://www.aras.com/en/resources/all/ (2019). Retrieved on 9.25.2020
4. Aras eBook: A guide to simulation for innovation, White Paper. https://www.aras.com/en/resources/all/ebook-guide-to-simulation-for-innovation-1 (2019). Retrieved on 9.25.2020
5. Aras eBook: How industry does quality: Lessons learned from the PLM-quality connection, White Paper. https://www.aras.com/en/resources/all/how-industry-does-quality (2019). Retrieved on 9.25.2020
6. Ahern, D.M., Clouse, A., Turner, R.: CMMI Distilled. A Practical Introduction to Integrated Process Improvement, 2. ed. Boston: Addison-Wesley, ISBN 0-201-73500-8 (2003)
7. Altium-Designer, Design-Variants, User Manual Version 20.1. https://www.altium.com/documentation/altium-designer/design-variants-ad. issue 4.14.2020. Retrieved on 6.16.2020
8. Anderl, R., Eigner, M., Sendler, U., & Stark, R.: Smart Engineering Interdisziplinäre Produktentstehung. Springer Berlin, Heidelberg: acatech DISKUSSION (April 2012)
9. Ashton, K.: That 'Internet of Things' Thing. Publisher RFID Journal Online. https://www.rfidiournal.com/that-internet-of-things-thing (2009)
10. Aurich, J.C., Koch, W., Kölsch, P., Herder, C. (Publisher): Entwicklung datenbasierter Produkt-Service Systeme – Ein Ansatz zur Realisierung verfügbarkeitsorientierter Geschäftsmodelle. Springer Vieweg, Berlin, ISSN 978-3-662-59642-5 (2019)
11. Becker, J., Knackstedt, R., Pöppelbuß, J.: Developing maturity models for IT management. Bus. Inf. Syst. Eng. **1**(3), 213–222. https://doi.org/10.1007/s12599-009-0044-5 (2009)
12. Bernard, Y.: Requirements management within a full model-based engineering approach. INCOSE Systems Engineering **15**(2), 119–139, Wiley Online Library. https://onlinelibrary.wiley.com/doi/epdf/10.1002/sys.20198. Retrieved on 4.24.2020 (2012)
13. Bitzer, M.,; Eigner, M., Faißt, K.-G., Muggeo, C., Eickhoff, T.: Framework of the evolution in virtual product modeling and model management towards digitized engineering. In: Proceedings of the ICED17/21st International Conference on Engineering Design, Vancouver, Canada, August 21–25, 2017, pages 379–388, The Design Society, Glasgow (2017)
14. Bleisinger, O., Kuhn, T., Eigner, M.: Autogenerierung von Simulationsmodellen mittels lernender KI-Verfahren auf Basis von PLM", ProduktDaten Journal 2020-01
15. Bley, K., Leyh, C., Schaffer, T.: Digitization of German Enterprises in the Production Sector – Do they know how "digitized "they are?", in Surfing the IT innovation wave: 22nd Americas Conference on Information Systems (AMCIS 2016) San Diego, California, USA, 11–14 August 2016, Curran Associates lnc, Red Hock, NY, MRP. 736–746.
16. Böhmann, T., Warg, M., Weiß, P. (Publisher): Service-orientierte Geschäftsmodelle erfolgreich umsetzen. Springer, Berlin (2013)
17. Bopp, R.: Design for X-Methoden, MRP. 195–203 in Forschungs- und Entwicklungsmanagement, Simultaneous Engineering, Projektmanagement, Produktplanung, Rapid Product Development, Publisher: Warschat, J., Bullinger, H.-J., Sringer Vieweg, Berlin, Heidelberg, 1997

18. Britt, L.: Generatives Design vs. Topologieoptimierung – Was ist was? In IT&Production, p. 88. 3/2020
19. Broy, M.: Cyberphysical Systems. Innovation durch software-intensive eingebettete Systeme. Springer, Berlin (2010)
20. Cadet, M., Meissner, H.: Cybertronische Systeme. In: Eigner, M., Koch, W., Muggeo, C. (eds.) Modellbasierter Entwicklungsprozess cybertronischer Systeme – Der PLM-unterstützte Referenzentwicklungsprozess für Produkte und Produktionssysteme. Springer Vieweg, Berlin, MRP. 19–22 (2017)
21. Capgemini Consult: Industry 4.0 – The Capgemini Consulting View. https://www.capgemini.com/consulting-nl/wp-content/uploads/sites/33/2017/08/industrie_4.0_0.pdf (2017). Retrieved on 4.23.2020
22. Chądzyński, P.: ECAD Connectors—Overview, Aras internal documents, July 2019
23. Chądzyński, P., Panthaki, M., Nicholich, M., Asuri, R.: Pervasive Simulation in a PLM Platform—The key to effective management of ever-increasing product complexity, 30th Annual Incose, International Symposium, Cape Town (July 18–23, 2020)
24. CIMdata: Extending the digital thread with integrated manufacturing process planning https://www.cimdata.com/de/component/docman/cat_view/43-complimentary-reports/45-commentaries (2016). Retrieved on 9.26.2020
25. Crnkovic, I., Asklund, U., Dahlqvist, A.P.: Implementing and Integrating Product Data Management and Software Configuration Management. Artech House, Boston (2003)
26. Crosby, P.B.: Quality is Free. The Art of Making Quality Certain. McGraw-Hill, New York (1979)
27. Dankwort, W., Podehl, G.: FIORES – ein europäisches Projekt für neue Arbeitsweisen im Aesthetic Design. VDI Berichte (1998)
28. Dori, D.: Model Based Systems Engineering with OPM and SysML. Springer Science+Business Media, New York (2016)
29. Eigner, M.: Modellbasierte Virtuelle Produktentwicklung auf einer Plattform für System Lifecycle Management. In: Sendler, U. (Publisher): Industrie 4.0 – Beherrschung der industriellen Komplexität mit SysLM. Springer, Berlin, Heidelberg 2013, MRP. 91–110
30. Eigner, M.: Das Industrial Internet. In Sendler, U.: Industrie 4.0 grenzenlos. ISBN: 978-3-662-48277-3 Springer, (2016)
31. Eigner, M., Dickopf, T., Apostolov, H.: Interdisziplinäre Konstruktionsmethoden und -prozesse zur Entwicklung cybertronischer Produkte – Teil 1. In: Konstruktion, 11-12-2018, VDI Fachmedien (2018)
32. Eigner, M., Dickopf, T., Apostolov, H.: Interdisziplinäre Konstruktionsmethoden und -prozesse zur Entwicklung cybertronischer Produkte – Teil 2. In: Konstruktion, 01-02-2019, VDI Fachmedien (2019)
33. Eigner, M., Detzner, A., Tharma, R., Schmidt, P.: "Definition des Digital Twin im Produktlebenszyklus" ZWF 114 (2019)
34. Eigner, M., Haesner, D.: Konfigurationsmanagement als integrierter Teil von PDM. EDM Report 3, 03.01.1998
35. Eigner, M., Koch, W., Muggeo, C. (eds.): Modellbasierter Entwicklungsprozess cybertronischer Systeme – Der PLM-unterstützte Referenzentwicklungsprozess für Produkte und Produktionssysteme. Springer Vieweg, Berlin (2017)
36. Eigner, M., Roubanov, D., Zafirov, R.: Modellbasierte virtuelle Produktentwicklung. Springer Vieweg, Berlin (2014)
37. Eigner, M., Stelzer, R.: Product Lifecycle Management – Ein Leitfaden für Product Development und Life Cycle Management, 2. ed. Springer Vieweg, Berlin (2009)

38. Europa Lehrmittel: Technisches Zeichnen, Technische Kommunikation, Grund- und Fachbildung Metall, 9 Aufl. (2006)
39. Floyd, S.: The impact of IOT on product design, ACE USA. https://cloudblogs.microsoft.com/industry-blog/manufacturing/2016/03/08/ace-2016-learn-about-the-impact-of-iot-on-industrial-product-design/ (2016)
40. Gausemeier, J., Bätzel, D., Orlik, L.: Potenzialfindung im Rahmen der strategischen Produkt- und Prozessplanung. ZWF 97(9), MRP. 453–458. https://doi.org/10.3139/104.100568(2002)
41. Gausemeier, J., Ovtcharova, J., Amshoff, B., Eckelt, D., Elstermann, M., Placzek, M., Wiederkehr, O.: Strategische Produktplanung. Adaptierbare Methoden, Prozesse und IT-Werkzeuge für die Planung der Marktleistungen von morgen. Heinz Nixdorf Institut. Paderborn. https://edocs.tib.eu/files/e01fb16/870185012.pdf (2016)
42. Gill, H.: A Continuing Vision: Cyberphysical Systems, Fourth Annual Carnegie Mellon Conference on the Electricity Industry (10–11 March 2008)
43. Gilz, T.: Requirements Engineering und Requirements Management. In: Eigner, M., Koch, W., Muggeo, C. (eds.) Modellbasierter Entwicklungsprozess cybertronischer Systeme – Der PLM-unterstützte Referenzentwicklungsprozess für Produkte und Produktionssysteme. Springer Vieweg, Berlin (2017)
44. Gilz, T.: PLM integrated interdisciplinary system models in the conceptual design phase based on model based systems engineering, PHD thesis, in Eigner M. (publisher) publication series VPE, volume 14 (2014)
45. Grieves, M.: Product Lifecycle Management: the new paradigm for enterprises. lnt. J. Product Development (Nos. 1/2) MRP. 71–84 (2005)
46. Hammerschmidt, C.: Das PLM von morgen. In Automotive IT, 17 Sep. 2018. https://www.automotiveit.eu/im-mittelpunkt/das-plm-von-morgen-178.html. Retrieved on 4.9.2020
47. Herrmann, A., Knauss, E., Weißbach, R.: Requirements Engineering und Projektmanagement. Springer, Berlin (2012)
48. Höhn, H., Sechser, B., Dussa-Zieger, K., Messnarz, R., Hindel, B.: Software Engineering nach Automotive SPICE. Entwicklungsprozess in der Praxis - Ein Continental-Projekt auf dem Weg zu Level 3. dpunkt Verlag, Heidelberg, ISBN 978-3-89864-578-2 (2009)
49. Humphrey, W. S. (1988) Characterizing the software process: a maturity framework. IEEE Softw. 5(2), MRP. 73–79. https://doi.org/10.1109/52.2014
50. Huxel, R.: ECAD design data management – an inventory, in Product Data Journal 2019-2
51. Hybertson, D.W.: Model-oriented Systems Engineering Science. A Unifying Framework for Traditional and Complex Systems. CRC Press, Boca Raton (2009)
52. INCOSE Systems Engineering Handbook: A Guide for System Life Cycle Processes and Activities. Wiley, New Jersey (2015)
53. ISO (2003), ISO/IEC 15504-2: Information Technology – Process Assessment – Part 2: Performing An Assessment
54. ISO/IEC/IEEE 42010:2011 Systems and Software Engineering – Architecture Description (2011)
55. ISO 9000:2005, Qualitätsmanagementsysteme. Grundlagen und Begriffe (2005)
56. Kampker, A., Schuh, G.: Strategie und Management produzierender Unternehmen. Springer, Berlin (2011)
57. Keßler, A.: Computer Aided Software Engineering (case). In: Eigner, M., Koch, W., Muggeo, C. (eds.) Modellbasierter Entwicklungsprozess cybertronischer Systeme – Der PLM-unterstützte Referenzentwicklungsprozess für Produkte und Produktionssysteme. Springer Vieweg, Berlin (2017)

58. Klingforth, B.: The necessity and benefits of ECAD-MCAD collaboration for PCB design, in Tech Design Forum, https://www.techdesignforums.com/practice/technique/ecad-mcad-collaboration-for-pcb-design/. Retrieved on 6.19.2020
59. Kohlberg, L., Lindquist, N.T.: Zur kognitiven Entwicklung des Kindes: Drei Aufsätze, Theorie, 1st edn. Suhrkamp, Frankfurt a. Main (1974)
60. Kuznets, S.: Economic Growth and Structure: Selected Essays, 1st edition. Norton (June 1 1965)
61. Lee, E.A.: Cyber Physical Systems: Design Challenges. Berkeley (California), USA, Electrical Engineering and Computer Sciences, University of California at Berkeley Technical Report No. UCB/EECS-2008-8. http://www.eecs.berkeley.edu/Pubs/TechRpts/2008/EECS-2008-8.html. Accessed 23 January 2008
62. Lee, E.A., Seshia, S.A.: Introduction to embedded systems. A cyber physical systems approach. http://leehia.org/releases/LeeSeshiaDigitalV108.pdf (2011). Retrieved on 4.23.2020
63. Littleavon, P.: Configuration Management – an Overview. Engineering Data Newsletter Volume 3 No. 9, 10.01.1995
64. Lock, M., Gaffney, S.: Weaving quality into the digital thread, Aberdeen, March 2020. https://www.aberdeen.com/
65. Maassen, A. et al.: Grundkurs SAP R/3®. Lern- und Arbeitsbuch mit durch-gehendem Fallbeispiel - Konzepte, Vorgehensoftwareeisen und Zusammenhänge mit Geschäftsprozessen. Vieweg & Sohn, Wiesbaden (2006)
66. Mahdavi, A.: Train AI-agents with Microsoft Project Bonsai, 14.7.2020. https://www.anylogic.com/blog/train-ai-agents-with-microsoft-project-bonsai/. Retrieved on 9.28.2020
67. Mahdavi, A., Wolfe-Adam, T.: Artificial intelligence and simulation in business. https://www.anylogic.com/resources/white-papers/artificial-intelligence-and-simulation-in-business/ (2019). Retrieved on 9.20.2020
68. Matt, C., Hess, T., Benlian, A.: Digital transformation strategies. Bus. Inf. Syst. Eng. **57**(5), MRP. 339–343 (2015)
69. Menges, M.: System Lifecycle Management als Backbone der Digitalisierung zur Verwaltung des Digital Threads, PHD Thesis, in: Eigner, Martin (Publisher), University of Kaiserslautern, publication series VPE, volume 28, Kaiserslautern, 2021 (submitted)
70. Menges, M., Roubanov, D., Ernst, J.: Produktionsplanung und Steuerung. In: Eigner, M., Koch, W., Muggeo, C. (eds.) Modellbasierter Entwicklungsprozess cybertronischer Systeme – Der PLM-unterstützte Referenzentwicklungsprozess für Produkte und Produktionssysteme. Springer Vieweg, Berlin (2017)
71. Mert, G., Herder, C.F., Menck, N., Aurich, J.C.: Innovative services for customized, availability-oriented business models for the capital goods industry. In: Proceedings of the 8th CIRP IPSS Conference on Product-Service Systems across Life Cycle, pages 501–506, Bergamo, Italy (20–21 June 2016)
72. Meudt, T., Wonnemann, A., Metternich, J.: Produktionsplanung und -steuerung (MRP) – ein Überblick der Literatur der unterschiedlichen Einteilung von MRP-Konzepten, September 4. 2017 – 2. akt. Version, Technische Universität Darmstadt, PTW/CIP
73. Mittal, S., Khan, M. A., Romero, D., Wuest, T.: A critical review of smart manufacturing & Industry 4.0 maturity models: Implications for small and medium-sized enterprises (SMEs). J. Manuf. Syst. 49, MRP. 194–214. https://doi.org/10.1016/j.jmsy.2018.10.005 (2018)
74. Nolan, Richard, L.: Managing the computer resource: a stage hypothesis. Commun. ACM **16**(7), MRP. 399–405. https://doi.org/10.1145/362280.362284 (1973)
75. Norris, M.: How to succeed at simulation data management. Paper submitted for NAFEMS (2020)

76. Panthaki, M., Mottola, E.: A coherent digital thread as enabler of systems thinking across the system lifecycle: Where's the simulation? in CAASE20, Virtual Conference, June 16th–18th
77. Pfenning, M.: Durchgängiges Engineering durch die Integration von PLM und MBSE, PHD Thesis. In: Eigner, M. (Publisher), University of Kaiserslautern, publication series VPE, Volume 20, Kaiserslautern 2017. 978-3-95974-060-9
78. Pfenning, M., Reichwein, A., Hooshmand, Y., Dickopf, T., Braun, S., Bleisinger, O., Psota, T., Masior, J., Roth, A., Muggeo, C., Hutsch, M.: Killing the PLM Monolith – the Emergence of cloud-native System Lifecycle Management (SysLM). Fraunhofer-Publica Verlag (2021) (submitted)
79. Pfenning, P.: Digitalisierung in der diskreten Fertigung – Ein holistischer Ansatz zur Reifegradeinordnung, wirtschaftlichen Bewertung und Umsetzung von Digitalisierungsstrategien, PHD thesis in: Eigner, Martin (publisher), University of Kaiserslautern, publication series VPE, Band 25, Kaiserslautern, 2021 (submitted)
80. Pfenning, P., Eibinger, H.A., Rohleder, C., Eigner, M.: "A Comprehensive Maturity Model for Assessing the Product Lifecycle", IFIP International Conference on Product Lifecycle Management PLM 2020: Product Lifecycle Management Enabling Smart X pp 514–526, (2020)
81. Pfenning, P., Eigner, M: A novel procedure model for developing individualized digitalization strategies. In Publication of Conference Design 2020, Dubrovnik (2020)
82. Pohl, K.: Requirements Engineering. Grundlagen Prinzipien, Techniken. Dpunkt, Heidelberg (2008)
83. Porter, M.E., Heppelmann, J.E.: How smart, connected products are transforming companies. Harvard Business Review, October 2015, MRP. 96–11
84. ProSTEP, Simulation Data Management (SimPDM) Recommendation, no. 2.0, Prostep iViP e. V., Darmstadt (November 2008)
85. PTC: PLM and ERP: Their respective roles in modern manufacturing, White Paper. https://www.ptc.com/en/resources/plm/white-paper/plm-and-erp-in-modern-manufacturing (2020). Retrieved on 9.26.2020
86. Rupp, C.: Requirements-Engineering und Management – Aus der Praxis von klassisch bis agil. Hanser, Munich (2014)
87. Schichtel, M.: Produktdatenmodellierung in der Praxis. Hanser, Munich (2002)
88. Schomburg, E.: Entwicklung eines betriebstypologischen Instrumentariums zur systematischen Ermittlung der Anforderungen an EDV-gestützte Produktionsplanungs- und -steuerungssysteme im Maschinenbau. Aachen, Techn. Hochschule., PHD Thesis. (1980)
89. Schönsleben, P.: Integrales Logistikmanagement. Operations und Supply Chain Management innerhalb des Unternehmens und unternehmens-übergreifend. Springer, Berlin (2011)
90. Schuh, G., Stich, V.: Produktionsplanung und -steuerung 1. Grundlagen der MRP. 4. ed. Springer, Berlin (2012)
91. Schumacher, A., Nemeth, T., Sihn, W.: Roadmapping towards industrial digitalization based on an Industry 4.0 maturity model for manufacturing enterprises. Procedia CIRP 79, MRP. 409–414. DOI: 10.10.2019 (2019)
92. Sendler, U.: Ganzheitliche Strategie: Systems Lifecycle Management, in PLM portal. https://www.plmportal.org/de/forschung-detail/ganzheitliche-strategie-systems-lifecycle-management-syslm.html (2013). Retrieved on 4.9.2020
93. Sendler, U.: PLM und die Zukunft, in PLMportal, U. Sendler https://www.plmportal.org/de/plm-und-die-zukunft.html (2012) Retrieved on 4.10.2020
94. Sendler, U. (Publisher): Industrie 4.0 – Beherrschung der industriellen Komplexität mit SysLM. Springer, Berlin (2013)
95. SIEMENS/PLM: NX PCB.xchange: ECAD/MCAD design interface, https://siemens.com/plm (2007). Retrieved on 9.26.2020

96. Steinmetz, C., Christ, A., Anderl, R.: Data Managment Based on Internet Technology Using Restful WEB Services, 10th International Workshop on Integrated Design Engineering IDE Workshop, Gommern (10th–12th September 2014)
97. Stöber, C., Gruber, G., Krehmer, H., StuMRPy, J., Westphal, C.: DFX 2009: Proceedings of the 20th Symposium on Design for X, Neukirchen/Erlangen, Germany, Editor H. Meerkamp, ISBN 978-3-9808539-8-9 (2009)
98. Ströer, F., Faißt, K.-G., Eickhoff, T., Apostolov, H., Sivasothy, P., Seewig, J.: Eigner; M.: Big Data in verfügbarkeitsorientierten Produkt-Service-Systemen am Beispiel einer Landmaschine. In: Schulze, S.-O., Tschirner, C., Kaffenberger, R., Ackva, S. (eds.) Tag des Systems Engineering, Paderborn, November 08-10, 2017, pp. 285–294. Hanser, Munich (2017)
99. Stuerken, M., Eigner, M.: Integrated model-based systems: Engineering solution to support digital transformation, ProStep IVIP Symposium, 9 and 10 April. ICS, Stuttgart (2019)
100. Tafvizi Zavareh, M., Sadaune, S., Siedler, C., Aurich, J., Zink, K.J., Eigner M.: A study on the socio-technical aspects of digitization technologies for future integrated engineering work systems. In Proceedings of NordDesign, ISBN: 978-91-7685-185-2. Linköpig (2018)
101. Tafvizi Zavareh, M: Systematik zur Reifegradmodell-basierten Validierung der Digitalisierung des Engineerings, PHD Thesis, in: Eigner, Martin (Publisher), University of Kaiserslautern, Schriftenreihe VPE, Volume 27, Kaiserslautern, 2021 (submitted)
102. Tao, F., Cheng, J., Qi, Q., Zhang, M., Zhang, H., Sui, F.: Digital Twin-driven Product Design, Manufacturing and Service with Big Data. Int. J. Adv. Manuf. Technol. **10** (2017)
103. Ulrich, K.T., Eppinger, S.D.: Product Design and Development. McGraw-Hill Education (ISE Editions), New York (1999)
104. UpKeep: Learning Center – Information for building effective maintenance programs. https://www.onupkeep.com/learning/maintenance-terms/maintenance-repair-overhaul. Retrieved on 9.26.2020
105. VDI/VDE 2206:2020-09 – Entwurf, Entwicklung cyber-physischer mechatronischer Systeme (CPMS), Beuth, https://www.beuth.de/de/technische-regel-entwurf/vdi-vde-2206/327616828 (2020)
106. VDI Verein Deutscher Ingenieure e. V.: Fertigungsmanagementsysteme – Manufacturing Execution Systems (MES), VDI 5600. Beuth Verlag (2007)
107. Zafirov, R., Roubanov, D.: Elektrik und Elektronik (E-CAD). In: Eigner, M., Koch, W., Muggeo, C. (eds.) Modellbasierter Entwicklungsprozess cybertronischer Systeme – Der PLM-unterstützte Referenzentwicklungsprozess für Produkte und Produktionssysteme. Springer Vieweg, Berlin (2017)
108. Zagel, M.: Übergreifendes Konzept zur Strukturierung variantenreicher Produkte und Vorgehensweise zur iterativen Produktstruktur-Optimierung, PHD Thesis, in: Eigner, Martin (Publisher), University of Kaiserslautern, Schriftenreihe VPE, Volume 1 (2006)

Bimodal SysLM Systems and Agile Implementation

5

Abstract

As already presented in the first chapter, the industry will experience a disruptive change process, which only agile and innovative providers will survive. Business and cooperation models will change and the entire product, and increasingly system structure, will be dominated by mechatronic and cybertronic, high-grade, networked system architectures. In order to satisfy these new requirements, the companies must have the competency to holistically orchestrate such interdisciplinary system architectures and the services built on these and to find new, intelligent forms of cooperation with suppliers to implement the system architecture. *Bimodal SysLM* systems must support these product and process changes with new, modular, federated and web-based technologies and architectures, through absolute openness and through new, customer-oriented business models. The objective is to create the prerequisites for agile implementation in addition to a drastic reduction in the total cost of ownership (→ TCO). After large PLM (It is the past, therefore PLM) projects in the past frequently failed due to their complexity, duration of implementation and also the permanent tracking and adjustment of the customer-specific customizing, (There are some PLM applications in industry that do not manage to adapt customizing between two PLM versions and are therefore further behind the current PLM version) bimodal SysLM systems, and agile implementation approaches are in demand today, where user stories can be implemented in short timeframes (sprints) based on interactive configuration instead of procedural customization. The objectives are implemented and refined in small, incremental steps, permanently supported by rapid prototyping and the feedback of system users.

M. Eigner, *System Lifecycle Management*, https://doi.org/10.1007/978-3-658-33874-9_5

5.1 Vision SysLM Versus Actual Status of Current PLM Implementations

System Lifecycle Management is at the center of the listed changes (cf. Chaps. 3 and 4) in the product and process world as the skeleton of a consistent, digitalized product lifecycle (PLC), which forms the basis for a digital product, production, service, and process model across all disciplines, distributed locations, and the entire supply chain. The roles of data and process integration between the product and production development, production as well as the service will be taken in as a continuous backbone on an SysLM-based engineering backbone (Chap. 4). The complexity of current SysLM architectures and implementation strategies is already very high and the further increase to be expected through cybertronic product and production systems will strengthen this trend. Traceability and reconfigurability across the entire product lifecycle, the disciplines and the supply chain according to ISO 9001, and other market segment specific standards must be guaranteed at all times in order to master the engineering process. The current introduction status of PLM stands in diametric opposition to these visionary objectives. The automotive industry is certainly a forerunner in implementation and has helped to shape many technological trends. However, here too, the PLM projects implemented to date are far removed from the upper target. Today's PLM implementations are often highly overcustomized due to excessive thoroughness and look something like in Fig. 5.1. This means a high initial investment; however, the permanent update of the customization when delivering a new PLM base version is far more cost-intensive in [1].

Fig. 5.1 Example of an "overcustomization"

Over 100 fragmented TDM/PDM/PLM and authoring systems as well as Excel and other "hidden DBs[1] and Shadow IT," partly historical, partly on the basis of M&A,[2] are typical for applications in large companies. Harmonization attempts fail due to user resistance and the lack of knowledge and disinterest of management for details in the application solution. A fully integrated engineering backbone as a basis for the engineering process was initially the vision; the reality still shows a focus in the phase of design/development and the management of CAx data. Requirements, system architectures, simulations, electronics, and in particular software are not generally integrated into the product and process structure.

5.2 Innovative Products/Systems and Business Models Present New Challenges for SysLM

Changes in the product and process world mean, on the one hand, new challenges for the products and service-oriented business models, and on the other hand, requirements on SysLM systems (cf. Sect. 4.3), for example, extended functions, easier intuitive operation and on new IT technologies, architectures, and standards (web services, microservices, cloud, OSLC/REST/RDF, multi-tenant databases, …). Digitalization of products, systems and engineering processes means a transformation process which rearranges the classic limits of a fragmented and competitive IT solutions. A consistent, lightweight and federated engineering backbone would take over the role of data and process integration between product development and production planning, production, as well as service. On one hand, this includes the support of the following extended functional SysLM areas of use (cf. Sect. 4), which are summarized in Fig. 5.2 [1]:

On the other hand, a purely functional extension to an SysLM solution would overwhelm many older PLM solutions. For this reason, approaches such as "bimodal IT" [3]or "digital PLM" (Accenture, 2016) demonstrated the necessity for a technological evolution in a similar way to SysLM. Transferring these bimodal approaches to SysLM results in the requirements on SysLM in Sect. 4.3. The significant points of a methodical application extension and a reduction of the total cost of ownership should be referred to once again here:

- From document- to model-based engineering in PDP[4] means support of the MBSE and System Thinking methodology,
- Depiction of hierarchical (standard BOMs of the hardware) to network (system architecture) and linear structures (software),

[1]DB Database.

[2]M&A Merger and Acquisition.

[3]Marc Halpern, Gartner Group, PDT 2016.

[4]PDP Product Development Process.

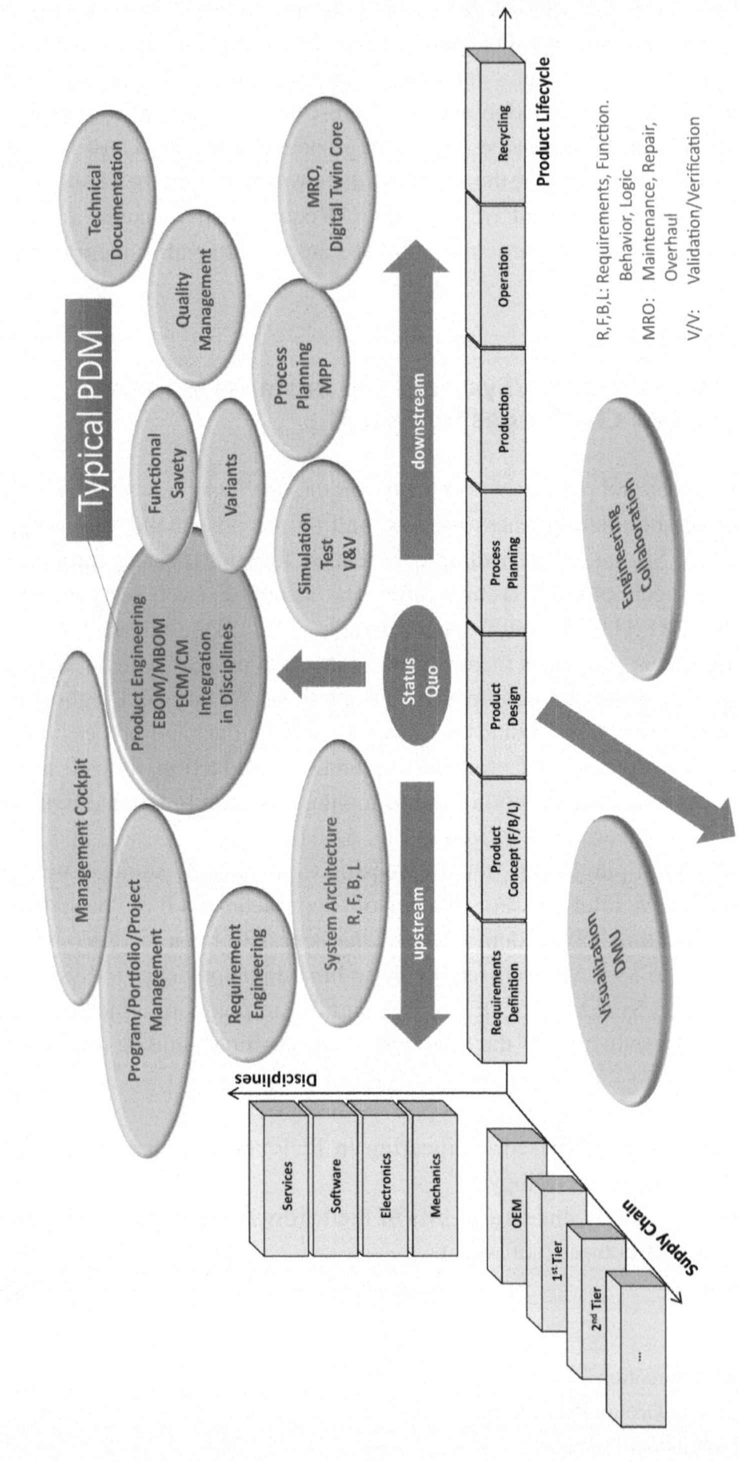

Fig. 5.2 Extended SysLM areas of use

- Support of the typical engineering processes, such as engineering change management (ECM) and configuration management (CM), standardized and coordinated for all disciplines,
- Derivation of digital twins in the development and operation phase,
- Technologies for information synchronization of the authoring systems to backbone and vice versa with modern web-based technologies (cf. Sect. 4.3),
- Agile introduction and operational adjustment through extensive interactive configuration instead of procedural customizing,
- Automatic adoption of all customizing when upgrading the base system through differential analysis of the underlying data and process models (→ model-based repository and low-code modeling engine), and
- New, flexible business model on the basis of transaction-orientated payment models instead of licenses.

The three final points in particular lead to a significant reduction in the TCO in procurement, implementation, and operation of an SysLM system. In addition, bimodal systems offer the optimum prerequisites for successful digitalization of the engineering processes and agile implementation in companies thanks to their agility and flexibility.

5.3 Reduction in the Total Cost of Ownership with Bimodal SysLM

A fundamental problem of all PLM[5] systems introduced up to now is the high proportion of customizing and implementation costs. For every one euro of investment, two to three euros of customizing and implementation costs are calculated on top. According to details from Unity AG, only 2% of the PLM introduction costs are spent on hardware, 25% goes on PLM software, and 71% of the total costs on customizing. Of this, approx. 75% alone goes on functional extensions and 13% on integrations and interfaces.

The most frequent factors of customizing concern.

- The functional extension of the system which is reflected in new classes, attributes, and methods and their integration in the "out of the box" basic functionality of the underlying SysLM system,
- The integration and extension of the typical product lifecycle-orientated processes, such as release, change, and configuration management. Of course, all new objects from the functional extensions along the product lifecycle must be implemented in this process in order to achieve continuous integration, and

[5] We are referring to the past here, hence PLM!

- The systematic integration of the various IT solutions on the different architecture levels as well as the integration of the IT solutions and the resulting data from suppliers and customers. This depends in particular on the openness of the interface, the guaranteed upward compatibility from the system supplier, and the use of international standards.

If the initial customizing has been completed in the first system implementation phase, each system upgrade again incurs expenses of 20–30% for the typical transfer of the customized portions of the operational solution.

In the following assumptions of a bimodal SysLM system that influence the TCO in particular:

- Subscription instead of licenses (roughly corresponds to the maintenance costs of standard market PLM systems),
- Simplified interactive customizing through repository technology and linking the data model with the user interface on the screen (→ low-code development) as well as,
- Automatic upgrade of the customer-specific customizing based on difference analysis of the repository integrated in the upgrade of the basic system included in the subscription price.

The TCO relationships presented in Fig. 5.3 for the one off and permanent costs result in both a bimodal SysLM and a standard solution (1000 seats, Project Management, Master Data, BOM, CAx management, ECM/CM, and integration to MRP) calculated for five years [1].

The immense cost benefit when upgrading the basic system and the customizing/configuration proportions is evident (Fig. 5.4).

This largely results from the fact that, in a bimodal SysLM system, the automatic tracking and testing of the customized solution proportions are contained parallel to the SysLM basis version in the subscription price. Technically, this process is through a differential analysis of the model-based repositories that identifies the customized part of the repository and transfers to the new SysLM base version. The only thing that remains for the customer is an integration test that incorporates the interface validation to authoring and other systems (MRP, MES, …). The typical customizing effort for various PLM/SysLM systems was analyzed and determined in an analysis by the institute VPE at the University of Kaiserslautern with a typical detailed customizing profile [1]. This included:

- Creating new attribute classes,
- Creating new object classes,
- Creating new relationship classes (→ relations),
- Which methods are standard (insert, delete, change, version, effectivity, …),
- Which methods can be inherited when copying an object class (→ object orientation),

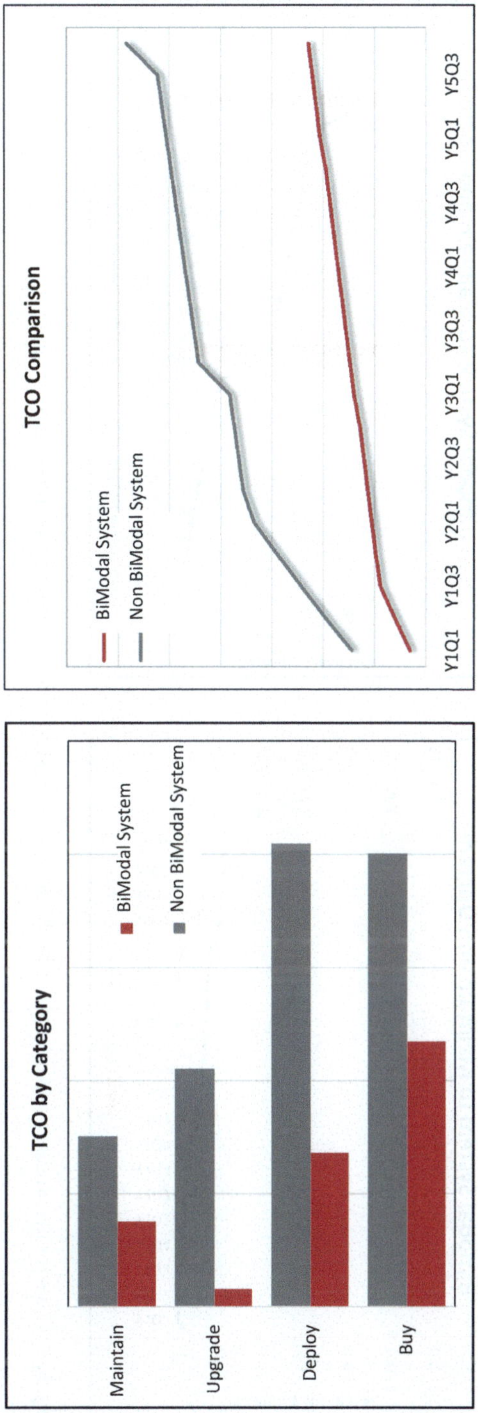

Fig. 5.3 TCO comparison of bimodal to non-bimodal system (overall representation)

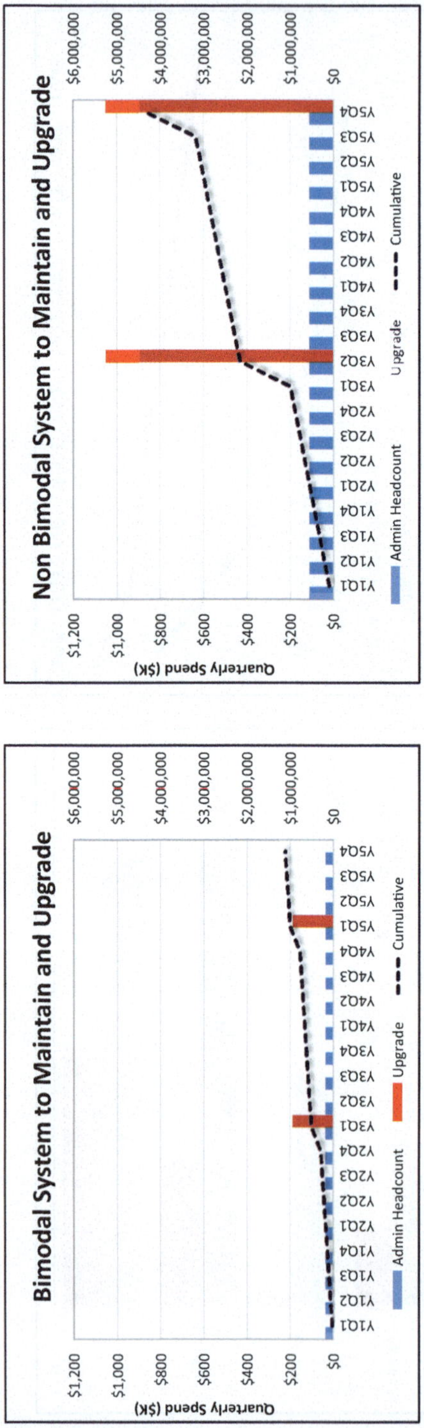

Fig. 5.4 Costs for maintenance and upgrade of the basic system (SysLM) and the customized proportions

- The data model of the SysLM system is linked with the user interface (→ automatic adaptation of the user interface if data model changes),
- Creating and integrating new menus,
- Creating and integrating new methods (→ for example format transfer from native into lightweight, before or after a process step of ECR),
- Creating a typical workflow (for example, ECM)
- Creating a typical lifecycle model,
- Existence and openness of integration interfaces,
- Typical deep integration of an authoring system (for example, M-CAD), and
- Typical simple integration of an authoring system (for example, office tools with drag and drop)

The selection is determined on the basis of pragmatic experience from a number of PLM/SysLM implementation projects. The very clear statement also results from this experience that the frequent promises of the PLM/SysLM providers of an "out of the box" (→ OOTB) solution in no way correspond to reality and represent an absolute exception. Therefore, every functional benchmark of a SysLM system should also be supplemented with the customizing capabilities. The importance of this benchmark is also shown by the determined result of the customizing effort of various SysLM/PLM systems. If the best system is normalized at the value 1, the worst system is at 2.8, i.e., 2.8 times more effort for the customization than the best system.

A further relevant criterion for the delivery and operation of SysLM systems is the understanding and preparedness of the supplier to deliver on the topic of DevOps.[6] DevOps currently still allows a wide bandwidth of interpretation, which often leads to misunderstandings. Fundamentally, there is agreement that DevOps allows for more efficient and smoother cooperation between development, delivery, and operations. John Willis, a veteran in the DevOps movement, describes the framework of the management strategy with five basic principles:

- Culture: Mutual trust, continuous flow of information, willingness to learn,
- Automation: Automation of certain work processes,
- Lean: Avoiding waste, generating value, transparency, and holistic process optimization,
- Measurement: Standard evaluation criteria (including beyond the application and its components) and
- Sharing: Willingness to share information, learn from one another, and proactively share recognitions.

[6]DevOps describes an approach to process improvement from the areas of software development and system administration. DevOps is an artificial word from the terms development (some also supplement delivery) and IT operations. https://de.wikipedia.org/wiki/DevOps, (retrieved on 08.18.2020).

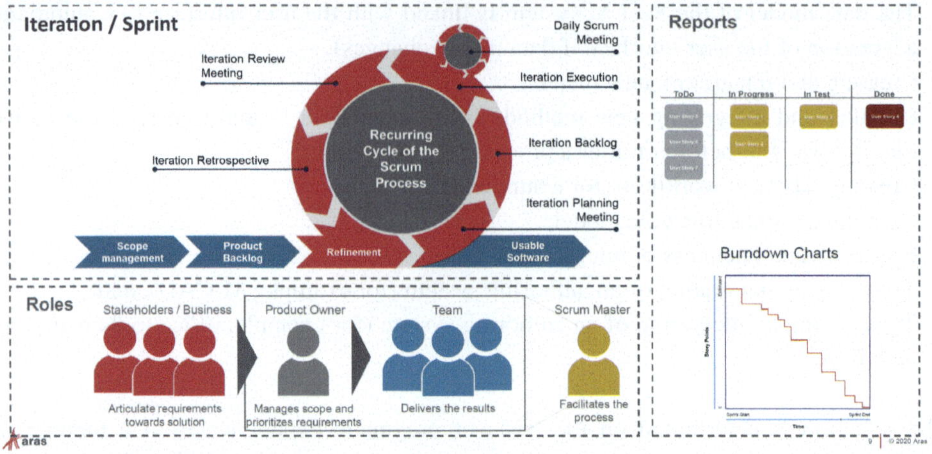

Fig. 5.5 Agile software development and implementation process. *Source* Aras

This includes the preparedness of the SysLM suppliers to support the continuous delivery process with a suitable development environment which is also installed in the customer environment.

5.4 The Agile Implementation of SysLM systems

All the technologies mentioned which are considered as components of a bimodal IT system, are, in turn, a prerequisite for agile software development and implementation strategies. Agile software development and the implementation of software solutions describe methods that increase the transparency and change speed and should lead to a faster deployment of the developed system in order to minimize risks and misdevelopments in the implementation process [2–4]. The foundation of this approach is to reduce the design phase to a minimum degree and to achieve executable software in the development process as early as possible. This is agreed regularly with the customer in short intervals. In this way, it should be possible to flexibly approach customer wishes in order to increase customer satisfaction overall.[7] Agile software development is characterized by self-organizing teams as well as an iterative and incremental approach [3].

Figure 5.5 shows an agile development and implementation process which fits perfectly with a bimodal SysLM.

[7]https://de.wikipedia.org/wiki/Agile_Softwareentwicklung, (retrieved on 08.20.2020).

The fundamental principles of an agile implementation method are:

- Information decisions are made by the entire team.
- The project is based on a repeating number of iterations or sprints that are in a cycle or released. The product always has several release cycles.
- Adaption means permanent changes.
- The product develops with customer feedback.
- Product features are added in every iteration by the user until acceptance of the function and operation is achieved.

Scrum is a framework that supports cooperation in teams. "Scrum" is well known in rugby teams and just as a rugby team trains for the big game, teams with Scrum are in a position to learn from experiences, organize themselves to resolve problems, and reflect on their successes and defeats in order to continuously improve. In Scrum, a project is developed in a series of iterations that are described as a sprint or iteration, in order to break down large, complex projects into manageable, individual steps. Sprint planning is the event in Scrum with which the sprint begins. In sprint planning, we define what can be delivered in the sprint and how it should be implemented. Scrum is mostly used by software developers; however, the principle and the recognitions gained are helpful for all kinds of teamwork. The framework includes meetings, tools, and roles, which together support the structures and manage the team work.[8] Some additional components of the method are described below:

- Product Backlog
  A well-prioritized agile product backlog not only simplifies release and iteration planning but also contains all the tasks for the team—including internal tasks that the customer never gets to see. The backlog helps to coordinate and prioritize the expectations of stakeholders and other teams. It is practically the initial requirements catalog which, however, can have flexible limits.
- Iteration/Sprint Backlog.
  These are the requirements that are relevant and to be implemented for a sprint.
- Iteration/Sprint Planning Meeting
  Planning of the next x-week phase of an iteration/sprint.
- Daily Scrum Meeting
  Daily exchange of the achieved implementation status.
- Iteration/Sprint Review
  Functional check of the development status of the project
- Iteration/Sprint Retrospective.
  Permanent methodological optimization of the approach.

[8]https://www.atlassian.com/de/agile/scrum, (retrieved on 08.22.2020).

- Roles
  Roles of those involved in a project with their functions according to Fig. 5.5 (→ Stakeholder, Product Owner, Team Member, and Scrum Master).

The project status is documented through so-called reports:

- Story Board
- The story board structures the user stories in the phases:
  - Yet to complete,
  - In process,
  - In test and
  - Complete.
- Burn-Down Chart
  The incomplete user stories are listed on the X-axis, the time on the Y-axis.

Below, the agile implementation of an SysLM system is presented using the example of a project at Microsoft (→ project OnePDM). Changing the concept to a standard platform on the basis of a standard market SysLM system for several business areas marks a cultural transformation for Microsoft IT applications. Previously, the business areas conceived or selected—and partly also still self-developed—and then implemented their own solutions.

5.4.1 Integration of Product Engineering and Supply Chains to support OnePDM[9]

PLM[10] supports complex processes in which numerous global teams are involved that conceive and develop new, innovative products. The processes extend from product presentation to discontinuation and include many company areas. PDM is an important subset of PLM and ensures that critical product data flows correctly from the engineering groups to the manufacturing teams, in order to manufacture high-quality mechatronic products. The integration of the supply chain in OnePDM depends on whether hardware product data can be managed and controlled in a system. The concept of OnePDM can be taken from Fig. 5.6.

---

[9]The contribution from Microsoft which forms the basis for this chapter, "Harnessing the power of agile development" (2016) can be found and downloaded at https://eigner-ec.com/.

[10]Microsoft uses the terms PLM and PDM in the same way as they were introduced in Sects. 2.2 and 2.3.

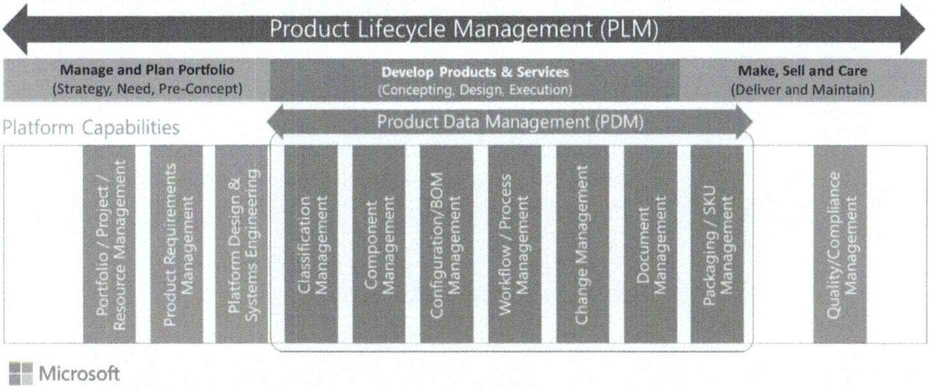

Fig. 5.6 Overall scope of OnePDM. *Source* Microsoft

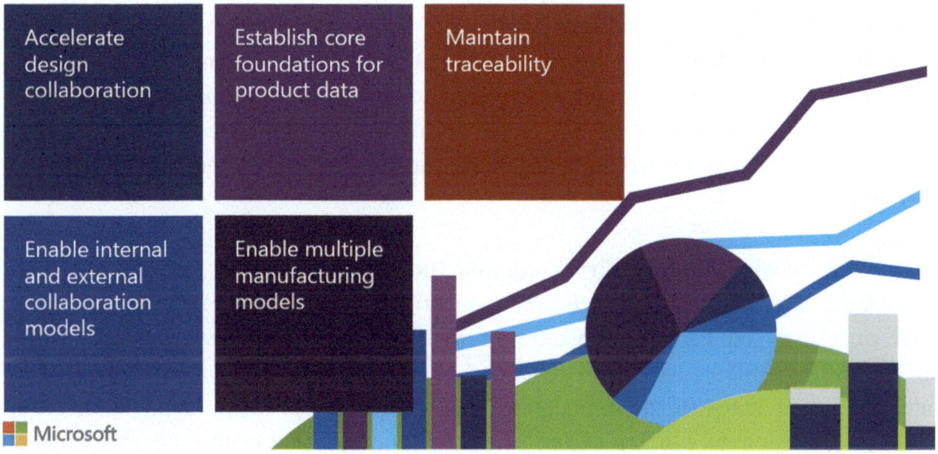

Fig. 5.7 Determining business objectives. *Source* Microsoft

Determining Business Objectives

OnePDM was developed after an analysis and evaluation of the actual status, in order to transfer and integrate several old systems in a modern solution that contains the extended functions for the management of products, descriptive documents, and for the optimization of engineering processes. Therefore, the following super-ordinate business objectives were defined early on (Fig. 5.7):

- Acceleration of the cooperation in the development process beyond the product lines,
- Creation of a joint database for product data (→ one single source of truth),
- Guarantee of traceability across the entire product lifecycle,
- To enable internal and external collaboration models for those involved in the engineering process, and
- To flexibly adapt own and external production.

Fig. 5.8 Observation of strengths and weaknesses within the application and maps with industry standards—and recommendations. *Source* Microsoft

Creating a Development Roadmap

A roadmap was developed that documents and prioritizes the implementation capability of the business objectives mentioned above. The objective was to achieve an initial minimum viable product[11] (MVP[12]). The roadmap also contains the iterative path of additional functions that were permanently added and improved after the rollout version of OnePDM MVP.

Analysis of Existing Business Processes and Industry Standards

Initially, the requirements of the relevant application groups that OnePDM will use in the future were analyzed. Parallel to this, the strengths, capabilities, and weaknesses of existing PDM/PLM systems were investigated. A significant criterion was to clearly understand the actual status of the applications and the processes of the application groups involved. The extent to which previously applied processes and methods corresponded to or deviated from industrial standards and recommendations should also be investigated. Wherever possible, development should be on the basis of best practice in the industry, rather than reinventing the wheel (Fig. 5.8).

In the scope of the project, Microsoft brought two completely different companies together in the project and simultaneously sought standards and recommendations that would be feasible and acceptable for all groups within the company. The market and the sector were analyzed to find best practice solutions which have proven that they can contribute to a process improvement.

[11]Product here refers to a released software solution in an iteration loop.

[12]A minimum viable product (MVP) is the initial minimal viable iteration of a product. https://de.wikipedia.org/wiki/Minimum_Viable_Product, (retrieved on 08.25.2020).

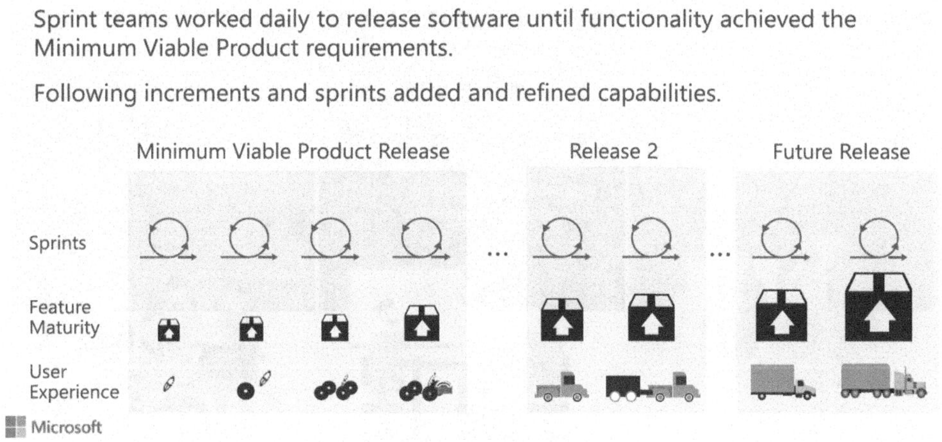

Fig. 5.9 Determining an agile OnePDM implementation strategy. *Source* Microsoft

Creating an Agile Methodology for OnePDM

The agile methodology helped to quickly publish a functioning OnePDM platform in which fast prototyping of sector-based processes was made possible. The application groups could quickly decide whether they could be accepted as a joint process or whether a certain process had to remain individually for Microsoft. End user representatives of the company groups and IT team members met daily in Scrum[13] meetings in order to discuss how the business process could function and to give feedback (Fig. 5.9).

The Agile Team Structure

Three different groups each play a key role in our program; they represent the primary governance, steering, and provision roles (Fig. 5.10).

- The governance board meets quarterly and includes executive sponsors who take on the organizational sponsoring. They inspect execution plans, the program roadmap, and "moving forwards" recommendations and evaluate their effects.
- The steering board meets weekly and is composed of management who enable decision-making and program consultancy of the key groups. They identify work group participants for implementation, integration, and provision and promote change management and communication activities for their respective company groups.

[13]Scrum is a process model of project and product management for software development in particular. It was originally developed in software technology; however, it is independent of this. Scrum is now used in many other areas. It is an implementation of lean development for project management, https://en.wikipedia.org/wiki/Scrum, (retrieved on 08.24.2020).

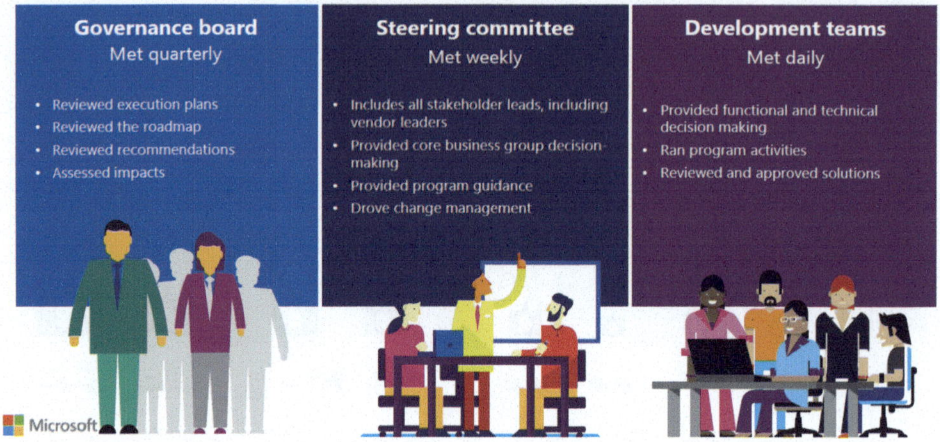

Fig. 5.10 Development of an agile control and management structure. *Source* Microsoft

- The development team and its partners met daily for functional and technical decisions that identify consortium auditors who are defined as part of the overall project. They also carried out program activities, from the collection of requirements to the testing and acceptance of solutions and supportive provisioning.

Agility Was Used for a Continuous Delivery Process

Agile development enables Microsoft to try out several configurations simultaneously. With agile development, it was possible to configure and create work software within one or two days, in order to receive immediate feedback from users. Agile development enables fast decision-making, taking measures, fast failures, and continuous function delivery that bring business benefits. The agile methodology was more efficient with resources than the traditional approach. It was known from the experiences of other IT teams that a robust, functional IT tool should be implemented according to a fast and iterative approach. Therefore, agile development methods were particularly attractive for publishing and releasing incremental software components. Feedback was obtained from users daily. This meant the OnePDM system with the initial core functions could be rapidly published and the stakeholders could continually improve the platform in an agile process. With agile methodology, development cycles (sprints) become faster and more iterative. After the core functionality was created and released, the core functions were improved with user feedback and tests and additional functions were configured with the help of daily builds in the system.[14] As a result, it can be determined that a deployable work platform was achieved in fewer than eight months (Fig. 5.11).

[14]In bimodal SysLM systems, the predominant part of customizing is carried out by interactive configuring of the repository, which supports the agile work technique extensively.

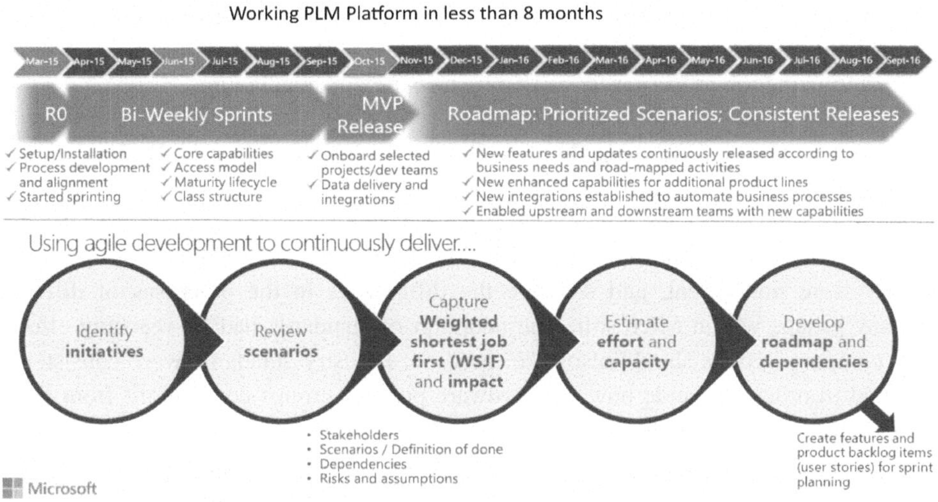

Fig. 5.11 Agile methodology enables the delivery of a deployable platform in fewer than eight months. *Source* Microsoft

Benefits of an Azure Certified PLM Platform Available on the Market

OnePDM uses a third-party platform certified by Azure (→ Aras) in order to be quickly ready for deployment, instead of creating a new system from the ground up [5]. This gave Microsoft an advantage within the industry for the development of the business processes and a standardized taxonomy in all Microsoft company groups. The platform functions and processes, which are immediately available for deployment, were used as a starting point and offer the additional advantage of receiving future innovations from a PLM sector leader. Many functions and part processes were immediately available with this initial situation. Thus, a standard market and tested standard system could be introduced that uses best practice possibilities in the sector.

5.4.2 Best Practice

Microsoft developed best practices in the development of the OnePDM platform, from which other organizations that consider an agile development in order to create or simplify a process-based PLM platform could profit [5]:

Document Your Processes

The guiding principle for OnePDM was the use of industry standard processes, unless a process was individually necessary for Microsoft, in that there was a business justification or an advantage was offered compared to the industry standard process. However, not all processes were well documented by Microsoft, which led to some reworking.

In the cases where the documentation was available, agile development sprints had much better stimuli to ensure that solutions were feasible for all groups. The greatest difficulties in optimizing the process were caused by missing documentation.

Understanding and Accepting That a Better Path Was Already Defined in the Industry

Microsoft recognized the necessity to take industry standards into consideration in order to optimize capabilities and competencies. The differences between Microsoft and the industry were transparent, and so were the differences in the processes of different company groups within Microsoft. The adoption of standards had far-reaching effects, as all company groups should also take these on. Industry standards were consistently evaluated in order to decide how the hardware product groups could profit from them. Through the testing and the understanding of existing, industry-typical best practice solutions, effective, justified decisions could be made as to when an out of the box process should be adopted or rejected. This resulted in a certain obligation to consider why some processes and definitions should be individual for Microsoft. Was it really that these processes offered an advantage, or because they had developed historically from solutions without awareness of the industry standard, or was it because of the users' inertia (→ under the motto: But we've always done it like this)? If individual processes offered an advantage, they were investigated in order to see how they fit with our superordinate objectives and could be integrated into the Microsoft standard.

Agile Dynamics Received through Participation

Agility requires daily cooperation in order to achieve the previously defined business objectives. In this way, teams can take features into consideration and test, validate, and retest features daily, and also quickly correct misdevelopments. For this reason, agile methodology is only effective if teams invest time in order to try out new functions and functions which are published in almost daily incremental builds. It is important to ensure the integration of management and the appropriate participation of all company groups. User feedback is important for the process and the dynamic immediately diminishes if user participation falls.

Treat Product Data as a Resource

The agile cooperation of several teams in a distributed organization is a challenge to recognize complex, interdisciplinary connections and derive resulting improvements from these. Agility and faster decision-making are supported in the product development, as all product data is accessible from an individual system. Users can find their data in a system through a physical presence or REST-based link, whereby high data transparency and time-saving is achieved that would have been applied to search for information across several systems. It is also important that management is convinced of the importance of the classified provision of development data (→ data is the oil of the future).

5.4.3 Benefits

Typical PLM implementations last 18 to 24 months with less complexity than at Microsoft. Agile development accelerates our program rollout considerably while simultaneously retaining high quality. Microsoft had their first program version with functioning software within six months. Within one year, Microsoft could replace the old systems and make additional functions available. Further advantages of agile development are [5]:

- Reduced development and provision times of the software. Code builds and environment provisions were reduced by 7200 min every three weeks to 10 min per day.
- Improved interoperability between engineering, production, and delivery reduced the release times through PLM in up and down stream systems.
- Similar functions were modeled, developed, and deployed 50% more quickly than in previous approaches. This increased the agility and reaction capability to the company's requirements.

OnePDM will save several million dollars (USD) in the first two operational years, as the operating costs of several older PDM systems will no longer be incurred. In addition to these cost savings, the business advantages of OnePDM include:

- The introduction of global parts and the promotion of reuse beyond hardware product lines.
- Lower resource requirement. The consolidation of the PDM systems and the increasing automation reduced the effort for engineering, support and supply chain activities.
- The availability of a much more agile base platform reduces the time needed for the incorporation of new groups in the company.

OnePDM offers all Microsoft company groups a standardized opportunity to develop, manage, and release their hardware products. Microsoft has implemented a platform which will continuously develop and extend the classic PLM functions to all company areas. Thus, Microsoft now has a platform that covers the entire product lifecycle and enables agile development. The platform will be improved consistently and step by step on the basis of the business requirements and simultaneously oriented toward the long-term objectives. Additional important advantages of the Microsoft PLM solution are lower costs and accelerated delivery timeframes.

References

1. Eigner, M.: TCO-REDUZIERUNG durch bimodale PLM-Systeme. Digital Engineering **6**, 62–64 (2018)

2. Beck, K.: Extreme Programming. Die revolutionäre Methode für Softwareentwicklung in kleinen Teams. Munich among others 2000, ISBN 3-8273-1709-6
3. Dogs, C.; Klimmer, T.: Agile Software-Entwicklung kompakt. Bonn (2005), ISBN 3-8266-1503-4.
4. Eckstein, J.: Agile Softwareentwicklung in großen Projekten. Heidelberg (2011), ISBN 978-3-89864-790-8.
5. Microsoft Case Study: Harnessing the power of agile development. https://www.aras.com/en/resources/all/microsoft-harnessing-the-power-of-agile-development (2016). Accessed 26 Sept 2020

Summary

6

The development of innovative, mechatronic, and cybertronic products, systems and connected services will lead to changes in several areas of the company. All areas are affected by this, however, in particular product development, production planning, production, logistics and service. For operational implementation, the planning and design methodologies of all disciplines and all functional and operational units must be put to the test. The aim is to transform this to a common, holistic, integrated, and interdisciplinary design process and IT solution approach (→ Engineering 4.0). New approaches of model-based system engineering and system thinking form the basis, which force in particular interdisciplinarity, digitalization, agility, and collaboration within the company and along the supply chain. The digitalization of products, services, and engineering processes introduces an important transformation process. It re-arranges the classic borders of a fragmented and competitive work and IT solution world. Here, a standard, lightweight, and federated engineering backbone on the basis of SysLM takes on the role of data and process integration along the product lifecycle (PLC). Characteristics of SysLM systems are functions, in connection and optimum orchestration with MRP and MES systems, that fully integrate the engineering, production planning, production and service processes along the PLC on the basis of Digital Models, the Digital Thread, and Digital Twin. In the operating phase of the product, the digital twin supports service-oriented business models. These changes are only possible through new IT technologies. Here, in particular, artificial intelligence, big data analytics, intelligent generative CAD systems with optimization for 3D printing, and linked data on the basis of REST/ RDF, new communication and database technologies and architectures as well as the cloud and the optimized, scalable application systems within it are major drivers.

In this book, the digitalization of engineering, embedding in the IOT/IOS environment, and the changed or further developed methods, processes, and IT tools were presented. These are the prerequisites for optimum implementation of the digitalization

M. Eigner, *System Lifecycle Management*, https://doi.org/10.1007/978-3-658-33874-9_6

strategy. Moreover, tangible examples of digitalized engineering processes in a standard SysLM system are presented. This should generate incentives and ideas to implement the vision of a digital transformation step-by-step, pragmatically and in an evolutionary manner in your company.

Without management standing behind the change and including and integrating motivated employees on the path of the digital transformation, the objective will not be achieved. In particular, the skeptics, who instinctively doubt such transformation, must be convinced. For this, we need creative, innovative, system thinking, and socially skilled employees at all levels of the company (Fig. 6.1) [1, 2, 3].

Beyond the technical aspects demonstrated in this book, people's skills and the challenges facing the employees are to be actively integrated. In addition, however, a corporate culture must also be common values, leadership, beliefs, and behaviors. These company-internal standards have an influence on how people behave and perform operations. In order to transform the digitalization of engineering, the DoD has proposed an approach to promote cultural change [4]. These efforts include training, strategic communication, leadership, and continuous improvements as shown in Fig. 6.2.

Moreover, the digitalization topics must be precisely determined and not be lofty, but rather capable of being pragmatically implemented. Last but not least, the added value both for the company and the employees must be defined and made transparent.

In this respect, the best of luck with the evolutionary design and implementation of your digitalization project.

Fig. 6.1 People's skills which need to be implemented in Engineering 4.0[1]

Fig. 6.2 Engineering digitalization culture and workflow enablers [4]

Reference

1. Seidenspinner, E.: Socio-technical system design with the help of MBSE in the digitalized product engineering, Master Degree Thesis, Institute VPE, University of Kaiserslautern, 2020
2. Tafvizi Zavareh, M., Sadaune, S., Siedler, C., Aurich, J. C., Zink, K., J., Eigner, M.: A Study on the socio-technical aspects of digitization technologies for future integrated engineering work systems. NordDesign, 2018.

3. Deuse, J., Busch, F., Weisner, K., Steffen, M.: Gestaltung sozio-technischer Arbeitssysteme für Industrie 4.0. In: Hirsch-Kreinsen, H., Ittermann, P., Niehaus, J.: (Hrsg.) Digitalisierung industrieller Arbeit, Seite 148–165, ISBN print: 978-3-8487-2225–9, ISBN online: 978-3-8452-6320-5, https://doi.org/10.5771/9783845263205-148, 2015
4. Department of Defense.: Digital Engineering Strategy, https://sercuarc.org/wp-content/uploads/2018/06/Digital-Engineering-Strategy_Approved.pdf, June 2018, Washington, DC.

Appendix

This chapter gives you the opportunity to delve a step deeper into a SysLM system. First of all, a typical command sequence from the area of product engineering is shown step by step in a dry exercise,[1] i.e., master data management, BOM, and M-CAD integration.[2] Parallel to this, you will also be given the opportunity to be able to download the SysLM system Aras Innovator[3] to access an even deeper level. The starting point of a SysLM introduction is always the application "product engineering", which essentially represents the typical functionality of a PDM system. But as a start, this is good, the application is self-enclosed, and most of the commands are repeated in other applications. Before you begin getting to know the command sequence, please look at Fig. A.1. Here, you can once again see the basic principle of a SysLM system: the complete part-centric networking of the information elements which arise in the individual product lifecycle phases. This mesh of networked information elements regarding building and maintaining all configuration statuses is the key task of an SysLM system and forms the basis for the digital thread.

In the following are some commands in the product engineering area of Aras Innovator (Fig. A.2, A.3, A.4, A.5, A.6, and A.7):

[1]Support is provided by Mr. Raoul Markus from XPLM during the creation of slides for the CAD coupling.

[2]The slide set is somewhat shortened here. You can download the long version here: https://eigner-ec.com/

[3]Aras Innovator can be downloaded from a free version of Microsoft SQL Server Express here.https://www.aras.com/en/support/downloads.

M. Eigner, *System Lifecycle Management*, https://doi.org/10.1007/978-3-658-33874-9

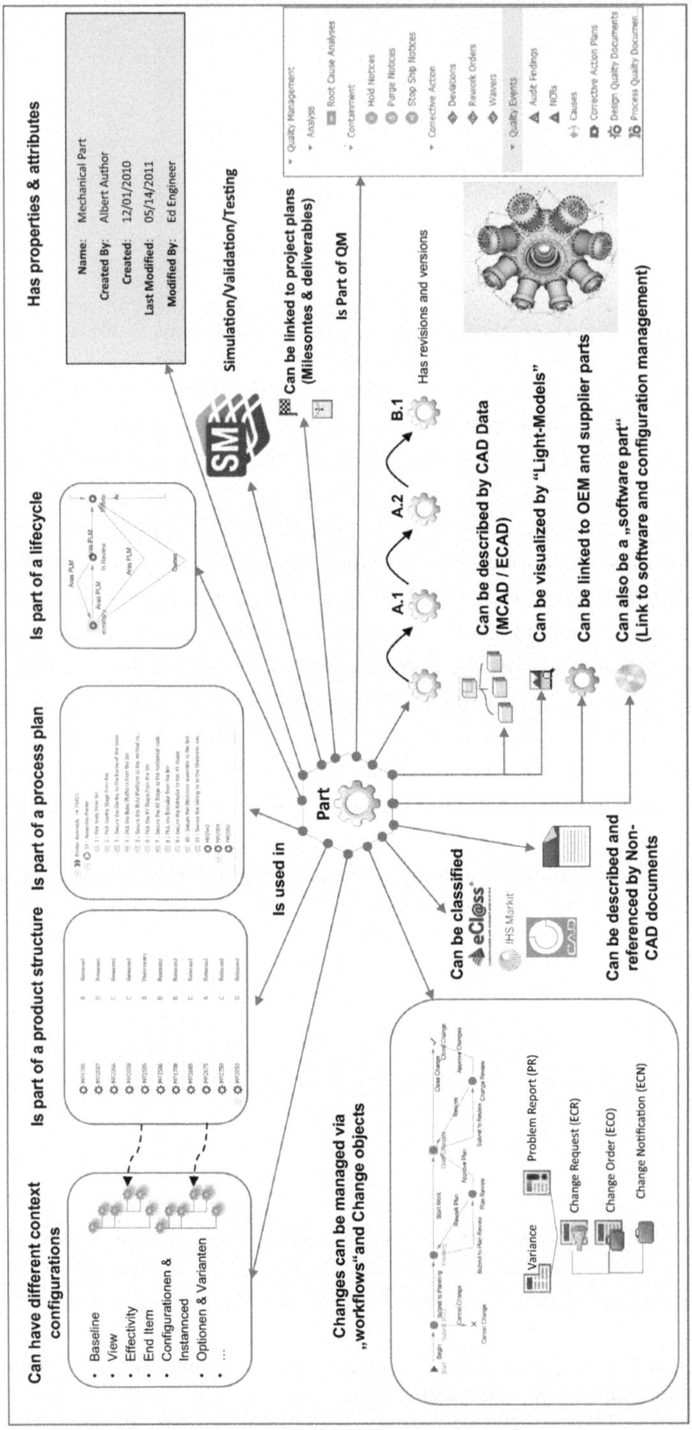

Fig. A.1 The part as a center of a fully operational network of information. *source* changed according to documents from Capgemini

Fig. A.2 Display of master data (parts) and selection of an assembly

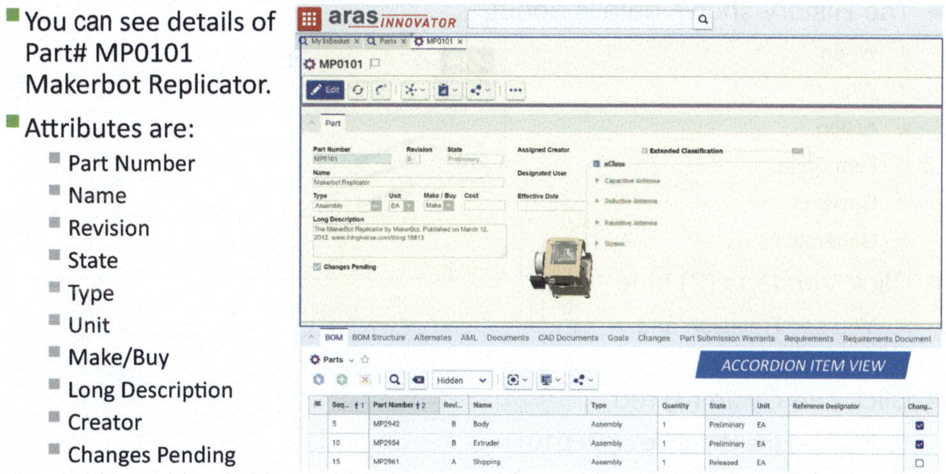

Fig. A.3 Display of the attributes and properties associated with a part (I)

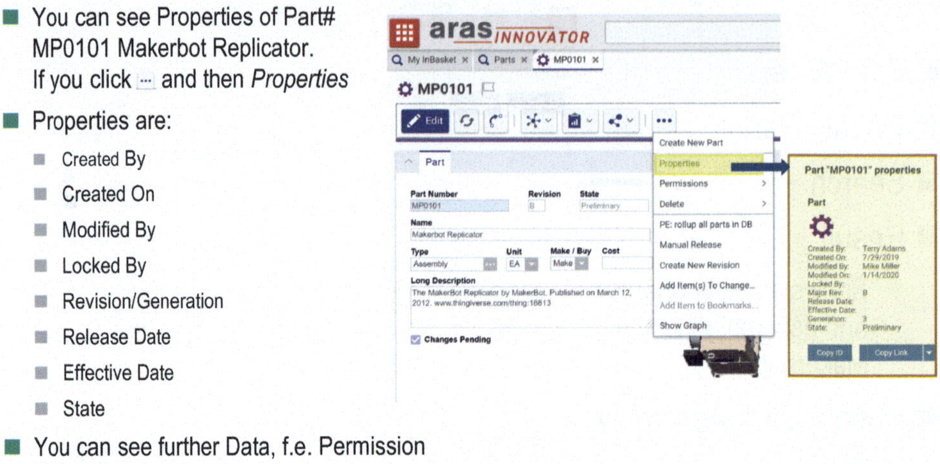

Fig. A.4 Display of attributes and properties associated with a part (II)

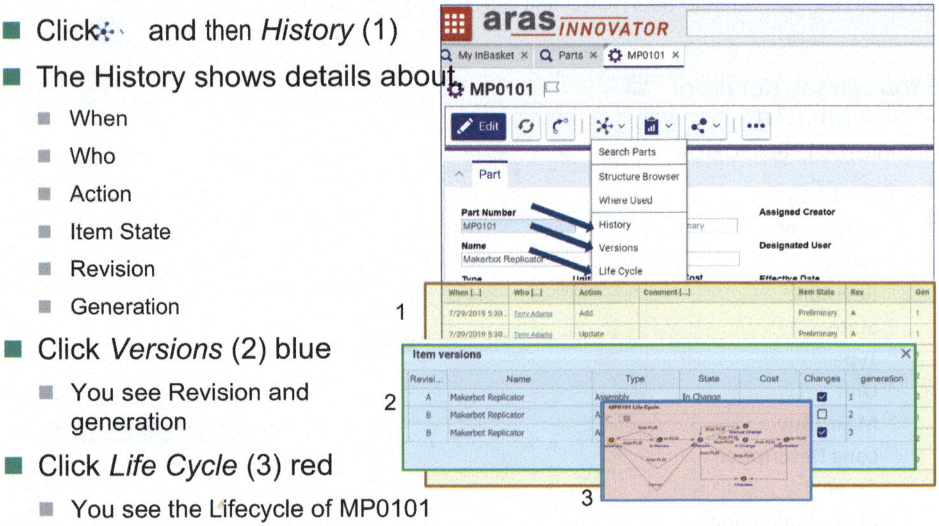

Fig. A.5 Display of history, revision/version, and lifecycle associated with a part

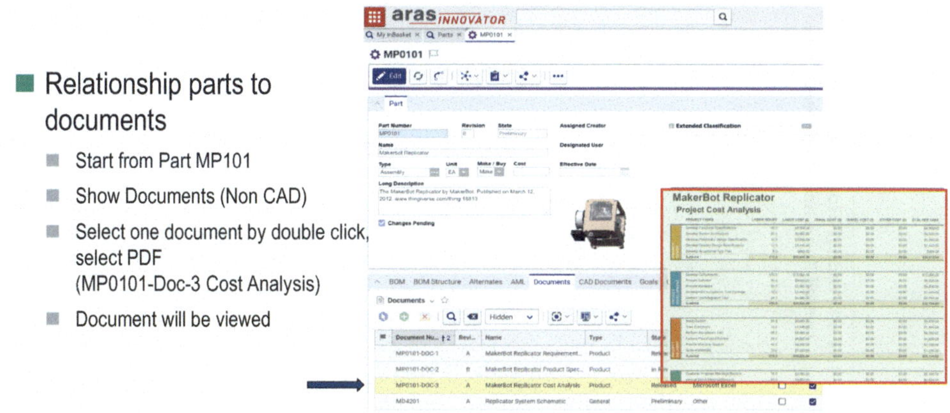

Fig. A.6 Display of non-CAD documents associated with a part

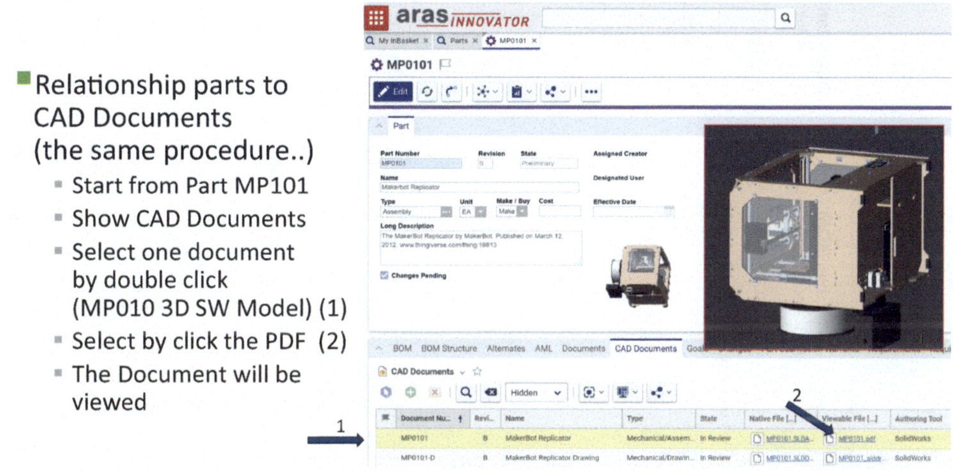

Fig. A.7 Display of CAD documents associated with a part and viewing

So far, some functions, directly related to the technical master record (→ part, documents, properties, and viewing), have been presented. The next point deals with the topic of the BOM (Fig. A.8). In order to prevent redundancy, only the two-level BOM is saved in a SysLM system. Quantitative (rollup) and multi-level BOM are generated from it on an ad hoc basis (Fig. A.9, A.10, A.11, A.12 and A.13).

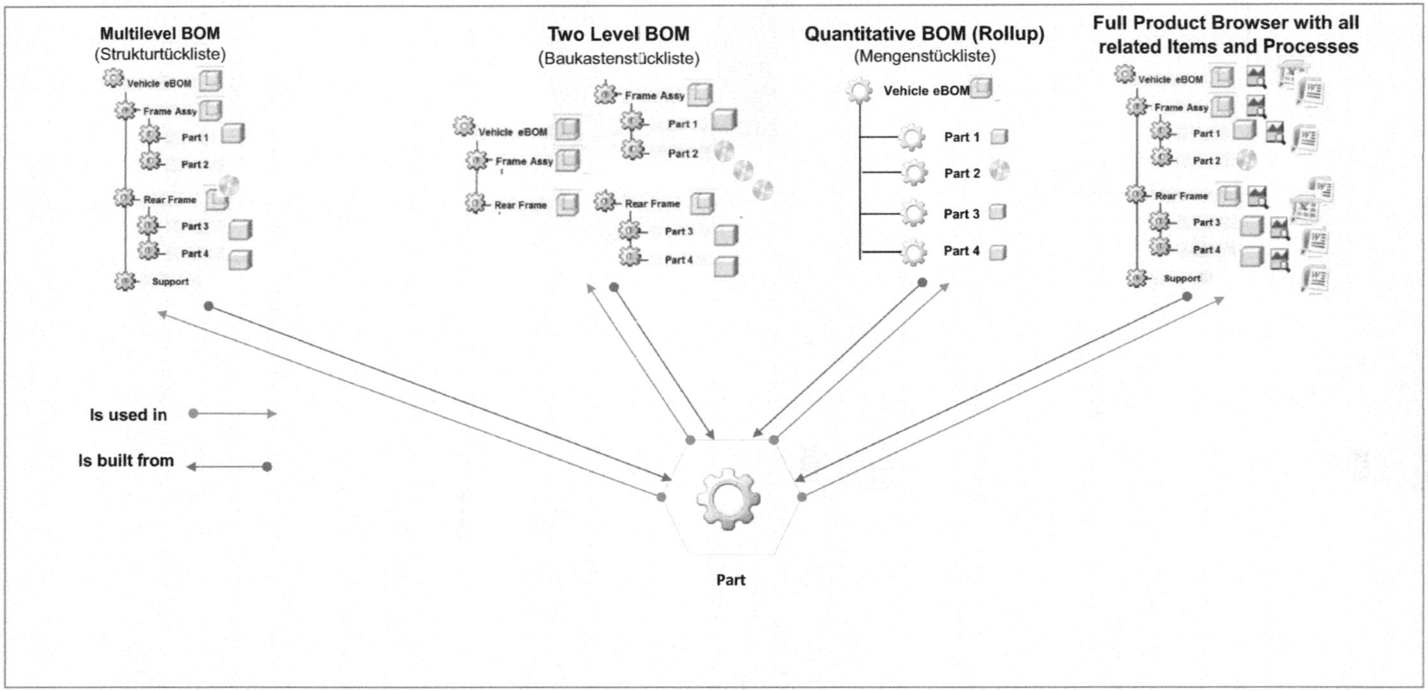

Fig. A.8 Overview of different BOM types

- Start from upper level MP0101
- Double Click MP2942 (1)
- The next level of assemblies and components will be shown
- Double Click MP2939 (2)
- The next level of assemblies and components will be shown

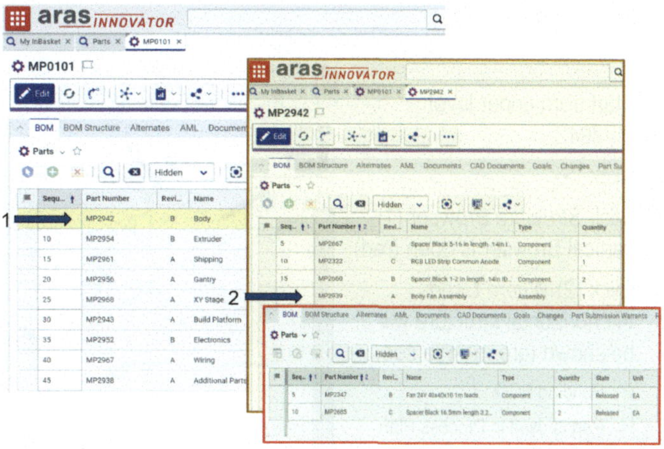

Fig. A.9 Display of two-level BOM

- Start from upper level MP2942
- Click BOM Structure (1)
- Click + (2) in the Browser and the next level will be open
- Click (3)
- Click *Multilevel BOM Report* Report in PDF, Excel or Word (4)

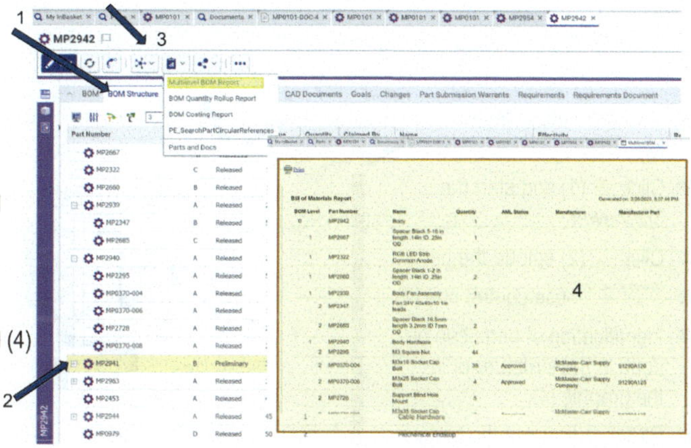

Fig. A.10 Display of the multi-level BOM and output in PDF, Word, or Excel

- Start from upper level MP2942
- Click
- Click *Structure Browser* (1), vertical Browser is open (red)
- Click (2) *Where Used* the horizontal structure will be shown (green). We see:
  - Pending ECO
  - Part is used in Process Plan
  - QM is telling us there is a CAPA activity and a Problem Report

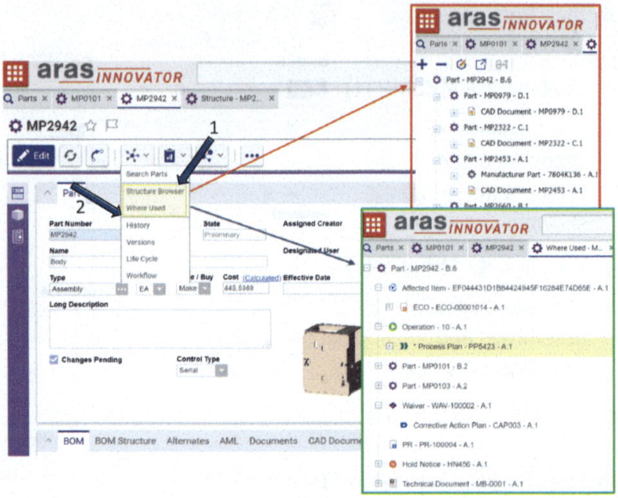

Fig. A.11 Use of the horizontal and vertical browser function

- Start from upper level MP0101
- Click (1) and start the 3D Viewer
- Click (2) eplode the parts
- use control box (3)
- Identification of parts can be done in the structure or in the graphic (4)
- Click (5) to cut the parts
  use control box (6)

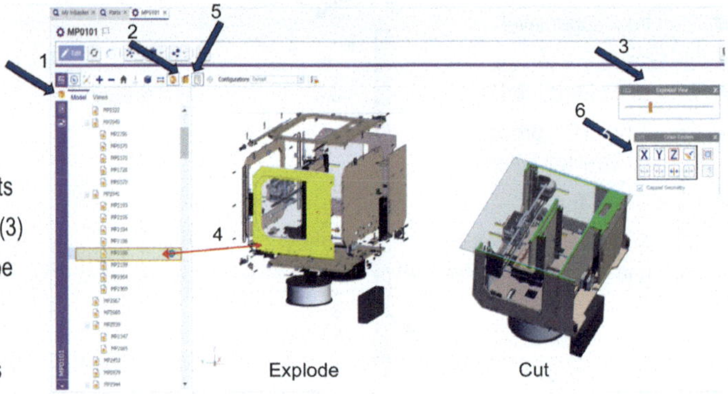

Fig. A.12 Exploded and sectional view in 3D dynamic viewer

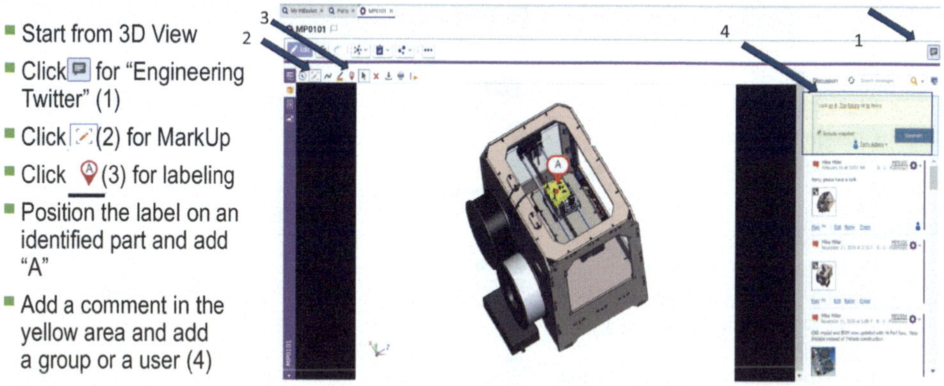

Fig. A.13 Use of the engineering collaboration tool (engineering twitter)

The next functions provide an insight into the engineering change management (cf. Sect. 4.2.2.1) and also the CAD integration (cf. Sect. 4.2.1.3), triggered by the situation that a CAD-designed part must be changed in the CAD system. A compatible change should be carried out which affects the part MP2966 and its CAD document. The part belongs to the assembly MP2954 (Fig. A.14, A.15, A.16, A.17, A.18, A.19, A.20, A.21, and A.22).

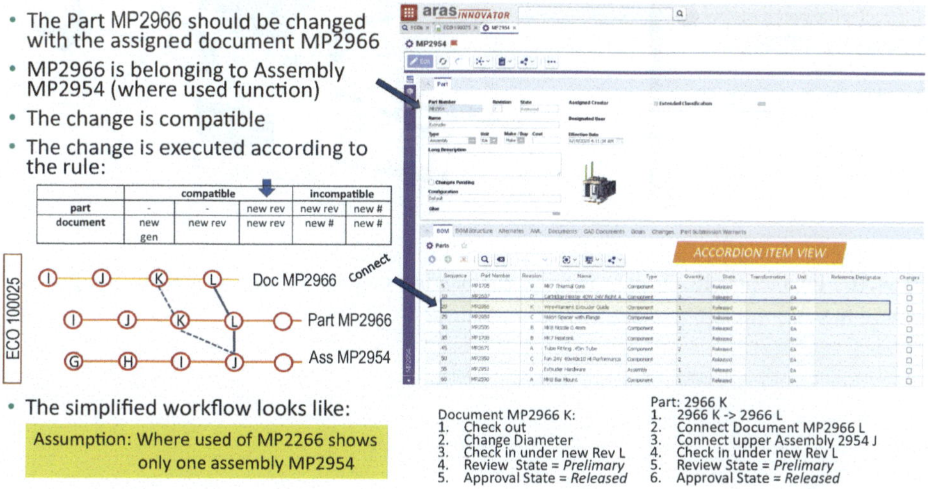

Fig. A.14 Pre-analysis of a compatible change to part MP2966

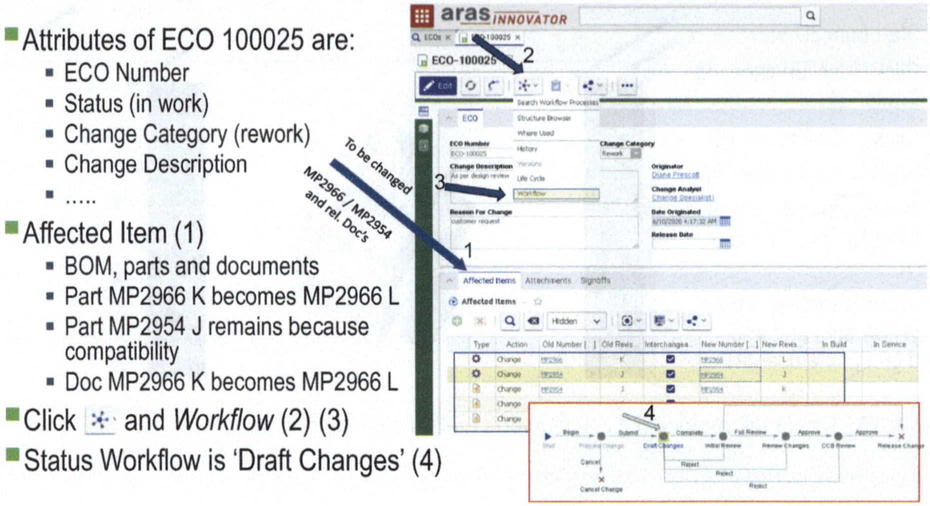

Fig. A.15 Issuing Engineering Change Order (ECO) and entering the affected items

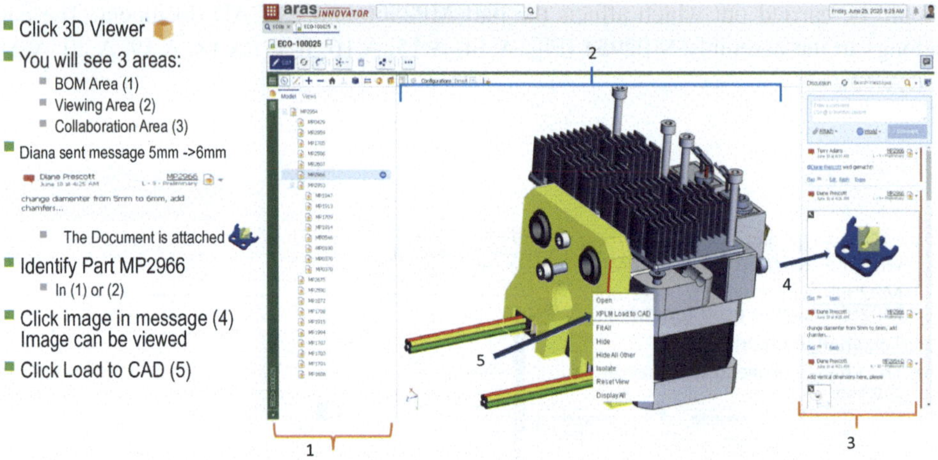

Fig. A.16 Interaction ECO, 3D viewing, engineering collaboration, and loading the part to be changed to CAD

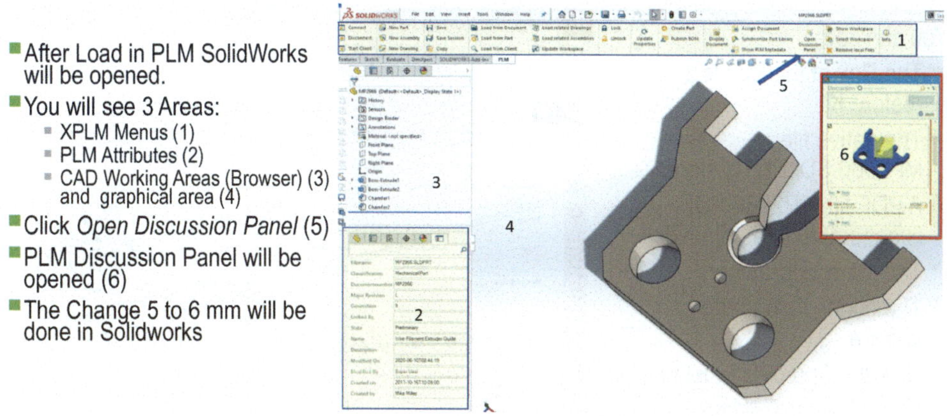

Fig. A.17 Graphical user interface of the CAD system SolidWorks and synchronization between SysLM and CAD

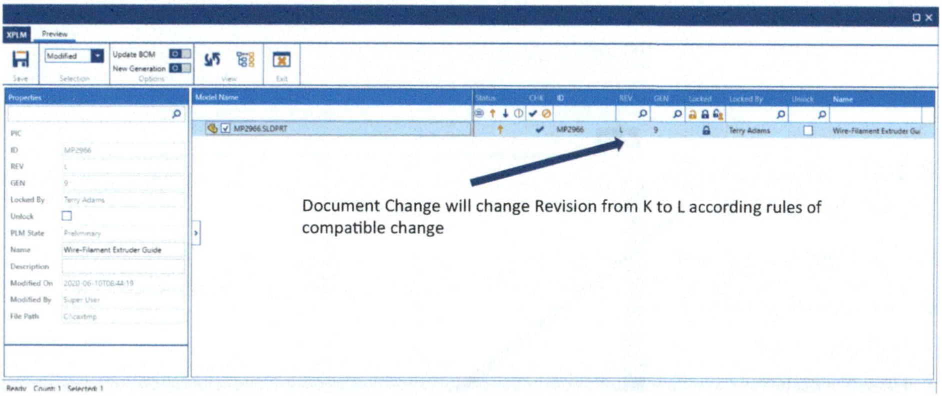

Fig. A.18 Return and check of the results in the XPLM integration tool

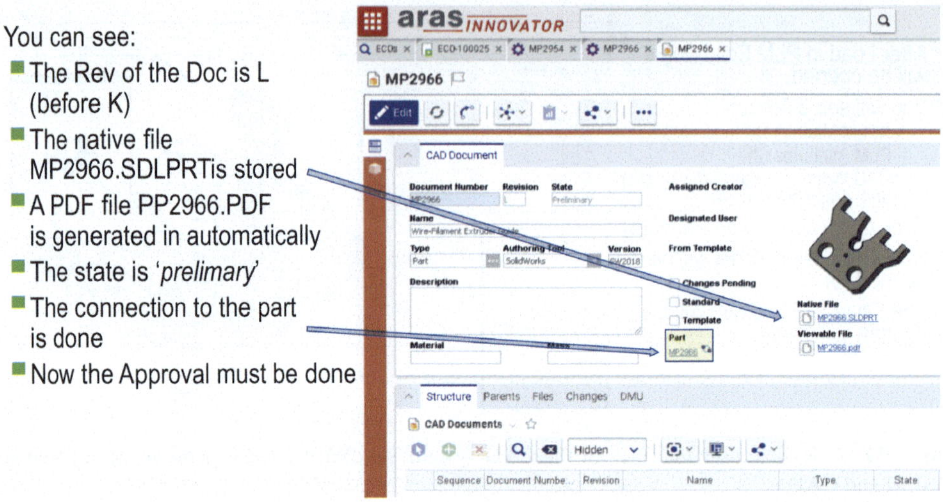

Fig. A.19 Check of the change execution in SysLM

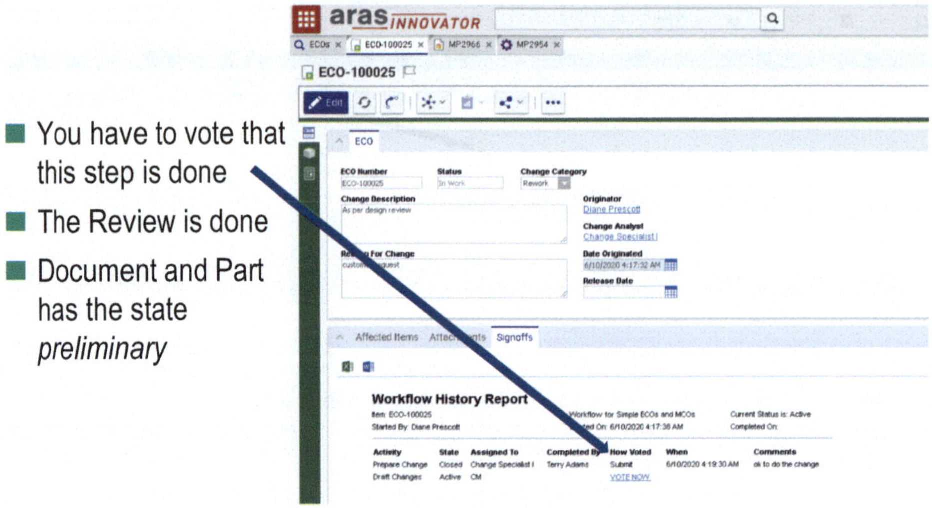

Fig. A.20 Review the change execution in SysLM and vote "Review done"

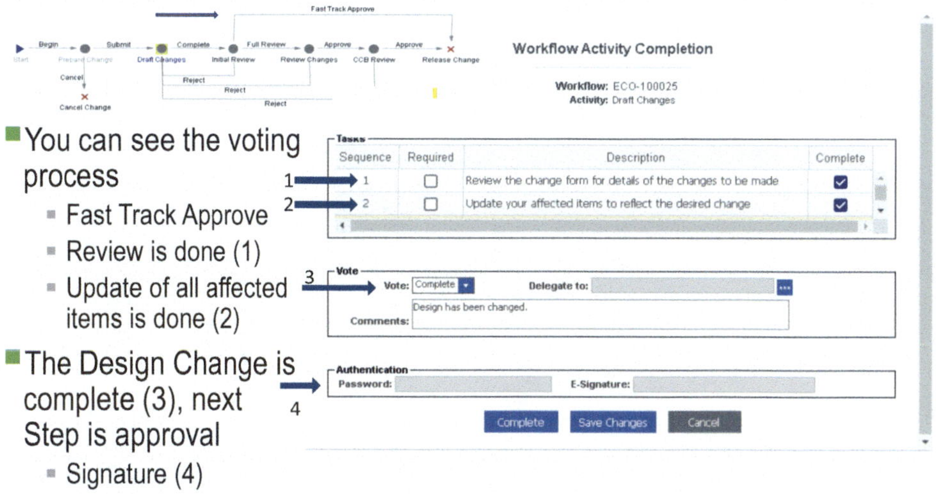

Fig. A.21 Legal signing off the change execution on the CAD and SysLM side

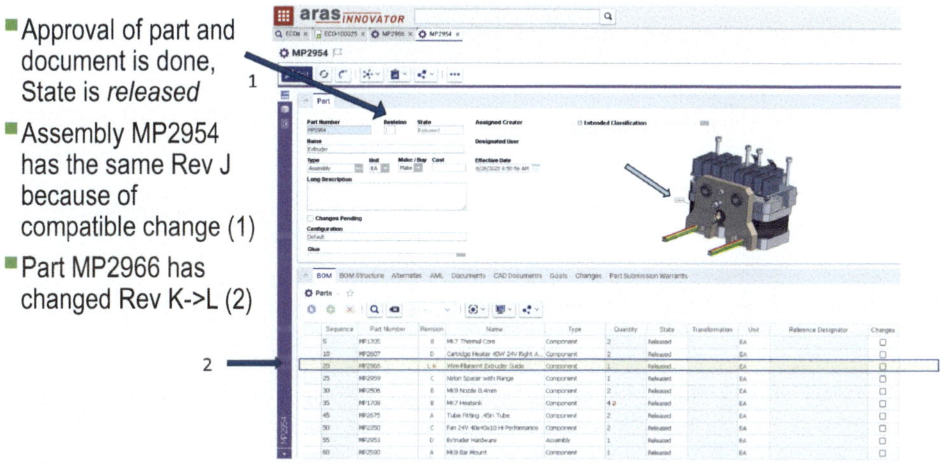

Fig. A.22 Showing the results of the executed ECO in SysLM on the part and document level

The manufacturer's authorised representative in the EU is Springer Nature Customer Service Centre GmbH, Europaplatz 3, 69115 Heidelberg, Germany. If you have any concerns regarding our products, please contact ProductSafety@springernature.com

Printed and bound by CPI Group (UK) Ltd, Croydon, CR0 4YY
30/09/2026
02225654-0017